SMART ACTUATION AND SENSING SYSTEMS - RECENT ADVANCES AND FUTURE CHALLENGES

SMART ACTUATION AND SENSING SYSTEMS - RECENT ADVANCES AND FUTURE CHALLENGES

Editor

Eden Slaver

scitus
academics

Smart Actuation and Sensing Systems - Recent Advances and Future Challenges
Edited by **Eden Slaver**

ISBN: 978-1-68117-604-8
Library of Congress Control Number: 2016946459

SCITUS Academics LLC,
616, Corporate Way, Suite 2, 4766,
Valley Cottage, NY 10989

www.scitusacademics.com

Notice

Printed in United States of America on Acid Free Paper

PREFACE

The development of integrated circuits, made possible by the enormous advances in semiconductor technology, resulted in an affordable microprocessor. Therefore, if it is possible to design an inexpensive silicone-based sensor, the general price of the control system can be reduced. We can have integrated sensors with electronics and transduction elements on a single silicone chip together. This complete system can be called a system-on-chip, the main objective of integrating the sensor and the electronics is usually to create an intelligent sensor that can be called an intelligent sensor. Intelligent sensors then have the ability to decide. A smart sensor physically consists of a transduction element, electronic signal conditioning and a controller/processor that supports some intelligence in a package. A smart sensor is a unit that takes input from the actual physical environment and uses calculated resources to perform predefined functions when specific input is detected and process data before it is transferred. Intelligent sensors enable more automated and accurate environmental data collection with less erroneous noise in the accurately recorded information. These units are used in a wide range of environments, including smart grids, battlefield recognition, exploration and excellent scientific applications, for monitoring and control mechanisms. The integrated actuator of all components, including motor, sensors, controller and communication unit, is a smart actuator. The robot connected to a smart actuator can be assembled or possibly disassembled and reset. As a result, industry demands for a smart sensor and actuator are increasing.

This book emphasizes the achievements of intelligent actors and intelligent sensors in combining researchers and scientists from different fields into one virtual room. The book thus reflects the multicultural nature of the field and enables the reader to taste and appreciate different points of view, different engineering methods and various tools that must be taken into account when designing and implementing intelligent actuation and sensing systems.

TABLE OF CONTENTS

Chapter 1 New Developments on Mini/Micro Shape Memory Actuators 1

Chapter 2 Piezoelectric Motors, an Overview 25

Chapter 3 Shape Memory Actuators for Medical Rehabilitation and Neuroscience 57

Chapter 4 A Technology for Soft and Wearable Generators 105

Chapter 5 Status and Perspectives of Multiferroic Magnetoelectric Composite Materials and Applications 123

Chapter 6 Position estimation of RFID-based sensors using SAW compressive receivers 177

Chapter 7 Diagnosis of carbonation induced corrosion initiation and progression in reinforced concrete structures using piezo-impedance transducers 199

Chapter 8 Base Metal Co-Fired Multilayer Piezoelectrics 227

Chapter 9 Simultaneous Smart Actuating-Sensing Devices Based on Conducting Polymers 263

Index 299

1

NEW DEVELOPMENTS ON MINI/MICRO SHAPE MEMORY ACTUATORS

Adelaide Nespoli, Carlo Alberto Biffi, Riccardo Casati, Francesca Passaretti, Ausonio Tuissiand Elena Villa

[1] Consiglio Nazionale delle Ricerche, Istituto per l'Energetica e le Interfasi (CNR-IENI), Lecco, Italy

1. INTRODUCTION

Shape Memory Alloys (SMAs) are functional materials, characterized bysome attractive features, such as pseudo-elasticity (PE)and shape memory effect (SME) [1].The latterconsist inthe capability of the material to recover high deformation values by heating and can be considered very suitable for actuation applications.

In the actuation field, the most common shape memory alloy is quasi-equiatomicNiTi system, commercially known as Nitinol®. Ti-rich NiTi compounds show characteristic transformation temperatures higher than the room one and they can recover high values of deformation. Moreover, these intermetallic compounds are widely used since they exhibit high thermal and mechanical cycling stability. Other NiTi-based alloys are also employed for this kind of applications. In particular, NiTiCu, with Cu substituting Ni in the 3-10at% range, or NiTiCosystem are used respectively when the application requires narrow thermal hysteresis or high stiffness.

In Figure 1, the comparison between differential scanning calorimetry (DSC) data derived fromNiTi wire (200 μm in diameter) in fully annealedcondition and after aging (500ºC for 10 minutes) are reported. As it can be seen, the fully annealed NiTi alloy goes into a one-step transformation both during cooling and heating. In this case, shape memory effect occurs by a martensitic transformation (MT) between a low-temperature monoclinic structure, B19', called martensite, and a high-temperature body-centered cubic parent phase, B2, called austenite. When the material is aged at specific temperatures, this transition may occur in two steps in association with a trigonal phase, called R-phase [1].

During heating, the phase transformation,namedinverse MT, is defined by austenitic start (As) and finish (Af) temperatures. Similarly, during cooling direct MT is defined by martensitic start (Ms) and finish (Mf) temperatures, respectively.

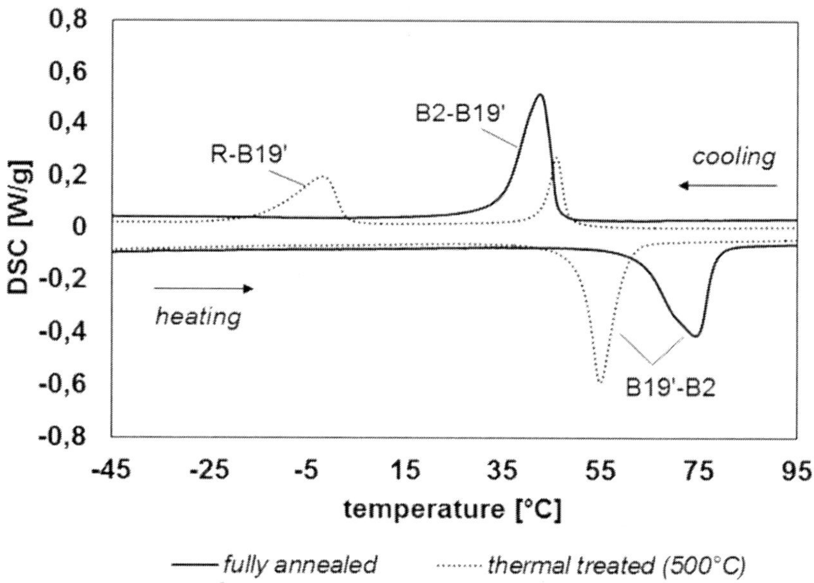

Figure 1. DSC scan of 200μm NiTi wirein the fully annealed state and after aging (500°C for 10 minutes).

Then, the deformation induced by a state of stress on a SMA element can be recovered heating the material over the austenitic characteristic temperatures: the shape recovery starts at As and finishes at Af.

The transition temperatures are strongly influenced by the applied load, since it stabilizes the martensitic structure causing an increase of all transition temperatures. This behavior of SMA is well described by the Clausius-Clapeyron law [1-5].

Furthermore, SMA actuators generally need atraining to stabilize their fatigue properties in terms of transformation temperatures and strain recovery. For example, a possible procedure consists in thethermal cycling undera constant load.Figure 2depicts the evolution of the temperature-stroke relationship during the first thermo-mechanical cycles. It is seen that the stabilization of the material behavior can be reached after few thermal loops.

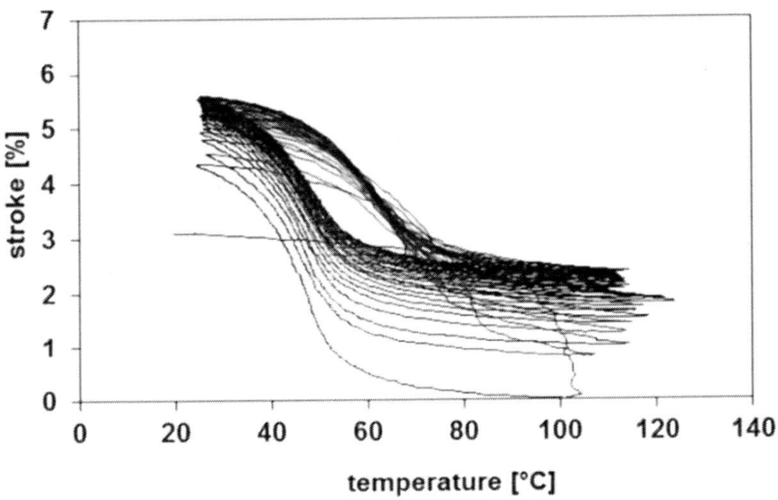

Figure 2. Example of thermo-mechanical hysteresis typical of a SMA material.

Since SMA can generate mechanical work, they are employed as actuators.The functional properties of SMA element used in actuators strictly depend onseveral factors, such as the chemical composition and the thermo-mechanical history[1,4,5].

Nowadays, the tendency is to produce ever smaller actuators because of the continued miniaturization of all the industrial products. In general, a mini-actuator may be composed by only a SMA elementor by a mini-modular device, in which the SMA element is embedded and acts as the core and active part.

In this chapter, the fundamental operating parameters, which affect the performance of shape memory thin wires, are reviewed. Then, the influence of electrical heating conditions performed by different waveforms on functional fatigue of NiTi micro-wires is also reported.

When the SMA element is embedded into a mini-modular mechanical device, its shape could be a serious problem as spaces are very restricted. To solve this drawbacks, a new SMA conformation suitable for the mini and micro scales is presented: it consists of a planar wavy formed NiTi wire, called snake-like arrangement.Currently, this configuration of the SMA wireis principally exploited in the textile and medical domains [6-8].n the micro scale, original works are reported by Mineta et al. [7-8], who produced different SMA snake elements by means of electrochemical etching for obtaining bending motion of active catheters. Khol et al. [9] used the micro snake geometry to activate a microgripper system.Moreover,Leester-Schadel et al. [10] adopted laser technology to produce micro SMA snake actuators and then used the batch fabrication process to obtain more articulated samples. In this chapterthe main mechanical performance of this unusual geometryand its exploitation in a mini-modular mechanical device are presented.

Another topicof this work is related to a current research on fabrication and characterization of SMA micro-snakes, by means of laser technology and following polishing processes. The reason why this non-traditional production technology has been chosen for this purpose is due to its flexibility in the machining of small features, high productivity and repeatability [11-12]. Considering the direction of miniaturization of the products during the recent years [13], in the last part of the chapter the rescalingofSMA mini-snakes down to micro-scale is presented. The attention is also focused on the evaluation of the capability of laser micro-cutting industrial process in the fabrication of micro-snakes from NiTi SMA sheets [14-17]. The evaluation of the thermo-mechanical properties of the produced actuator in correspondence of the different fabrications stepsis then presented and discussed in order to describe the behavior of the micro-snake-like element for actuation.

2. SMALINEAR ACTUATORS

The first case study is the application of a simple straight wire as base element for actuation systems [18].The most important functional properties, which characterize shape memory wire for actuators, are the capability to recover high strain values, the characteristic transformation temperatures and the thermo-mechanical cycling stability [1]. These characteristics of SMA can be tailored acting on chemical composition, thermal and mechanical processing, shape setting treatment and training [1, 20-25].

In recent years, the good ductility of NiTi intermetallic has been exploited for the production of few tens micron wires [21, 26, 27].

Thanks to their small dimensions, such thin wires permit to reach very short cooling time but, obviously, they can be employed just in all those applications in which high loads are not required. The fabrication of these products has led to the miniaturization of SMA components for actuators that are nowadays employed for new kind of devices such as optical image stabilizer and autofocus for small camera [26]

During working life, the SMA wire is subjected to thermo-mechanical cycling and accumulates plastic deformation, elongating irreversibly and reducing its diameter (see Figure3). This behavior has to be considered in actuator design. Another important point to consider is the unsteadiness of the functional properties of shape memory wires during the working life. The plastic deformation induced during the thermo-mechanical cycling leads to a change of microstructure of SMA wire. Lattice defects, dislocations and nano-scaled precipitates are introduced into the matrix and they cause significant changes in transformation temperature and in the capability of the material to recover a deformation [28-30], as depicted in Figure 4.

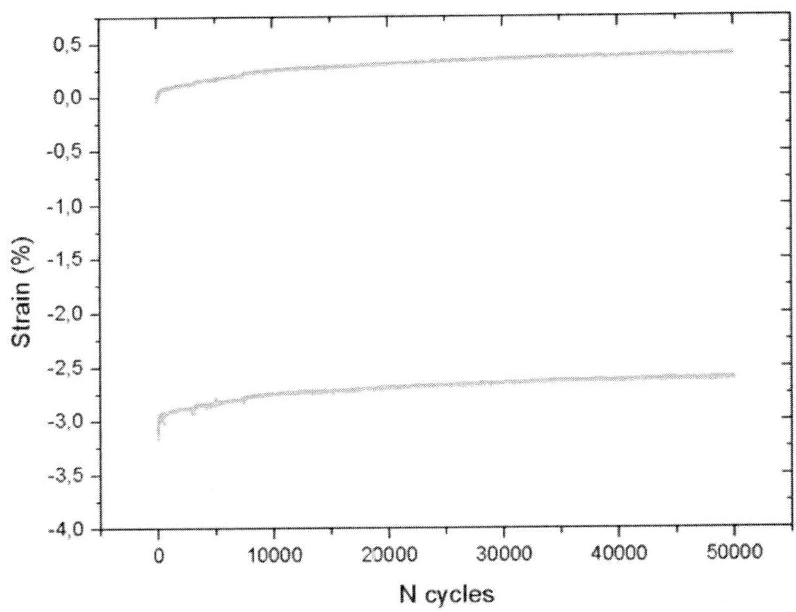

Figure 3. Thermal cycling of a 80μm wire under constant stress (200MPa). The heating was performed by electrical pulse, and stroke was limited at 3%.

In most cases, SMA wires used as actuator, are heated by means of an electrical pulse (Joule effect) and cooled by natural air convection. So, it is very easy to reach short actuation time (heating time) but the reset time (cooling time) cannot be substantially controlled. Slight improvement in the reduction of reset time is achieved positioning the shape memory wire horizontally [19], as shown in Figure 5. In this figure, it can be noticed that after heating (at 200s) the strain is maintained to two constant values for both the two configurations till time is near to 230s and 250s for horizontal and vertical position respectively. After these points, the two wires start the cooling process which results to be faster for the horizontally positioned wire than the vertical one. As an example, it can be seen that at 275s the horizontally positioned wire recovers half of its maximum strain while at the same time the vertical positioned wire has just started the cooling route.

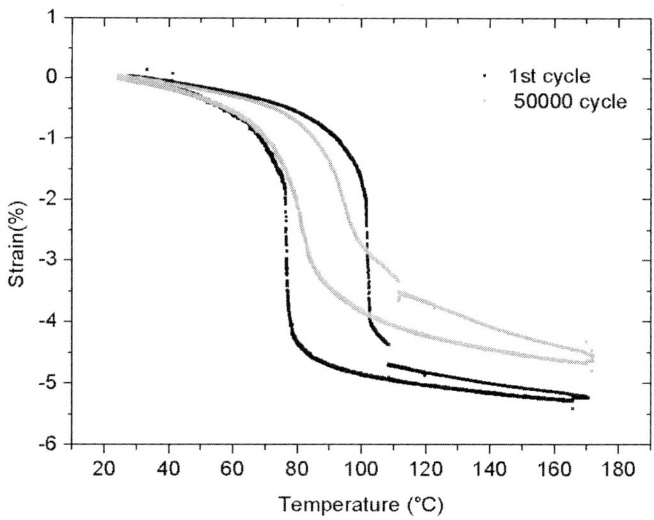

Figure 4. Thermal loopunder constant stress (200MPa). The heating was performed by a thermal chamber. The test was carried out on a 80μm wire before and after $5x10^4$ thermal cycles under constant stress (200MPa).

As previously stated, the heating process of a SMA actuator can be easily obtained by Joule effect. In this case, it has to be considered that the shape of the electrical pulse used for the actuation strongly affects the functional properties of SMA [27]. As an example, Figure 6 shows the time required by the wires to recover the 3% of deformation (actuation time) when it is heated by two different current pulses. At the 1st cycle ramp and step electrical pulse employed to heat two 80μm wires were designed to have the same electrical efficiency but the actuation time is 400ms by step current pulse while it is 623ms by ramp. After 5 10^3 cycles the wire heated by step pulse employs 618ms, so 218ms more than the time employed at the first cycle. As opposite, the ramp pulse leads to an increase of just 22ms. After 5 10^4 cycles the actuation time related to the wire heated by step is even higher than the one related to wire heated by ramp. This behavior leads to a drastic decrease of the step heating method efficiency (from 0.91% to 0.55%), vice versa the efficiency of the ramp heating method does not change so much (from 0.91% to 0.75%).

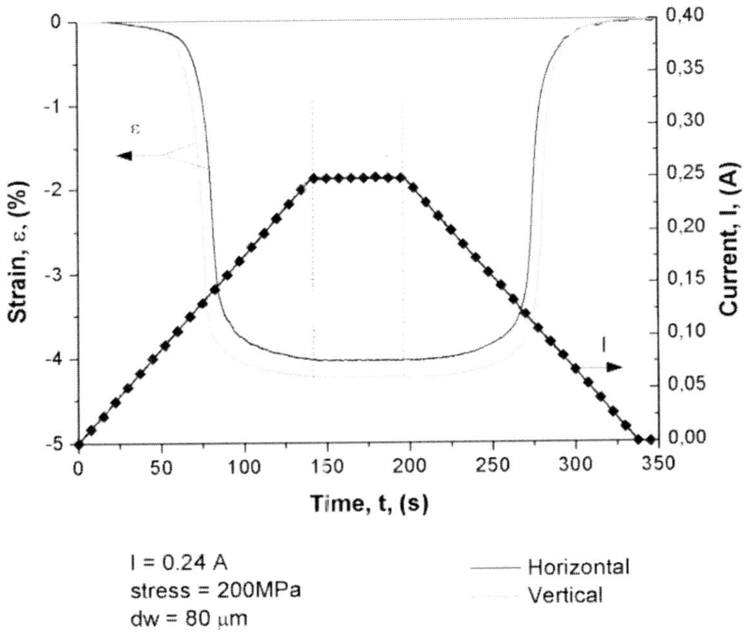

Figure 5. Comparison between the mechanical performance of horizontal and vertical configuration of 80μm SMA wire used as actuator [**19**].

Then, in order to achieve fatigue performance acceptable for the specific device, the right electrical pulse has to be chosen considering the number of cycles that the SMA actuator has been designed to work.

Recently, the effect of drawing procedure on functional fatigue of thin NiTi wires has also been investigated [21]. Basically, 80μm NiTi wires were produced through two different drawing procedures reaching the same final cold working level before shape setting. These two processes differ for the number of drawing steps carried out to reach a certain cold working level before each heat treatment. After the last thermal treatment the specimen that underwent to a less number of drawing steps shows a narrow thermal hysteresis, even after thermo-mechanical cycling (see Figure 7). It means that adopting a severe drawing procedure leads to improve the actuation performance.

3. SNAKE-LIKE SMA ACTUATORS

The main focus of miniaturization is to develop devices with high mechanical performance in very limited spaces. In the SMA field, the mini-actuator may be composed by only the SMA element or by a mini-modular device in which the SMA element is embedded and acts as the core part. In the latter case, some issues typical of SMA miniaturization should be taken into a serious consideration: first of all, the fixing points of the SMA element and the electrical contacts should be adequately designed to prevent abrupt rupture, and then the use of a wire geometry generally implies the employment of additional mechanical parts which may affect the long run performance of the SMA element.

The mechanical work resulting from the SMA shape recovery is used to produce linear or rotational motion in smart actuators. To attain high strokes in limited space, designers developed several miniature devices which include SMA elements with a straight or a helical spring shape [31]. These two shapes present two opposite mechanical behaviors: the straight conformation allows high forces but smart wire arrangements should be employed in order to achieve also high strokes. As an example, Jansen et al. used two sets of seven pulleys to arrange a 1000mm long SMA wire with a diameter of 0.200mm, and attain 3% of stroke under a constant load of about 130MPa [32]. The use of additional mechanical components may affect the overall performance of the SMA element, leading to friction which couldinfluence the final functional parameters of the device.As opposite, the helical spring shape guarantees high strokes without the use of any further mechanical part but fairly small forces can be attained especially when embedded in miniature devices.

Hence, straight and helical spring conformations represent two opposite limits in the force versus stroke plane.In order to answer both to the need of stroke and force between these two limits, new SMA geometries should be studied.

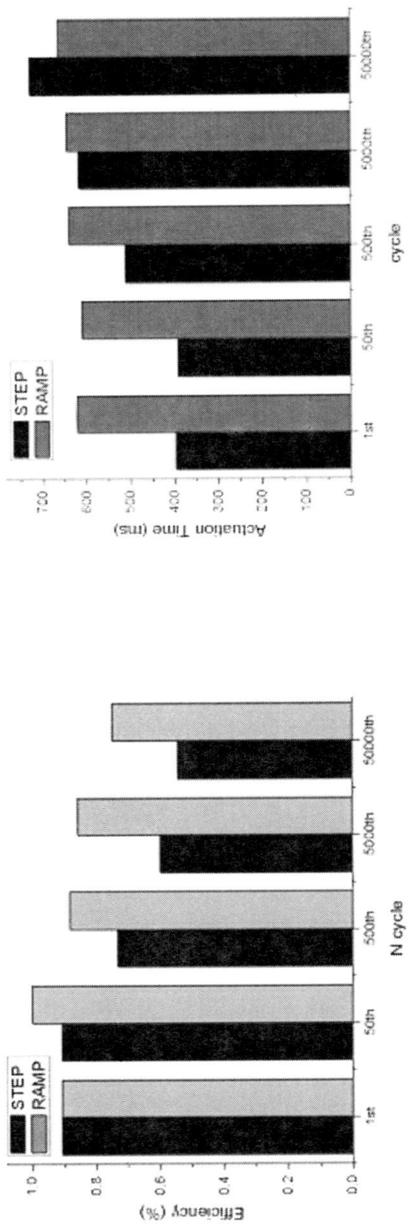

Figure 6. Actuation time tests performed by electrical heating (on the left) and by efficiencies (on the right).

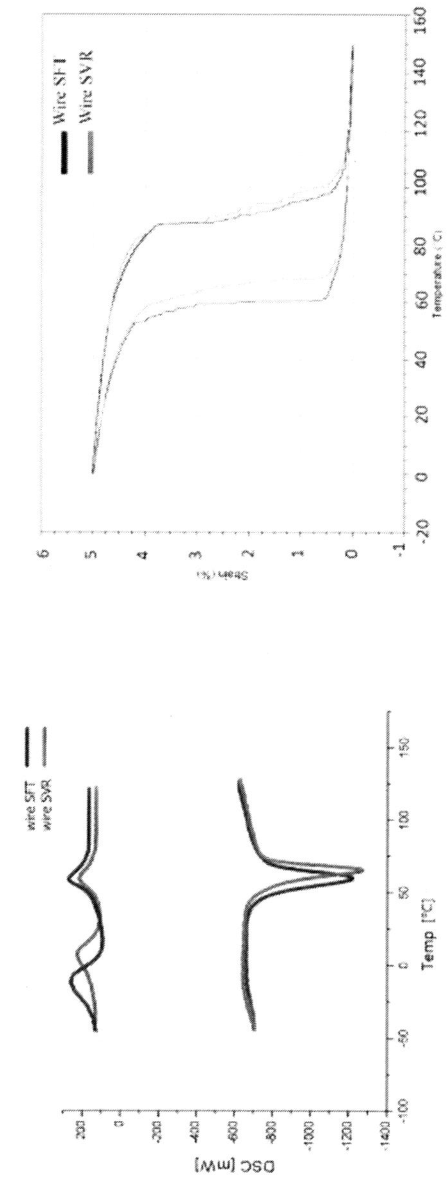

Figure 7. DSC scans, on the left, and thermal loop under constant tensile stress, on the right, of two 80μm NiTi wires obtained by a severe (SVR) and soft (SFT) drawing procedure.

In a recent work, a new SMA conformation has been proposed [33]; it consists in a planar wavy formed NiTi wire, named snake-like arrangement.To understand the mechanical performance of this unusual conformation as respect to other shapes, three NiTi samples with different geometry, but with identical elemental composition, thermal history and power consumption were testedby means of a standard tensile test. In this investigation the NiTi wire was formed to have the straight conformation, the helical spring and the snake-like arrangements, and austenite and martensite mechanical responses were assessed separately. As can be seen in Figure 8 the snake-like shape has a mechanical behavior between the wire and the helical spring ones.

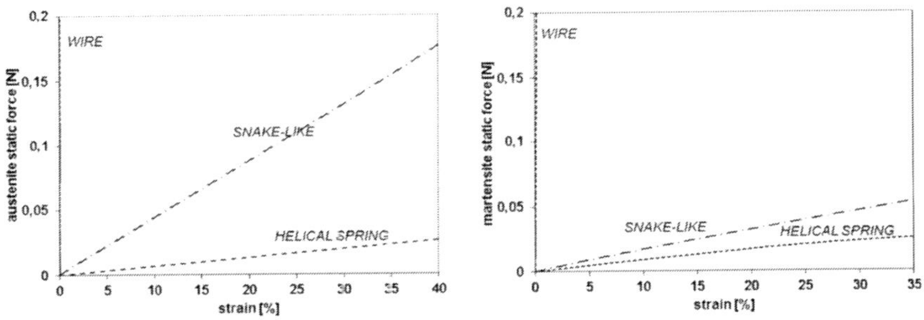

Figure 8. Comparison between austenite and martensite mechanical behavior of different NiTi wire geometries for actuators.

The snake-like conformationis characterized by four main parameters: the number of curvatures, N, the curvature radius, R, the distance between two consecutive curvatures, D, and the height, H.A snake-like specimencanbe prepared by constraining a SMA wire in the snake-like arrangement by using a drilled aluminum bar and a series of nails; heating at high temperature is finally used to fix the shape.Figure 9reports a snake-like sample prepared from a commercial NiTi wire having the diameter of 0.2mm,formed at 500°C during ten minutes and quenched in water at room temperature, with N=4, H=5.27mm±0.07mm, R=0.54mm±0.03mm and D=0.82mm±0.06mm.

As previously stated, during the ten minutes formation process the curved parts of the snake element are in contact with metal devices. This fact does not modify the calorimetric properties of the SMA element in all its length, as can be seen in Figure 10 where the overlapping of the DSC scans relative to the straight and the curved segments is clearly visible.

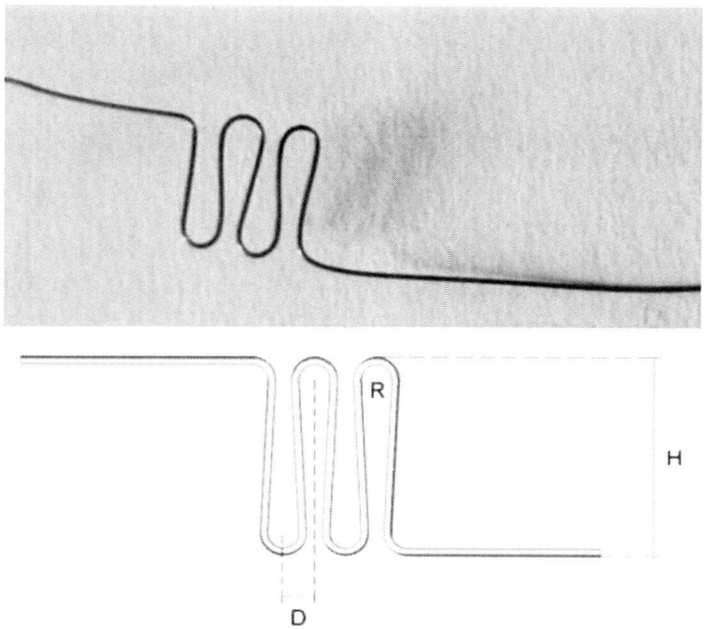

Figure 9. Snake-like NiTi wire with N=4, H=5.27mm±0.07mm, R=0.54mm±0.03mm and D=0.82mm±0.06mm.

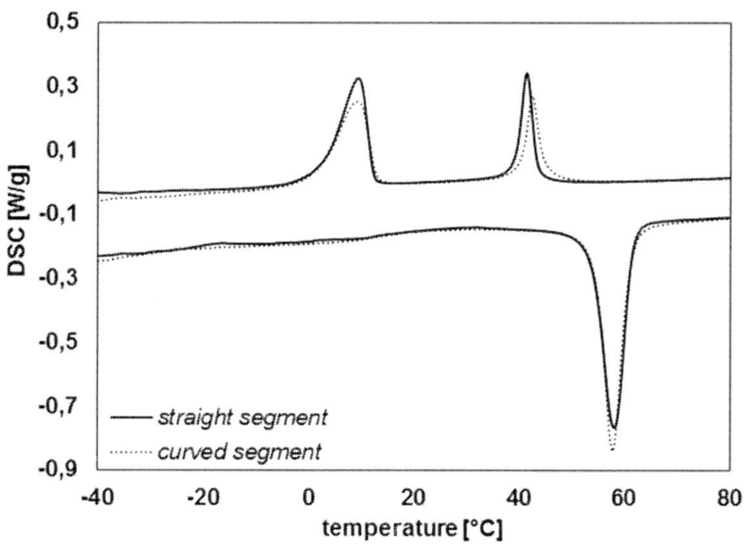

Figure 10. DSC scans of the straight and the curved part of the snake-like NiTi wire.

For small deformation values(see Figure 11), this snake-like shape shows an almost complete recovery of the deformation both for austenite and for martensite.

The relative higher output stroke performed by the snake-like NiTi wire as respect to the straight wire conformation is substantially due to the bend/unbend movement of the curvatures. In order to avoid plastic deformation and to achieve simultaneously a significant output stroke, the snake sample, formed as previously explained, should be stressed with a load near 0.1N, as forces lower and higher than 0.1N do not guarantee a proper functioning of this snake element: in fact, lowerforces do not induce martensite formation, as opposite higher forces cause plastic deformations. As an example, in Figure 12the curves derived from five thermal cycles of the snake-like NiTi sample under two constant loads, 0.1N and 0.2N is depicted. By comparing the two trends, it can be seen that in both cases the snake sample lose part of its shape recovery during cycling; this is much more visible for the specimen stressed with 0.2N.

Fatigue life was studied under the constant load of 0.1N; the snake-like sample was electrically activated with 0.6A with a maximum power consumption of 1W and a work frequency of 0.1Hz. Fracture takes place at the apex of that curvature most far from the applied load and from Figure 13it can be observed that it occurs near the $7{,}5 \times 10^4$cycle. The fractured section was analyzed through SEM observation, see the inset of Figure 13; it can be seen a region with the striation of the crack propagation departing from a nucleation site placed on the inner surface of the snake and a region characterized by dimples, typical of ductile materials.

After the cycling test, the DSC scan shows a feeble shifting of the transformation temperatures towards higher values; a better definition of the rhombohedral phase during heating is also visible (see Figure 14).

When embedded in mini-modular device, the snake-like wire does not lose its mechanical performance. As an example, in Figure 15 a miniature rotational actuator composed by two mutual antagonist NiTi snake-like wire is reported [34]. The device exerts a stroke of about 120° under a constant torque of 0.1Nmm, as shown in Figure 16; it is activated by 0.6A with a maximumpower consumption of 1W and a work frequency of 0.05Hz.Due to the particular construction, plastic deformation of the SMA element is prevented as the SMA elements never exceed 6mm of deformation.

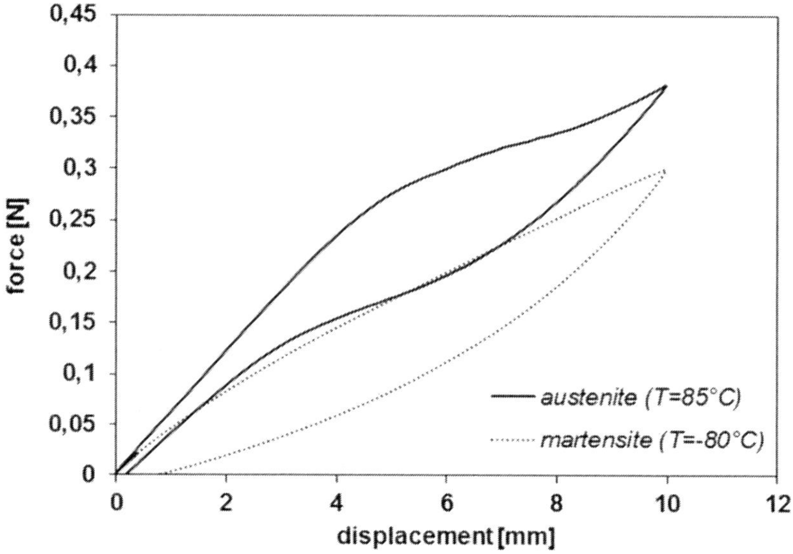

Figure 11. Force versus displacement of austenite and martensite of the snake-like NiTi wire.

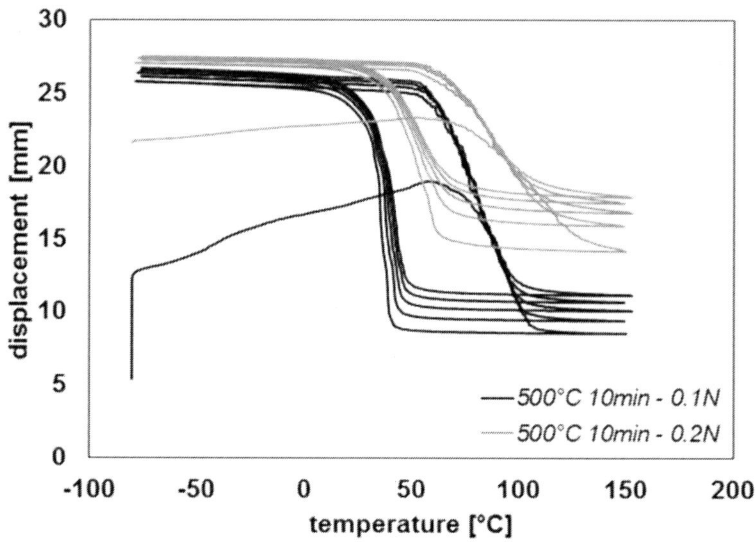

Figure 12. Hysteresis curves of the snake-like NiTi sample treated at 500°C for 10 minutes under 0.1N and 0.2N constant load.

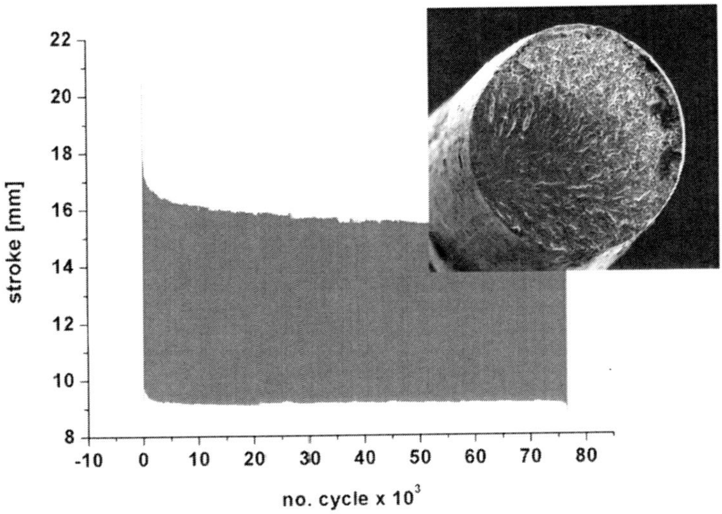

Figure 13. Snake-like NiTi sample fatigue test result under 0.1N load.

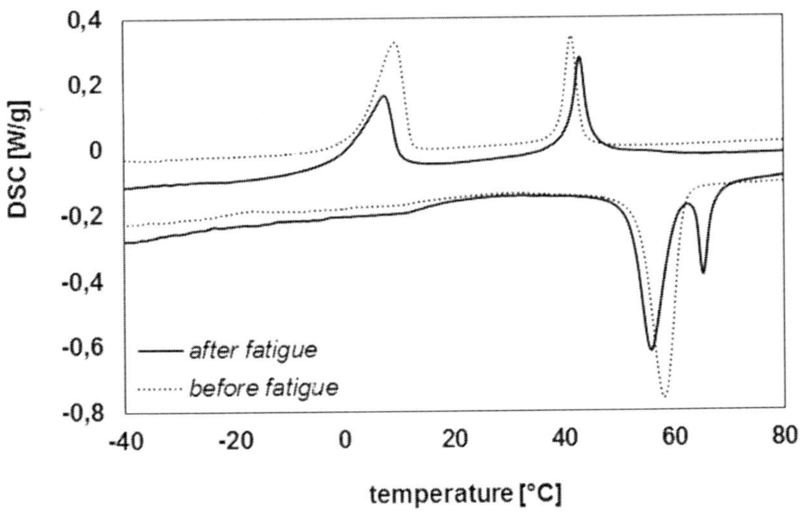

Figure 14. Snake-like NiTi sample DSC scan before and after the fatigue test

Figure 15. Mini rotational actuator activated by two antagonists snake-like NiTi wires.

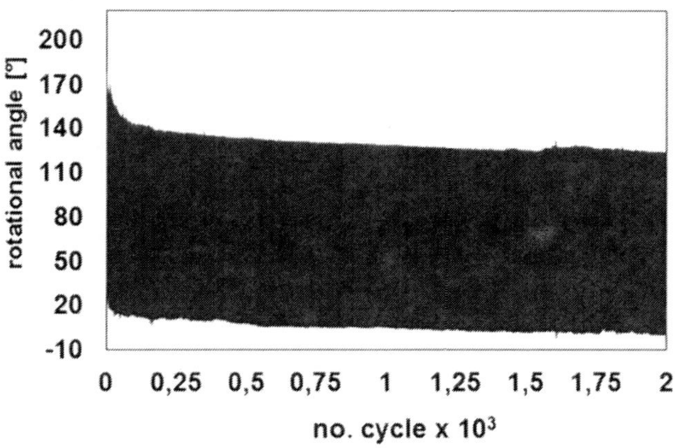

Figure 16. Fatigue test results of the mini rotational actuatorunder 0.1Nmm constant torque.

3.1. Micro Snake-Like Shape Memory Alloy Elements

In the last years the trend of miniaturization of components and products has been evidently introduced in several industrial fields, such as biomedical, electronics, aerospace and mechanics [35-36]. SMAs, as it already happens in biomedical applications [14,15,37-40], can be machined using laser machining with appreciated and high qualitative results [12]. The decision to investigate this method for the production of SMA actuators is considered an interesting topic because it shows some relevant advantages in comparison with other technologies, such as high productivity, high quality and repeatability,when applied in the industrial world.

Hence, the snake geometry can be easily scale down to the micro scale by the employment of laser technology with appreciated and high qualitative results. Because laser material processing is a thermal process, in which an heat source is used to remove a certain volume of material from the work piece, some thermal damages are evidently obtained, such as melted material, heat affected zone and oxides. To remove these defects, chemical etching and electro-chemical polishing have to be performed on the snake SMA samples [37,41]. Figure 17 depicts SEM images of the snake NiTi sample after laser, chemical etching and electro-chemical processes; top and bottom surfaces as well as snake sample section are shown. It can be seen that laser damages are mainly visible on the bottom surface and that the chemical etching remove the great part of defects. Moreover, a significant loss of mass and consequent reduction of the dimensions of the micro-snake can be evidently observed from SEM pictures due to the material removal of the post-processing.The snake NiTi sample geometry and calorimetric properties change after each process, as can be seen in Figure 18 and in Figure 19respectively.In particular, the geometrical dimension are referred to a micro-snake obtained starting from a 120μm NiTi sheet. As concern the DSC data, it can be observed that the laser machining produces a refinement of the rhombohedral phase peak both on cooling and heating while the martensite peak barely change as respect to the unworked 120mm thick sheet, being almost spread and invisible. The laser process also yields a conspicuous diminution of the rhombohedral phase transformation temperatures and of the martensite starting temperature while the other characteristic transformation temperatures do not have an analogous variation being almost unchanged.Chemical etching and electro-polishing steps do not visibly alter any phase transformation temperature both on cooling and on heating, even if some little change due probably to the hydrogen introduction in the chemical process, is visible [45,46].

The fabrication process previously described can be used to produce snake SMA sample of different length. As an example, Figure 20 depicts the SEM image of a micro-snake NiTi sample with 19 curvatures (i.e. 3.8mm of length) obtained starting from a 120μm NiTi sheet treated at 400°C for 15 minutes and quenched in water.

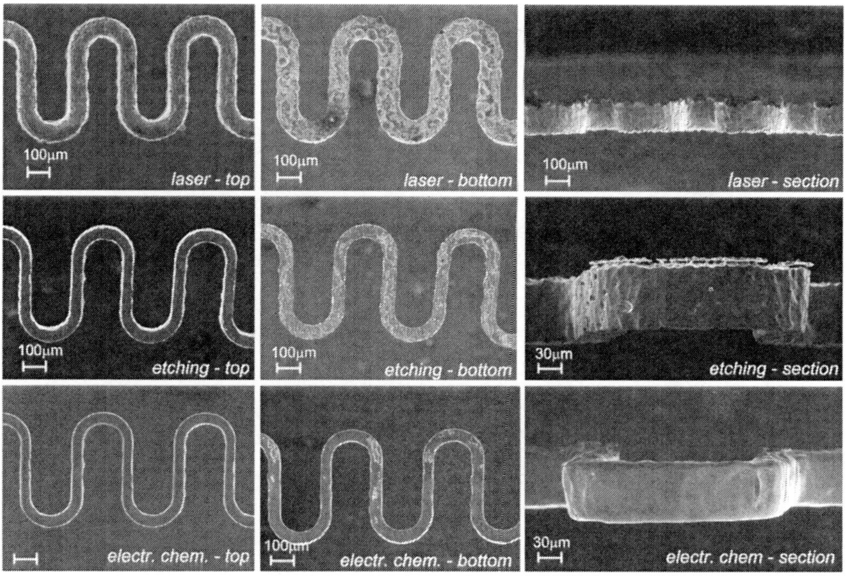

Figure 17. SEM images of the snake NiTi sample surfaces and section after laser, chemical etching and electro-polishing processes.

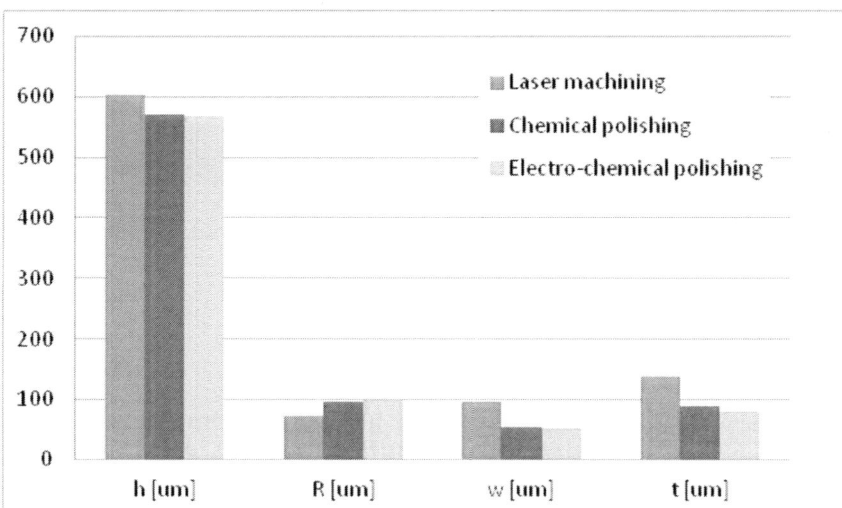

Figure 18. Variation of the geometrical parameters of the micro-snake during the fabrication process (h: high, R: radius of curvatures, w: width, t: thickness).

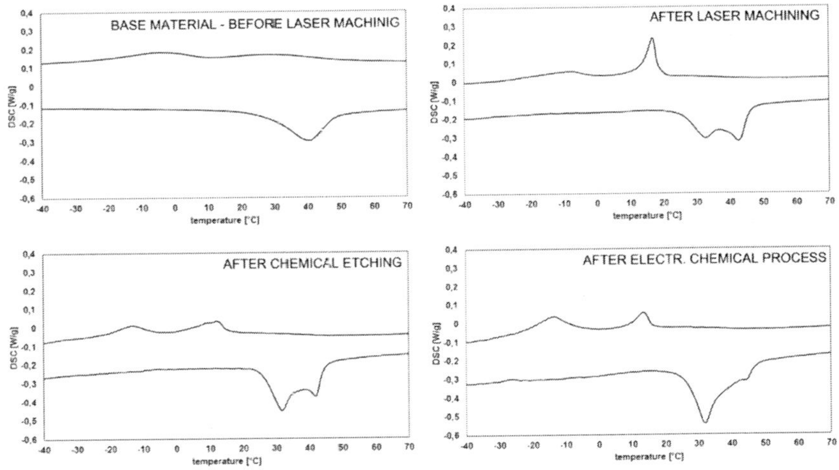

Figure 19. DSC scans before laser machining and in correspondence of each fabrication process.

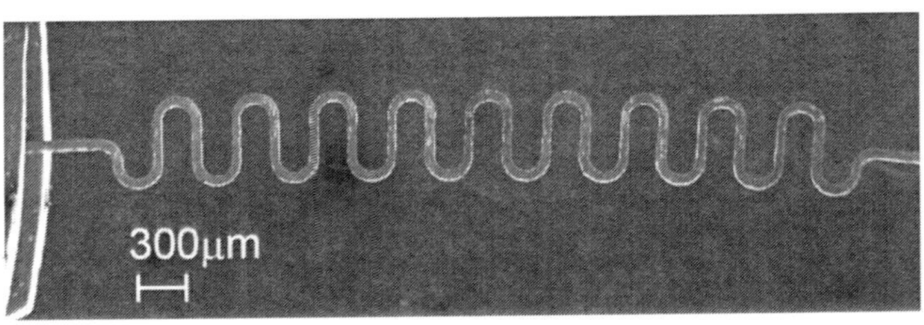

Figure 20. SEM image of a micro-snake NiTi sample obtained starting from a 120µm NiTi sheet.

Strain-recovery and thermo-mechanical cycling tests demonstrate the high performance of this micro-snake sample. As an example, under the constant load of 16mN, it reaches about 40% of recovery with a cycling stabilization within few hundreds of cycles, when heated by electrical current (130mA) with a maximum power consumption of 0.3W and a frequency of 0.15Hz.

4. CONCLUSIONS

In this work, new developments on SMA actuators in the mini/micro scale are presented: the attention was focused on straight NiTi thin wires and mini/micro snake-like NiTi samples.

Several working parameters have proved to strongly affect the functional and fatigue properties of SMA actuator, such as the electrical pulse performed to heat a wire, the different drawing procedures carried out to produce the wire or the wire arrangement into the device.

Then, the functional parameters of a snake-like NiTi wire demonstrate that this kind of geometry may be a good alternative to the most common geometries. It can generate a shape recovery in a very large range of deformation comparing to the straight SMA wire. This point is very beneficial for practical uses since the new proposed SMA conformation just occupies small space to generate the same displacement as the straight one, and it demonstrates to maintain its general mechanical performance when embedded in a mini-modular mechanical device.Finally, the presented fabrication process adopted for the manufacturing of micro-actuators looks to be very attractive (laser process can be performed on a large family of materials with high repeatability). The thermo-mechanical behavior of the micro-snake actuator can be considered interesting in the range of low applied load valueswith a good cycling stability.

ACKNOWLEDGEMENT

Authors would like to thank Prof. BarbaraPrevitali,head of Sitec-Laboratory for Industrial Laser Applications of Mechanical Department of Politecnico di Milano, for her precious support on the laser processing.

REFERENCES

1. Otsuka. Kazuhiro, M. Clarence, Wayman, 1998ShapeMemoryMaterials. Cambridge University Press. 284 p.

2. Hiroyasu Funakubo1984Shape memory alloysLondon: Gordon and Breach Science Publishers.

3. Miller DA, Lagoudas DC2001Influence of cold work and heat treatment on the shape memory effect and plastic strain development of NiTiMater. sci. eng. A. 30816175

4. T. Fukuda, T. Kakeshita, M. Kitayama, T. Saburi, 1995Effect of aging on martensitic transformation in a shape memory Ti-40.5Ni-10Cu alloyJ. phys. IV. 571722

5. S. Degeratu, P. Rotaru, Gh. Manolea, H. O. Manolea, A. Rotaru, 2009Thermal characteristics of Ni-Ti SMA (shape memory alloy) actuatorsJ. therm. anal. calorim. 97695700

6. E. Villa, S. Arnaboldi, A. Tuissi, M. Giacomelli, E. Turco, 2009Mechanical analysis of hybrid textile composites with NiTi wiresJ.mater.eng.perform. 18(5-6):517 EOF521 EOF

7. T. Mineta, T. Mitsui, Y. Watanabe, S. Kobayashi, Y. Haga, M. Esashi, 2001Batch fabricated flat meandering shape memory alloy actuator for active catheterSens.actuators A. 8811220

8. T. Mineta, T. Mitsui, Y. Watanabe, S. Kobayashi, Y. Haga, M. Esashi, 2002An active guide with shape memory alloy bending actuator fabricated by room temperature process. Sens.actuators A. 97-98:632 EOF

9. M. Kohl, B. Krevet, E. Just, 2002SMA microgripper systemSens. and actuators A. 97-98:646 EOF

10. M. Leester-Schadel, B. Hoxhold, C. Lesche, S. Demming, S. Buttgenbach, 2008Micro actuators on the basis of thin SMA foils.Microsyst. technol.14697704

11. F. John, Dave. F. Ready, Larson, 2001LIA Handbook of Laser Materials ProcessingLaser Institute of America, Magnolia Publishing, Inc.715

12. A. K. Dubey, V. Yadava, 2008Laser beam machining-A reviewInt. J.mac.tools manuf.48609628

13. Gower MC2000Industrial applications of laser micromachining.Opt. Express 75657

14. C. Li, S. Nikumb, F. Wong, 2006An optimal process of femtosecond laser cutting of NiTi shape memory alloy for fabrication of miniature devicesOpt. and lasers in eng.4410781087

15. C. Momma, U. Knop, S. Nolte, 1999Laser cutting of slotted tube coronary stents-state of the art and future developments.Prog. in biomed.res. 3944

16. S. Buttgenbach, S. Butefish, M. Leester-Schadel, A. Wogersien, 2001Shape Memory Microactuators.Microsys.technol.7165170

17. J. Meijer, K. Du, A. Gillner, D. Hoffmann, Masuzawa. T. Kovalenko, A. Ostendorf, R. Poprawe, W. Schulz, 2002Laser machining by short and ultrashort pulses, state of the art and new opportunities in the age of the photonsCIRP Ann.Manufact. Technol.51531550

18. Saikrishna CN, RamaiahKV, AllamPrabhuS, Bhaumik SK2009On stability of NiTi wire during thermo-mechanical cyclingBull. mater. sci.32343352

19. C. Zanotti, P. Giuliani, S. Arnaboldi, A. Tuissi, 2010Analysis of wire position and operating conditions of functioning of NiTi wires for shape memory actuatorsProc. SMST 2010.688696

20. J. Frenzel, E. P. George, A. Dlouhy, C. H. Somsen, M. F. X. Wagner, G. Eggeler, 2010Influence of Ni on martensitic phase transformations in NiTi shape memory alloys. Actamater.5834443458

21. Casati R, Tuissi A, Belochapkine S, DickinsonC, TofailSAM Thin NiTi wires with reduced thermal hysteresis for shape memory actuators. doi:10.1142/S17936047125000991250009

22. Miller DA, Lagoudas DC2001Influence of cold work and heat treatment on the shape memory effect and plastic strain development of NiTiMat. Sci. Eng. A.308161175

23. TodorokiT, TamuraH1987Effect of heat treatment after cold working on the phase transformation in TiNi alloy. Tran. of the jpn.Inst. of met.288394

24. C. P. Frick, A. M. Ortega, J. Tyber, A. E. M. Maksound, H. J. Maier, Y. Liu, K. Gall, 2005Thermal processing of polycrystalline NiTi shape memory alloys.Mat. sci. eng. A.4053449

25. ContardoL, GueninG1990Training and two way memory effect in Cu-Zn-Al alloy. Acta metall. mater.3812671272

26. http://www.cambridgemechatronics.com/

27. R. Casati, A. Tuissi, (201, 2011Effect of current pulses on fatigue of thin NiTi wires for shape memory actuators. Proc. SMST 2011. Submitted

28. M. F. X. Wagner, S. R. Dey, H. Gugel, J. Frenzel, Ch. Somsen, G. Eggeler, 2010Effect of low-temperature precipitation on the transformation characteristics of Ni-rich NiTi shape memory alloys during thermal cycling. Intermet.1811721179

29. T. Simon, A. Kroger, C. Somsen, A. Dlouhy, G. Eggeler, 2010On the multiplication of dislocations during martensitic transformation in NiTi shape memory alloys. Acta mater.5818501860

30. A. Ibarra, Juan. J. San, E. H. Bocanegra, M. L. Nò, (200, 2007Evolution of microstructure and thermomechanical properties during superelastic compression cycling in Cu-Al-Ni single crystal.Acta Mater.5547894798

31. A. Nespoli, S. Besseghini, S. Pittaccio, E. Villa, S. Viscuso, 2010The high potential of shape memory alloys in developing miniature mechanical devices: A review on shape memory alloy mini-actuatorsSens. and actuators A.158149160

32. S. Jansen, J. Breidert, E. G. Welp, 2004Positioning actuator based on shape memory wires. Actuators 2004. 9th International Conference on New Actuators.

33. A. Nespoli, E. Villa, S. Besseghini, 2011Thermo-mechanical properties of snake-like NiTi wiresand their use in miniature devices. J. therm.analcalorim. DOIs10973-011-1324-0

34. A. Nespoli, E. Bassani, S. Besseghini, E. Villa, 2010Rotational mini-actuator activated by two antagonist shape memory alloy wiresPhis. Procedia10182188

35. HsuT2002Miniaturization- A paradigm shift in advanced manufacturing and education. Proc. IEEE/ASME

36. K. Feldmann, J. Franke, F. Schüßler, 2010Development of micro assembly processes for further miniaturization in electronics production. CIRP Annals-Manufact.technol.5914

37. KathuriaYP2005Laser microprocessing of metallic stent for medical therapy. J. of Mater. Process.Technol.170545550

38. H. Meng, J. Liao, Y. Zhou, Q. Zhang, 2009Laser micro-processing of cardiovascular stent with fiber laser cutting system.Opt. And laser technol.41300302

39. B. Previtali, S. Arnaboldi, P. Bassani, Lecis. N. Biffi, A. Tuissi, M. Carnevale, Conte. A. Lo, 2010Microcutting of NiTiCu alloy with pulsed fiber laser.978-0-79183-877-8Order I844CD

40. C. A. Biffi, P. Bassani, A. Tuissi, M. Carnevale, N. Lecis, Conte. A. Lo, B. Previtali, 2011FlexuralVibrationSuppression of Glass Fiber/CuZnAl SMA Composite.Funct.Mater.Lett., DOIS1793604712500142

41. A. Raval, A. Choubey, C. Engineer, D. Kothwala, 2004Development and assessment of 316LVM cardiovascular stents. Mater.sci. and eng. A.386331343

42. S. S. Kudesia, P. Solana, W. Rodden, D. P. Hand, J. Jones, 2005Appropriate Regimes of Laser Drilling Models Containing Melt Eject MechanismsJournal of Laser Applications14159164

43. B. T. Rao, R. Kaul, P. Tiwari, A. K. Nath, 2005Inert gas cutting of titanium sheet with pulsed mode CO2 laser. Opt. and Lasers in Eng.4313301348

44. Lecis. N. Biffi, B. Previtali, M. Vedani, G. Vimercati, 2010Fiber laser microdrilling and its effect on material microstructure. The int.j.of adv. manuf. Technol.51983994

45. F. M. Mazzolai, A. Biscarini, B. Coluzzi, G. Mazzolai, E. Villa, A. Tuissi, 2007Low frequency internal friction of hydrogen-free and hydrogen-doped NiTi alloys.Actamater.554243425

46. A. Biscarini, B. Coluzzi, G. Mazzolai, F. M. Mazzolai, A. Tuissi, 2003Mechanical spectroscopy of H-free and H-doped NiTiCu shape memory. J. of alloy and compd.356669672

2

PIEZOELECTRIC MOTORS, AN OVERVIEW

Karl Spanner and Burhanettin Koc *

Physik Instrumente GmbH & Co. KG, Auf der Roemerstrasse, 1, Karlsruhe 76228, Germany

ABSTRACT

Piezoelectric motors are used in many industrial and commercial applications. Various piezoelectric motors are available in the market. All of the piezoelectric motors use the inverse piezoelectric effect, where microscopically small oscillatory motions are converted into continuous or stepping rotary or linear motions. Methods of obtaining long moving distance have various drive and functional principles that make these motors categorized into three groups: resonance-drive (piezoelectric ultrasonic motors), inertia-drive, and piezo-walk-drive. In this review, a comprehensive summary of piezoelectric motors, with their classification from initial idea to recent progress, is presented. This review also includes some of the industrial and commercial applications of piezoelectric motors that are presently available in the market as actuators.

KEYWORDS

piezoelectric; inverse piezoelectric effect; friction coupling; motor

1. INTRODUCTION

Generation of electrical charge on certain materials in response to applied mechanical stress is known as the piezoelectric effect. The same materials have the ability to convert electrical energy into mechanical motion directly and that is known as the inverse piezoelectric effect. The range of mechanical movement generated on a piezoelectric element is very small, from 1 μm to 100 μm. This small movement is the key for obtaining extremely high precision positioning in the order of 1.0 nm. Piezoelectric motors are the devices that generate unlimited rotary or linear movements by harvesting this very small motion.

By the different drive and functional principles for generating unlimited rotary or linear movement, piezoelectric motors can be classified into three categories: (1) resonance drive (Ultrasonic motor), where vibration at ultrasonic frequency range on an oscillating element is transferred to a moving element through frictional coupling; (2) inertia-drives, where a movement is generated by speed dependent friction coefficient; and (3) piezo-walk-drives, where various piezo actuators are used for alternatively clamping and shifting a moving element. The actuators in these motors operate in a quasi-static mode.

2. EARLY STRUCTURES OF PIEZOELECTRIC MOTORS

Since the discovery of piezoelectricity, there have been several attempts to obtain longer mechanical motion using the inverse piezoelectric effect [1,2,3,4]. In a US patent titled "Converting electrical oscillations into mechanical movement" filed by Meissner in 1927 [1]. The structure of the motor described in this patent consists of a piezoelectric plate with a center-attached shaft and asymmetrically-attached two lever arms. When a driving signal is applied through two conductive plates to top and bottom surfaces of the piezoelectric plate, torsional vibration modes are excited due to asymmetrically-attached lever arms. It seems that the conductive plates are not only used for applying a driving oscillatory signal, but also to create a frictional coupling. Generated vibration on the piezoelectric plate at a suitable torsional mode causes the piezoelectric plate and the center shaft to rotate (Figure 1).

The piezoelectric motor developed by Williams and Brown in 1942 [2] proposed a structure to convert gyratory (hula-hoop) motion generated by the arrangement of multiple piezoelectric elements into a rotary motion through a gear structure. In the same patent, a structure to convert gyratory motion to a rotary motion by frictional coupling was also proposed. The necessary pre-stressing force to increase frictional coupling was applied through magnetic attraction force (Figure 2).

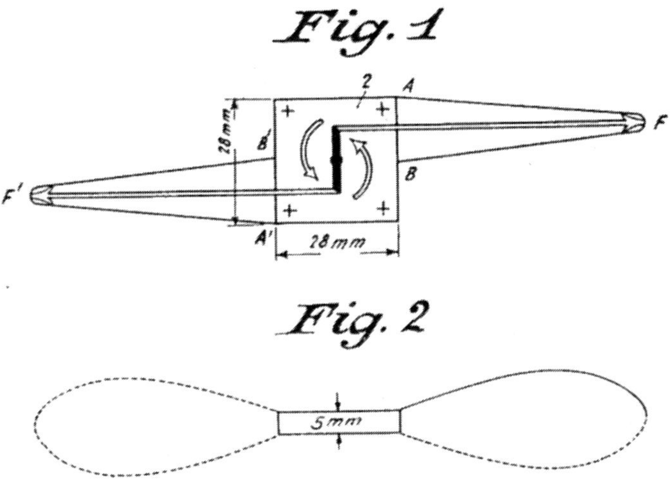

Figure 1. Piezoelectric oscillating element to obtain mechanical movement as described in the US Patent invented by Meissner [1].

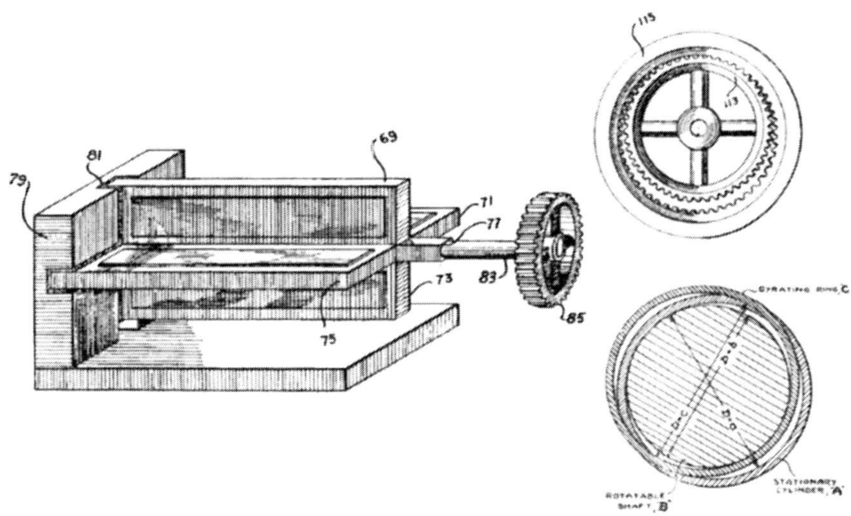

Figure 2. Piezoelectric motor proposed by Williams *et al.* in 1942 [2].

A systematic study of piezoelectric ultrasonic motors in the former Soviet Union can be seen after the introduction of a piezoelectric motor by Lavrinenko in 1965 [3]. This motor consists of a piezoelectric plate pressed against a rotor and operates unidirectionally (Figure 3). At about the same time, another unidirectional piezoelectric motor structure using a cylindrical piezoelectric element and a spinning rod with attached coupler was proposed by Tehon in the USA [4]. Following these initial ideas, several other motor structures have been proposed by Wischnewskiy [5]. Most of these early structures are unidirectional, which means only one mode on a vibrator is excited. One, or multiple, elastic vibration couplers oriented obliquely and attached either on a spinning element or on a vibrating element transfer the vibration into a rotating element. In the early structures, not only rectangular piezoelectric plates, of which longitudinal modes were excited, but also circular piezoelectric plates were used to convert radial vibratory motion into tangential motion via angularly-attached lamina (Figure 4) [5].

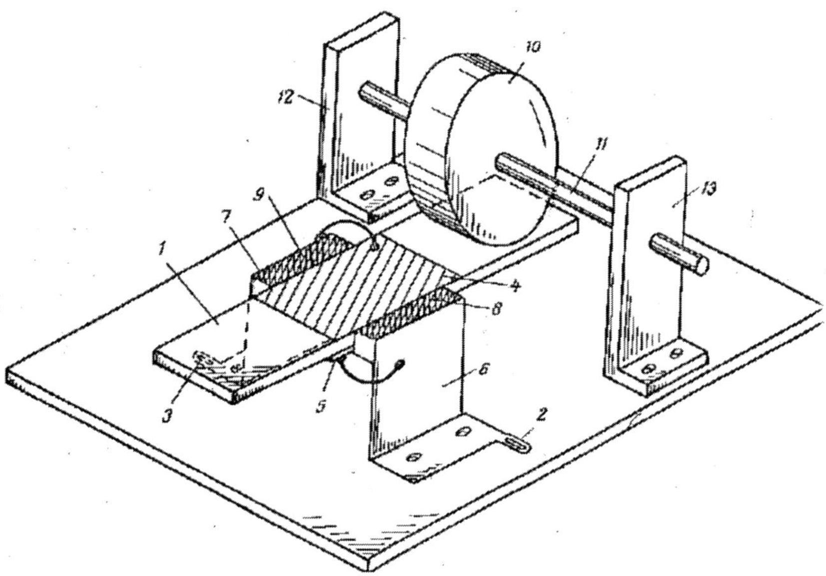

Figure 3. Piezoelectric motor introduced by V. Lavrinenko in 1965 [3].

Further investigations of piezoelectric motors led to a recognition of fundamental design principles that following structures could be operated bidirectional [6,7,8]. As a result of these studies by various people, a book on piezoelectric motor was written in 1981 by Bansevicius and Ragulsky

[9]. After the translation of this book into English language in 1988 [10], piezoelectric motors studied in former Soviet Union were recognized more.

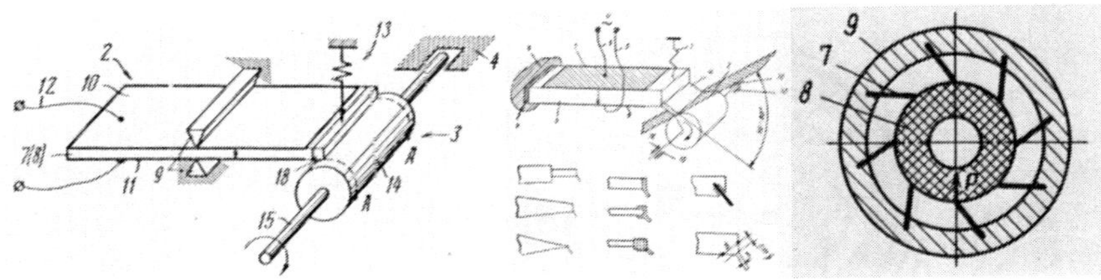

Figure 4. Early structure of various unidirectional piezoelectric motors proposed by Wischnewskiy [5].

In the meantime, to fulfill the needs for precise positioning, other piezoelectric devices were proposed using step motion of multiple piezoelectric elements. A structure called an incremental feeding mechanism proposed by Stibitz in 1962 was the first structure that could be considered to function according to piezo-walk and inertia drive principles [11]. Even though the design is using three magnetostrictive actuators that are driven simultaneously; in the same structure, there is a description for the possibility of using piezoelectric actuators to make a small incremental motion. The magnitude of the driving signals for each actuator has a typical characteristic of a slow increase and a fast decrease with a certain phase difference between them (Figure 5).

3. RESONANCE-DRIVE (PIEZOELECTRIC ULTRASONIC MOTORS)

Among the piezoelectric motors, resonance-drive (or ultrasonic) motors have been the most studied actuators [9,10,12,13,14,15]. Any stator element in a piezoelectric ultrasonic motor is a composite structure, in which piezoelectric and elastic elements are combined in various forms. These elastic elements can have different functions. They can simply act as a friction tip or vibration coupler to convert generated displacement from one direction to another.

When we were classifying resonance-drive type motors we tried to find answers to the following statements:

- Number of vibrating elements that generate needed microscopic motion,
- Number of mechanical modes that are excited at the operating frequency of the vibrating element,
- Number of driving signals applied to a vibrating element,
- Full or partial drive of a piezoelectric element in a vibrator, and
- Unidirectional or bidirectional motion, if bidirectional; the method of changing direction such as frequency, phase, or excited part.

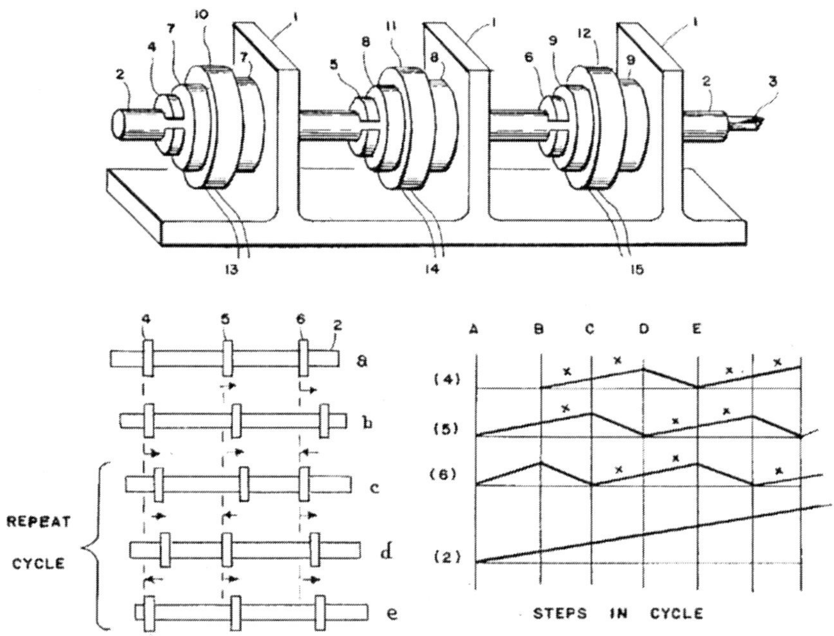

Figure 5. Precision positioning device proposed by Stibitz in 1962, which could be considered as the first structure functioning according to piezo-walk and inertia-drive principles [11].

We tabulated the answers of the above statements in Table 1.

Since all of the ultrasonic motors operate by exciting one or more than one mechanical resonance modes of their stator elements, these motors can be categorized according to the number of resonance modes excited on a stator element. Actually, any resonance type piezoelectric motor can be grouped into two. The generated motion is either oblique or elliptical, and these motions can be generated in various ways. On a stator element, if only one mechanical resonance mode at the operating frequency is excited, then this motor can be called a single-mode excitation type. If more than one mechanical resonance mode at the motor operating frequency is excited, then the motor can be called a multi-mode excitation type. In order to generate an elliptical motion on the vibrating element, there needs to be motion in two orthogonal directions with a certain phase difference.

Table 1. Classification of resonance-drive type piezoelectric (ultrasonic) motors.

Single-Mode Excitation Type		Multi-Mode Excitation Type		
Oblique motion		**Elliptical motion:** By exciting two orthogonal mechanical resonance modes at the same time.		
One Driving Source		One Driving Source		Multiple Driving Sources
Piezo-element is fully driven	Piezo-element is partially driven	Piezo-element is fully driven	Piezo-element is partially driven	Piezo-element is fully driven
Unidirectional	Bidirectional: Driven part controls the direction	Bidirectional: Frequency change controls the direction	Bidirectional: Driven part controls the direction	Bidirectional: Phase change between the driving sources controls the direction (traveling wave-motor)

3.1. Single-Mode Excitation Type

3.1.1. Piezoelectric Element Is Fully Driven

In these types of motors, the motion generated on a vibrating element is in the oblique direction. The vibrating element is excited by one driving source and the piezoelectric element in the motor is fully driven. Exciting only one resonance mode is enough to generate a motion. One can orient either vibrating element or moving element with a tilt angle so that this direction of microscopic motion at the contact point to have normal and tangential components at the interface. The other method to obtain oblique motion at the interface is the use of an attached vibration coupler or lamina [5,8,10,16,17]. Even though the motor is unidirectional, one can use two of these vibrating elements to obtain bidirectional motion [18,19].

3.1.2. Piezoelectric Element Is Partially Driven

When a motor is single-mode excitation type and bidirectional, the moving direction can be changed by partial driving of a piezoelectric element. A well–known piezoelectric ultrasonic motor fitting to this category was developed at Physik Instrumente (PI) [20, 21, 22]. The motor is manufactured under a trade name of "PIline®". The vibrating element in this motor is a rectangular type piezoelectric plate with a certain length to width ratio. Electrodes on one surface of the plate are segmented into two sections in width direction and a friction tip is attached at the middle of one side surface (Figure 6). When one half of the plate is driven with an electrical signal, a planar (or radial) mode resonance frequency of the driven section is excited. The friction tip and unexcited other half of the plate function like a perturbation mass. As a result, an oblique motion at the tip of the friction tip causes the sliding element to make a linear or rotary motion. A recent structure, where a linked twin square plate vibrator proposed by Yokoyama, functions according to the same principle [23].

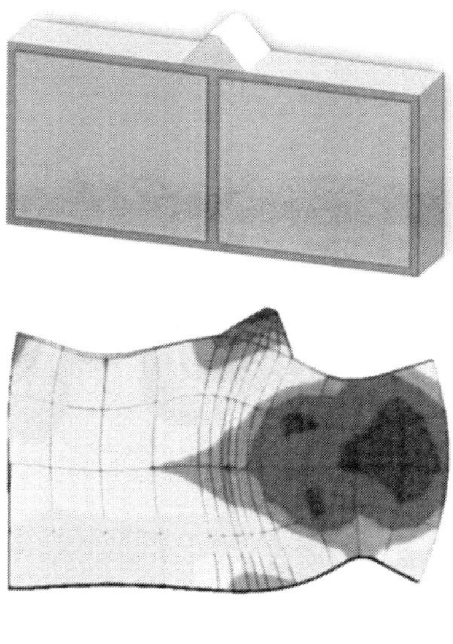

Figure 6. Single-mode excitation type ultrasonic motor from Physik Instrumente (PI). The excited single mode of the stator makes the pushing point to have an oblique impact to a sliding element [22].

The single-mode excitation type resonant piezoelectric motor (PIline®) with its simple structure and fast dynamic response has been used in various applications. One such application is in a theodolite. A theodolite

[24] is a precision instrument on which a movable telescope is mounted for measuring angles in the horizontal and vertical planes. Two pairs of PIline® motors are integrated into a theodolite structure, which can provide rotary motions in two perpendicular axes. The PIline® motors have been also used in various medical devices such as high-resolution optical coherence tomography (OCT) scanners and controllers for laser beam scanners used for eye surgery [22].

The same principle of partial driving was applied to a cylindrical piezoelectric element and again by exciting the segmented electrodes on the piezoelectric cylindrical element partially, an oblique motion on friction tips was obtained (Figure 7). Changing the driven part changes the direction of the oblique motion and, thus, the motor direction.

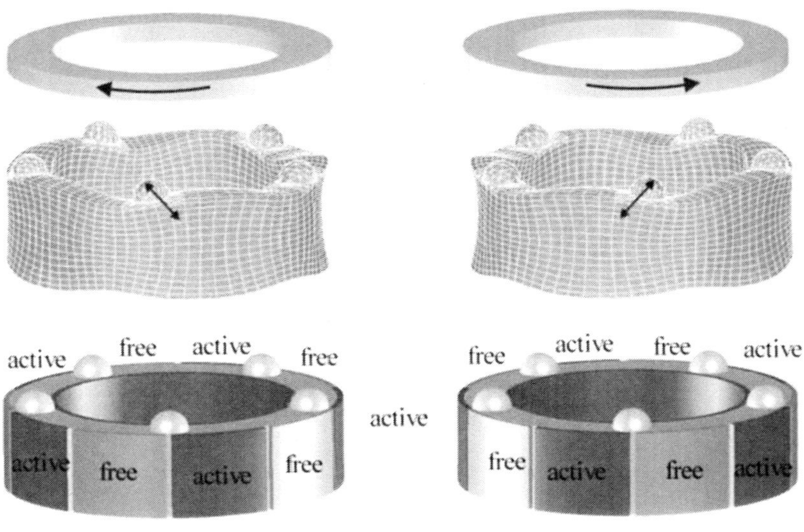

Figure 7. Principle of single mode excitation type resonance (ultrasonic) motor using a cylindrical piezoelectric element developed at PI [22].

Another well-known rotary version of a single-mode excitation type ultrasonic motor was developed by Takano *et al.* [25] using a piezoelectric ring with 12 segmented electrodes. Driving only six of the 12 electrodes on a circular piezoelectric element at one time excites a single flexural mode on the stator element. This flexure deformation of the stator element causes three of the corresponding friction tips (projections) to make an oblique motion. The oblique motion on the friction tips causes the rotor to spin.

A motor proposed by He [26] is also fitting to this category, where a piezoelectric tube with four side electrodes with end teeth is used as the vibrator of this motor. An electrical signal is applied in between one pair of surface electrodes (from half of the surface) to the other pair of surface electrodes. This driving excites the single bending mode. Specially-located teeth apply an oblique impact to the surface of spinning elements at both ends, thus causing the rotor to spin.

3.2. Multi-Mode Excitation-Type Motors

Unlike single-mode excitation type ultrasonic motors, where the motion at contact point is in oblique direction, the motion at the contact point of multi-mode excitation type motors is elliptical. The elliptical motion is generated by exciting two orthogonal mechanical resonance modes. These orthogonal modes can be excited by one or multiple driving sources. A piezoelectric element (or elements) in a vibrator can be fully or partially driven. One can group multi-mode excitation-type motors according to the number of driving sources and control methods used for direction change.

- **One Driving Source**
    - Piezoelectric element is fully driven
        - Frequency change controls the direction
    - Piezoelectric element is partially driven
        - Driven part controls the direction
- **Multiple Driving Sources**
    - Piezoelectric element is fully driven
        - Phase change between the driving sources controls the direction

3.2.1. One Driving Source (Frequency Change Controls the Direction)

In a multi-mode excitation-type ultrasonic motor, if all of the piezoelectric volume in a stator element is electrically driven with one AC source, and if more than one mechanical resonance modes are excited, the direction of the sliding or rotating element can only be changed when the frequency of the input electrical signal is tuned to a different orthogonal mechanical resonance mode. A good example fitting to this category is the motor developed by Elliptec [27]. A recent motor structure that uses the same operating principle was proposed by Xing *et al.* [28]. This motor is using a metal tube, on which a piezoelectric plate is attached on to the flattened side surface. The main advantage of these motors is the use of only two wires for driving. However, the direction dependent speed-driving voltage dependency could increase the complexity of the control electronics. The initial idea fitting to this type of motor structure was proposed by Uchino in 1988 [29].

3.2.2. One Driving Source (Driven Part Controls the Direction)

In these multi-mode excitation-type motors, piezoelectric element in a vibrator is partially driven with one driving source. Early structures of these motors were proposed by Wischnewskiy [6] and Bansiavichus [7]. Electrodes on one surface of a rectangular-shaped piezoelectric plate are divided into four equal small rectangles as shown in Figure 8. Orthogonal longitudinal and bending modes are excited at the same time by exciting only two electrodes that are located diagonally. The direction of the motor can be controlled by driving the other diagonally-located electrodes.

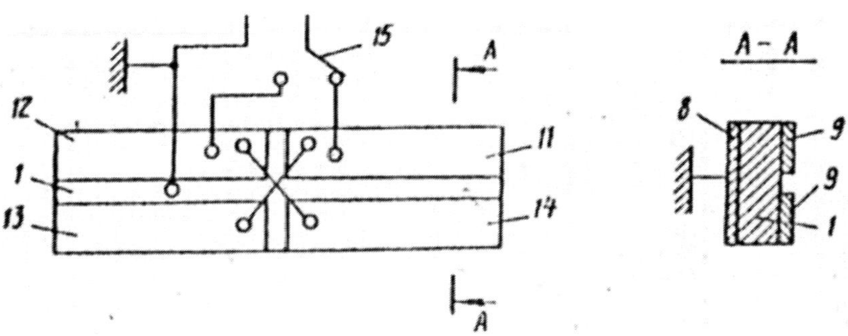

Figure 8. Bidirectional piezoelectric motor designed by Bansiavichus [7].

The ultrasonic motor from Nanomotion Ltd. also uses the same first longitudinal and the second flexural modes in width direction. The pusher in this motor is located at the end of the piezoelectric plate [30].

Piezoelectric motors can be made using non-magnetic materials and piezoelectric elements do not generate magnetic fields. Thus, these motors can be made non-magnetic and magnetic-insensitive, which is an advantage in medical applications, such as for magnetic resonance tomography (MRT). Piezoelectric motors manufactured by Nanomotion are used in surgical robots [31].

Another multi-mode excitation type motor using piezoelectric plate-type stator element was developed by Samsung Electro-Mechanics and this motor has been used for auto-focus applications in phone cameras [32,33]. Even though the basic operating principle is the same as the other multi-mode excitation-type motors, dilatational and thickness bending modes of a piezoelectric plate were combined and the miniature stator element was directly manufactured in multilayer form. On such a small actuator, which has a dimensions of 2.5 × 2 × 0.6 mm, two cylindrical friction elements are attached into two indented surfaces (Figure 9). This motor is probably a good example to show that ultrasonic motors can be manufactured in large quantities with a similar cost to their electromagnetic counterparts. Indeed, the manufacturing process of this motor is very similar to the multi-layer ceramic capacitor (MLCC) manufacturing method.

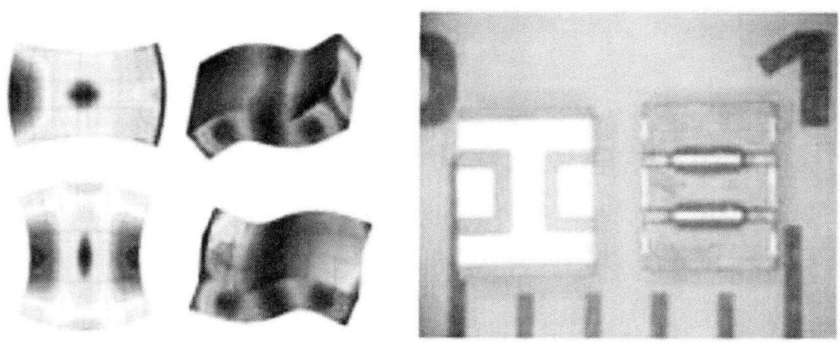

Figure 9. Ultrasonic motor developed by Samsung Electro-Mechanic. Orthogonal dilatational and bending mode shapes (**left**), and a stator element manufactured in multilayer form (**right**) [33].

An ultrasonic motor developed at the Penn-State University is another example for the multi-mode excitation type rotary motors [34]. The two orthogonal modes in this motor are the two bending modes of a metal cylinder. The stator of the motor consists of a metal cylinder on which two piezoelectric plates are attached on flattened surfaces. When one piezoelectric plate is electrically driven, a wobbling motion on the cylinder is generated because two bending modes of the cylinder are excited at the same time. This motion causes the rotor to spin. When the other piezoelectric plate is electrically driven, the direction of wobbling motion is changed (Figure 10).

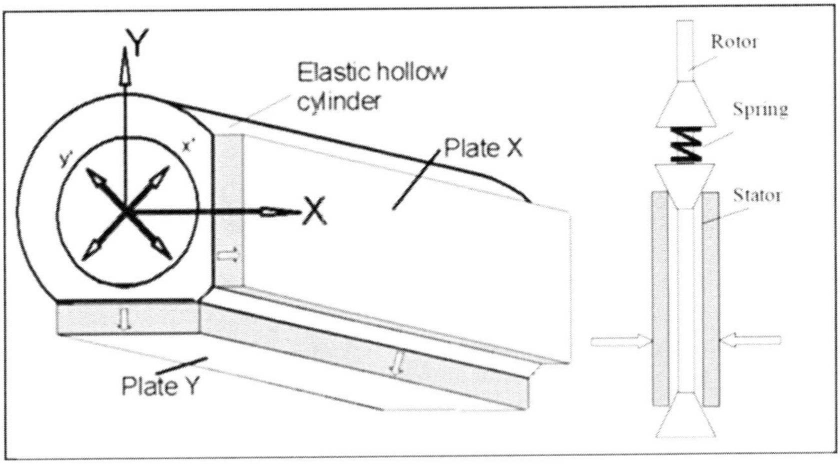

Figure 10. Metal tube piezoelectric ultrasonic motor [34].

3.2.3. Multiple Driving Sources (Phase Change between the Driving Sources Controls the Direction)

To generate an elliptical motion on a vibrating stator element, it would be useless to drive this element with two sources with a phase difference between them. Unless the vibrating stator element has two orthogonal modes and the resonance frequencies of these modes are equal or close enough. In case of these two orthogonal modes resonance frequencies are not equal; only one mode can be excited efficiently. Ultrasonic motors that are known in literature as travelling wave types could be considered as a multi-mode excitation type. A flexural vibration mode on a circular piezoelectric ring or disk has symmetry in every direction. Segmented electrodes on a piezoelectric element divide the surface into two groups. These two electrode groups are excited with two signals and each of these signals excite two identical flexural modes. Since there is a 90 degrees phase difference between the driving signals, each point on the stator surface makes an elliptical motion. The resulting motion is nothing but a travelling wave on this circular stator element. Ultrasonic motors using circular stator elements and operate the abovementioned principle found the first commercial applications in 80 s in Japan (Figure 11) [35,36]. A famous application of traveling wave type ultrasonic motor is for automatic focus systems in cameras used by Canon [37,38,39].

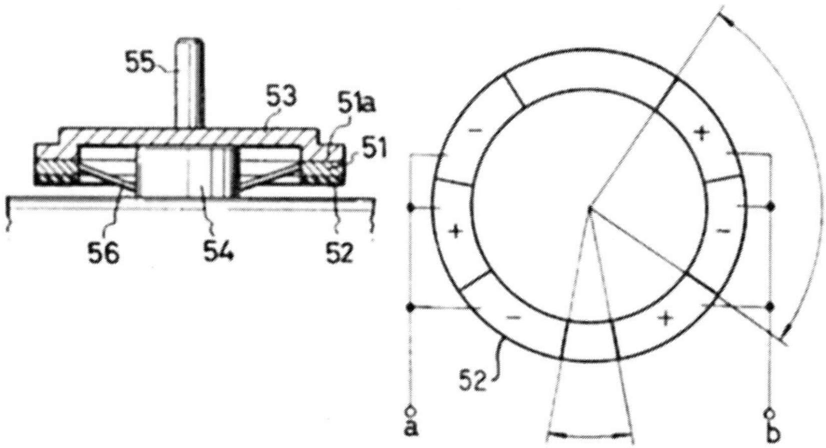

Figure 11. First commercialized ultrasonic motor proposed by Sashida [35].

The motor proposed by Wischnewskiy is an example of a multi-mode excitation-type motor, where a piezoelectric plate with segmented electrodes and corresponding orthogonal resonance modes are driven with two phase signals [40]. On the piezoelectric plate the larger electrode area in the middle is responsible for generating longitudinal mode, and four small electrode areas are responsible for exciting bending modes. Each input signal excites the corresponding orthogonal mode and phase difference of the two signals control the direction of the elliptical motion at the friction tip and, thus, the direction of the slider (Figure 12).

When multiple piezoelectric elements are combined in a stator element, motion generated by each vibrator should be orthogonal [41]. If the motions generated by these vibrators are identical then the vibrators need to be oriented orthogonally [42] (Figure 13). Integrating two vibrating elements by orienting them 90 degrees to each other was used by various researchers working on piezoelectric motors [43,44,45].

Externally hammering an elastic ring element by obliquely-oriented multiple longitudinal actuators to generate a travelling wave on an elastic ring could also be considered as a multi-mode excitation-type motor [46].

Since hammering points cause two identical Eigen modes to be excited at the same time, the phase difference causes contact points to make elliptical motions.

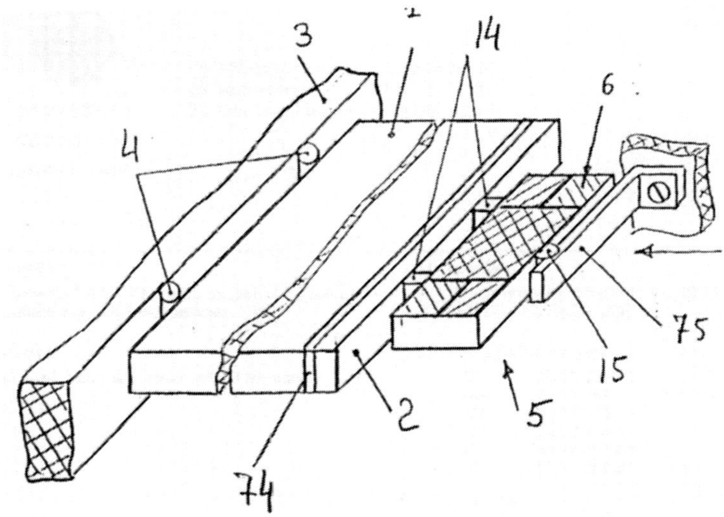

Figure 12. Two-phase motor from Physik Instrumente (PI) [40].

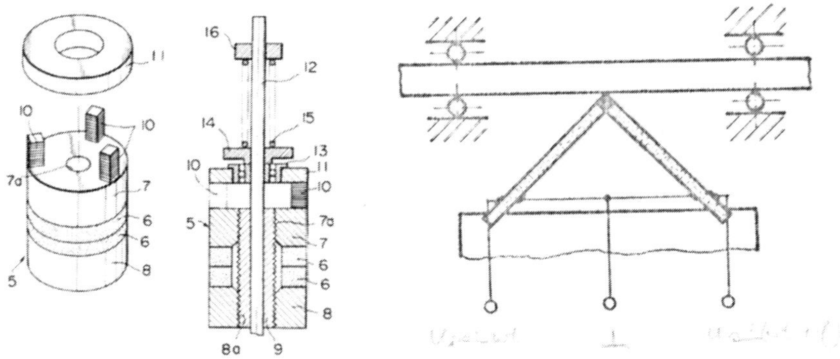

Figure 13. Ultrasonic motors in which multiple piezoelectric elements are used in a stator element. Torsional-Longitudinal (**left**) [41] and Longitudinal-Longitudinal (**right**) [42].

The ultrasonic motors using symmetric cylindrical, circular, or flat square structures, and using two identical orthogonal modes, should be considered as multi-mode excitation-types. Motors using piezoelectric cylinders with four segmented electrodes on outside surface and uniform metallization to its inner surface are excited with four phase signals (sine, cosine, -sine, -cosine). In this structure, two bending modes that are orthogonal are excited and the phase difference causes the hula hoop motion [47]. The same principle was applied to a squiggle motor manufactured by New Scale [48], where a metal tube with attached four piezoelectric plates and treaded at the inner side made a screw-type spinning element to cause a linear motion.

A cylindrical structure with embedded piezoelectric ring elements at a suitable position is commonly used as a stator for resonance-type piezoelectric motors. All of these motors are using two orthogonal and identical first bending modes including the structure proposed by Kurosawa in 1989 [49]. Commercially manufactured second-generation ultrasonic motor manufactured by Canon for lens positioning applications also has the same structure [39], where the piezoelectric ring element and the mechanical part in the motor is manufactured with a highly-specialized manufacturing technique [50].

A hula-hoop motion can be obtained on a piezoelectric square or circular plate by dividing one surface electrode into four equal segments and driving with four phase signals [37]. The same hula-hoop motion can be generated by using symmetric circular or square plates when both the top and bottom surfaces have two identical conductive electrodes that are dividing the piezoelectric element into two equal regions. If a signal is applied in between two divided surface electrodes, in-plane modes can be generated. Since electrode pairs are rotated by 90 degrees on the other surface, the same in-plane modes can be excited in orthogonal direction. When one source drives the electrodes on the top surface, and the other drives the electrodes on the bottom surface, these identical modes in the orthogonal direction are excited at the same time. When the two signals have 90 degrees out of phase, the resulting motion is nothing but a hula-hoop motion [51]. Any identical in-plane motion can be controlled independently, which means the shape of the microscopic elliptical motion at the contact point is fully controllable by magnitude and phase differences of the driving signals (Figure 14).

4. INERTIA-DRIVE-TYPE PIEZOELECTRIC MOTORS

The first practical inertia-drive (stick-slip) structure was proposed by Pohl in 1986 [52], where a piezoelectric cylinder is embedded into a four-bar mechanism. One side of the four-bar was attached to a base and a sliding mass is attached at the other parallel bar. When the actuator is driven with a saw-tooth waveform signal, the attached mass moves together with the bar during the slow expanding period of the saw-tooth signal (sticking) and left behind during the fast contraction period of the signal (slipping). Repeating the cyclic motions make the attached mass move continuously. When the electric field makes the piezoelectric element expand quickly and contract slowly, the attached mass moves in the opposite direction. This mechanism was applied for a precise multi-degrees of motion positioning device applications for an atomic force microscope.

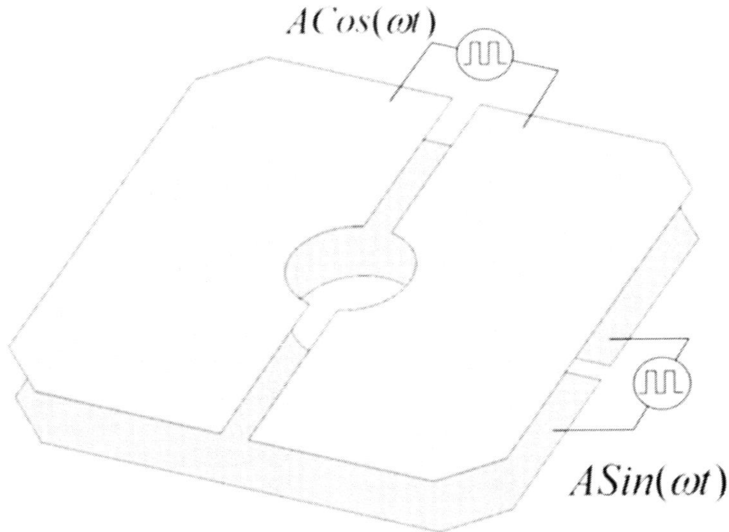

Figure 14. Generating identical orthogonal modes by driving surface electrodes [51].

About the same time, another impact drive mechanism was proposed by Higuchi [53]. A moving mass is placed on a base plate and held by frictional force acting between the base plate and the moving mass. A piezoelectric element is attached on one side of the moving mass and a weight is attached at the other side of the piezoelectric element. When an electric field is applied on the piezoelectric element, a rapid expansion of the piezoelectric element creates an acceleration causing the moving mass to win the static friction (slip) so both masses move in opposite directions. During the slow contraction time of the piezoelectric element, the acceleration on the moving mass cannot win against the static friction so only the weight moves (stick). At the end of one cyclic motion, a microscopic movement on the moving mass is obtained (Figure 15).

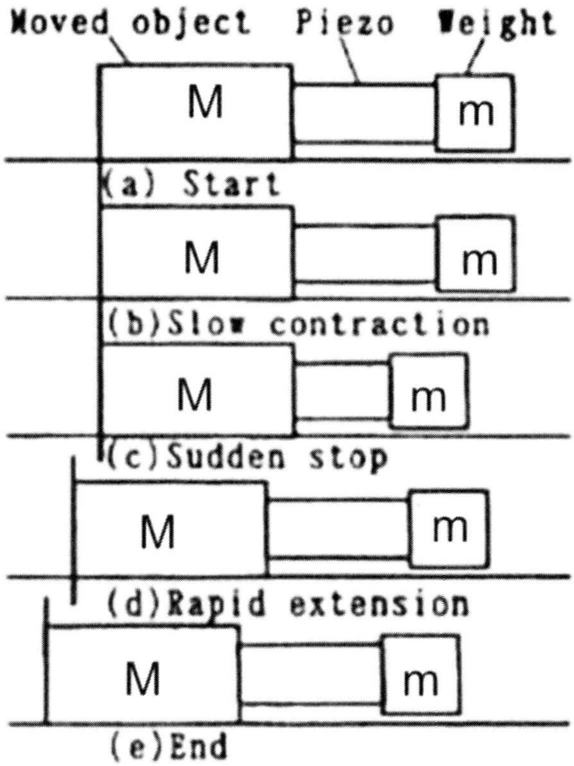

Figure 15. Operating principal of impact-drive proposed by Higuchi in 1986 [53].

In the later version of the impact-drive mechanism, a piezoelectric multilayer actuator is attached in between a body mass and a rod, on which a sliding element (inertia mass) is attached with a spring force creating a frictional coupling between the rod and the inertia mass. Even if the actuator is generating a displacement in the longitudinal direction, this motion on the rod is actually in a tangential direction to the attached moving inertia mass. This motor had been commercialized by Konica Minolta to use for camera applications, such as auto-focus, zoom, and image stabilization functions [54,55].

Almost all of the inertia-drive type motors are using the above described principle with various ways of generating tangential displacement at the interface in between a stationary and a moving element. A displacement in the tangential direction can be generated in two ways. One is direct generation by using longitudinal [53,54,55], transverse [56] or shear [57] deformations of the piezoelectric material. The other way is by converting generated displacement from piezoelectric element into tangential direction at the interface by using flexure or leverage mechanisms [58,59,60,61]. Presently, there are companies producing high-precision stages using inertia drive-type motors [58,59,60,61,62,63,64].

There are applications that use inertia drive motors for microscopy and biotechnology research such as cell manipulation. Piezoelectric inertia drive motors have been used for both optical and scanning probe microscopes successfully in vacuum environments [65,66].

Recently-proposed inertia-drive motor structures can operate at resonance frequency [67,68,69,70]. Saw-tooth-like motion to generate stick and slip actions is created by exciting two resonance modes at two different frequencies. The second-mode resonance frequency of the vibrating element in these motors is two times higher than the first-mode resonance frequency. These two resonance modes are excited by either a square wave with a duty ratio smaller than 50% or adding two sine signals. The resulting tangential motion is fast acceleration to slip by overcoming friction force and slow deceleration to stick or *vice versa*.

5. PIEZO-WALK-DRIVE-TYPE PIEZOELECTRIC MOTORS

For positioning in a long range of motion, a structure proposed by Brisbane in 1965 operates based on the piezo walk-drive principle [71]. In this structure, two piezoelectric disc shaped plates are attached to a hollow cylinder at both ends. While the piezoelectric disks function as clamping actuators, the hollow cylinder acts as a moving actuator. The whole structure is placed in a retaining cylinder as shown in Figure 16. One period of motion sequence of this motor is as follows; assuming one of the clamping actuators (4) is in the clamp position and the other (5) is in the release position, the step is the moving actuator (6) expanding and making a microscopic movement. Following that the clamping actuator (5), that is in the release position, clamps and then the other clamping actuator (4) releases. At this time, the moving actuator (6) contracts and makes another microscopic movement. One period of motion sequence is completed when the clamping actuator (4) clamps and then the other clamping actuator (5) releases.

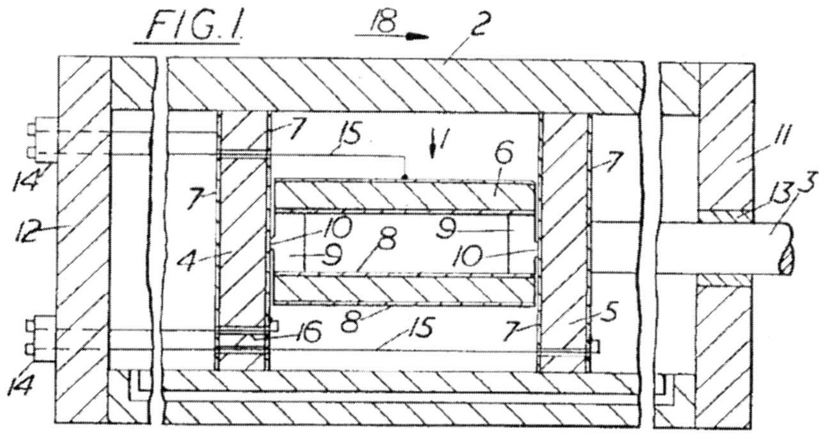

Figure 16. A piezo walk-drive mechanism uses clamping and moving principal, proposed by Brisbane in 1965 [71].

A similar structure, functioning with the same principle, was proposed by Galutva *et al.* in 1970 [72]. In this structure, three multilayer piezoelectric actuators were placed inside two parallel guiding plates. The actuator in the middle makes a microscopic stepping motion parallel to the guiding plates and the other end actuators make the structure to be clamped inside the two guiding plates simultaneously. In this structure, wedge-shaped elements eliminate manufacturing tolerances and apply pre-stressing forces on the piezoelectric actuators.

A well-known piezo-motor in this category is the Inchworm® motor proposed by Burleigh Instruments, Inc. in 1974 [73]. Multiple piezoelectric actuators make synchronous motion by repeating "release-clamp-move and clamp-release-move" steps to produce a movement.

A piezoelectric motor using four legs that are pressed against a slider is developed by Marth *et al.* in 1995 [74]. Each drive element is a piezoelectric bimorph that is electrically isolated and can be controlled independently. Deformation in length and bending direction of the bimorph element simultaneously causes the pressed slider to make a linear motion. A piezoelectric stepper motor from the Swedish company Piezo Motor Uppsala AB operates according to the same principle [75].

The manufactured piezo-walk drive motor at PI has integrated clamping and moving actuators (Figure 17). Shear actuators [76] that are responsible to make micro-stepping are attached at the tip of the longitudinal actuators that are responsible for clamping action. Excitation sequence is the same as the other piezo-walk drive motors, assuming actuator pairs (A) are in release and pairs (B) are in clamping condition. The sequence could continue as follow: Clamp (A) - release (B) - move (A) - clamp (B) - release (A) - move (B).

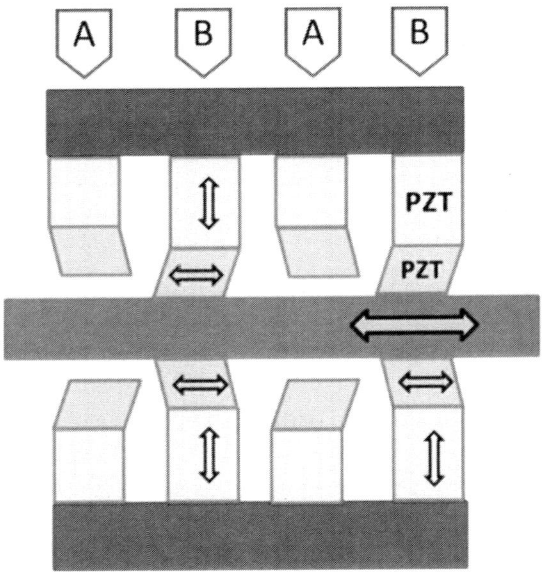

Figure 17. Piezo-walk drive motor at PI [22].

The high force and high stiffness of the piezo-walk drive motor with its high resolution (< 1 nm in open-loop), self-locking at rest, vacuum compatibility, and non-magnetic features, make this motor useful in applications in research, semiconductor processing, such as lithography, and hexapod robots used for computer-aided surgery. Nextline® piezo-walk drive motors developed and manufactured at PI also have two modes of operation; analogue mode for short travel range and step mode for long travel range [22].

Recently, we can see that particle accelerator devices also benefit from non-magnetic and compactness features of piezoelectric motors. Particle accelerators, such as a cyclotron, are devices producing radioisotopes that are used in medical therapy, imaging, and research in various industrial and medical applications. In addition to the electric and magnetic field generating systems in a closed environment (chamber), there are also mechanical devices in the chamber that need to be moved or adjusted at a desired position so that charged particles are incident thereon. A US patent was granted for a particle accelerator in which piezoelectric motors are used to move or adjust mechanical devices in it [77].

6. DISCUSSIONS

In piezoelectric walk-drive-type motors, the actuators are responsible for clamping and releasing actions simultaneously, and at least one actuator is responsible for the moving action. In a piezoelectric inertia-drive-type motor, there is only tangential back and forth motions at the interface of sliding and fixed elements within one period of time. In one direction of the tangential motion, inertia force acting on the sliding element is smaller than the friction force so the sliding element sticks to the fixed element. In the other direction of tangential motion, inertia force action on the sliding element is larger than the friction force, so the sliding element slips and makes a microscopic step.

In a single mode excitation type piezoelectric ultrasonic motor, there is an oblique motion at the interface. Since this oblique motion has components in tangential and normal directions, a net microscopic motion is obtained. In a multi-mode excitation-type piezoelectric ultrasonic motor, there is an elliptical motion at the interface, which means there are displacements in two orthogonal directions at the contact point. This elliptical motion causes the sliding element to make a microscopic motion.

Even though there are many piezoelectric motor designs functioning above mentioned operating principles, the number of commercially successful inventions is limited. This indicates that the described operating principles in those inventions were not sufficient to make commercially viable motors. There are contradictory conditions in piezoelectric motors that need to be carefully analyzed for commercially successful designs.

Holding of displacement generating or vibrating element: a vibrating element in a resonance-drive-type piezoelectric motor need to be held as tightly as possible so that a precise motion without any hysteresis can be obtained. However, tight holding of a vibrating element may suppress useful displacements or vibrations. At the same time, to generate enough vibration with sufficient magnitude, the vibrating element needs to move freely so that vibration on a stator element can be transferred to a sliding or rotating element. Therefore, holding of a vibrating element in a resonance-drive type piezoelectric motor could create a contradictory condition and increases complexity of the motor structure. If a stator element allowing tight holding means that there are nodal positions on it. In this case, it could be considered as a good design.

Friction contact: on a contact surface for low wearing, a smooth and slippery surface is desirable. At the same time, for an efficient motion transfer from vibrating to sliding element, a high friction contact is necessary. Thus, these two parameters, slippery surface for low wear and friction contact, work against each other. Moreover, the nonlinear nature of tribology makes the contact mechanics to be even more challenging.

Pre-stress on vibrating element: when a piezoelectric motor is integrated in a system, in order to increase force transfer, the vibrating element needs to be pre-stressed against the sliding element. While pre-stress force may disturb the guiding mechanism of the slider, the vibration of the stator element could initiate undesirable vibrations on the integrated system.

Additionally, temperature dependency of piezoelectric elements used in a motor requires the drive electronics to have special features.

In short, success of a piezoelectric motor is hidden in the details. Aforementioned issues require know-how from different engineering disciplines to be used for the design of piezoelectric motors. One needs to analyze these issues and apply them properly in a design for commercially viable piezoelectric motors.

Indeed, modeling of contact mechanics and control of piezoelectric motors to clarify the nature of the operating principal and to optimize performance has always been an important study area in the field of piezoelectric motors [78,79,80,81,82,83,84,85,86,87].

Distinguishing features of piezoelectric motors can be understood better if multiple degrees of motion for a system are needed in a compact structure. For example, one can get two directional motions just by putting one linear motor on top of another. Even if one motor has to carry the other motor plus the load, the compact size and light weight of one motor would not change the other motor´s characteristics [22].

In the literature, there are many piezoelectric motor structures for obtaining motion with multiple degrees of freedom. Regardless of the motor type used in these structures, obtaining single axis of motion functions according to the above-described operating principles; resonance, inertia, or piezo-walk drive [22,57,88].

Spherical ultrasonic motor proposed by Toyama has the ability to make motion around a sphere almost in any direction [89]. One can constrain the sphere and make the three vibrating ring element to move on the sphere. An attractive application would be a robot joint with motion with multiple degrees of freedom.

7. CONCLUSIONS

Fast response time, compact size, and self-locking at the rest position make piezoelectric motors suitable candidates for focusing, zooming, and optical image stabilization in cameras. Many camera manufacturers have piezoelectric motor solutions especially inertia- and resonant-drive types for lens moving mechanisms [38,90,91,92].

Piezoelectric motors have been integrated into various fields as compact, fast responding, and magnetic insensitive, vacuum compatible, and high-precision actuators. As systems in medical and micro robotic fields requiring more complex and multi-tasking actuations, the need for these motors would be in demand [77,93,94,95,96]. Research on more reliable friction contact materials, non-lead piezoelectric materials, and making motors to operate at lower voltage with improved driving voltage-speed non-linearity can create advantages for finding new applications.

CONFLICTS OF INTEREST

The authors declare no conflicts of interest.

REFERENCES

1. Meissner, A. Converting electrical oscillations into mechanical movement. US Patent 1,804,838 A, 12 May 1931.

2. Williams, A.L.W.; Brown, W.J. Piezoelectric Motor. US Patent 2,439,499, 13 April 1948.

3. Lavrinenko, V.; Nekrasov, M. Electrical Motor. USSR Patent 217,509, 10 May 1965. (In Russian).

4. Tehon, S.W. Electromechanical transducer motors. US Patent 3,211,931, 10 December 1965.

5. Wischnewskiy, V.; Kavertsev, V.L.; Kartashev, I.A.; Lavrinenko, V.; Nekrasov, M.; Prez, A.A. Piezoelectric motor structures. US Patent 4,019,073, 12 August 1975.

6. Wischnewskiy, V.; Gultajeva, L.; Kartashev, I.; Lavrinenko, V. Piezoelectric motor. USSR Patent 851,560, 16 February 1976. (In Russian).

7. Bansevicius, R. Piezoelectric motor. USSR Patent 693,493, 18 July 1977. (In Russian).

8. Vasiliev, P.E.; Vasilievich, A. Vibration Motor Control. GB Patent 2,020,857, 24 April 1979.

9. Bansevicius, R.; Ragulskis, K. Vibromotors; MOKSLAS Publishing House: Vilnius, Lithuania, 1981. (In Russian)

10. Ragulskis, K.; Bansevicius, R.; Barauskas, R.; Kulvietis, G. Vibro-motors for Precision Micro-robots, Application of Vibration Series; Rivin, E., Ed.; Hemisphere Publishing Co.: Detroit, MI, USA, 1988.

11. Stibitz, G. Incremental feed mechanisms. US Patent 3,138,749, 5 March 1962.

12. Ueha, U.; Tomikawa, Y. Ultrasonic Motors: Theory and Applications; Clarendon: Oxford, UK, 1993.

13. Sashida, T.; Kenjo, T. An Introduction to Ultrasonic Motors; Clarendon: Oxford, UK, 1993.

14. Uchino, K. Piezoelectric Actuators and Ultrasonic Motors; Kluwer: Boston, MA, USA, 1997.

15. Zhao, C. Ultrasonic Motors: Technologies and Applications; Science Press: Beijing, China; Springer Verlag: Berlin, Germany, 2011.

16. Kumada, A. Ultrasonic motor Using Bending, Longitudinal and Torsional Vibrations. US Patent 4,642,509, 21 February 1986.

17. Kumada, A. Piezo Electric Motor. US Patent 4,947,076, 16 December 1988.

18. Barth, H.V. Ultrasonic Drive Motor. IBM Tech. Discl. Bull. 1973, 16, 2263.

19. Zhelyaskov, V.R.; Petrenko, S. Piezoelectric rotary motor with high rotation speed and bi-directional operation. US Patent 8,183,742, 1 September 2010.

20. Wishnewskiy, W.; Kovalev, S.; Vyshnevskyy, A. A New Ultrasonic Piezoelectric Actuator for Nanopositioning. In Proceedings of the 9th International Conference on New Actuators, Bremen, Germany, 14–16 June 2004; pp. 118–122.

21. Spanner, K.; Wishnewskiy, O.; Vyshnevskyy, W. New Linear Ultrasonic Micro motors for Precision Mechatronic Systems. In Proceedings of the 10th International Conference on New Actuators, Bremen, Germany, 14–16 June 2006; pp. 439–443.

22. Piezo Technology. Available online: http://www.physikinstrumente.com/technology.html (accessed on 24 November 2015).

23. Yokoyama, K.; Tamura, H.; Masuda, K.; Takano, T. Single-Phase Drive Ultrasonic Linear Motor Using a Linked Twin Square Plate Vibrator. Jpn. J. Appl. Phys. 2013, 52, 07HE03.

24. Theodolite. Available online: https://en.wikipedia.org/wiki/Theodolite (accessed on 30 November 2015).

25. Takano, T.; Tomikawa, Y.; Kusakabe, C. Same Phase Drive-Type Ultrasonic Motors Using Two Degenerate Bending Vibration Modes of a Disk. IEEE Trans. Ultrason. Ferroelect. Freq. 1992, 39, 180–185.

26. He, S.; Chiarot, P.R.; Park, S. A single Vibration Mode Tubular Piezoelectric Ultrasonic Motor. IEEE Trans. UFFC 2011,58, 1049–1061.

27. Magnussen, B.; Varadi, P.C.; Hagemann, B.; Schuler, D.A.; Davidson, E.M. Vibratory motors and methods of making and using same. US Patent 7,368,853, 22 October 2004.

28. Xing, Z.; He, B.; Xu, K.; Wang, J.; Dong, S. A Miniature Cylindrical Piezoelectric Motor with an Asymmetric Vibrator.IEEE UFFC 2013, 60, 1498–1504.

29. Uchino, K.; Ohnishi, K. Linear motor. US Patent 4,857,791, 3 June 1988.

30. Zumeris, J. Ceramic motor. US Patent 5,453,653, 3 August 1993.

31. Sutherland, G.R.; Louw, D.F.; McBeth, P.B.; Fielding, T.; Gregoris, D.J. Microsurgical robot system. US Patent 8,005,571, 3 July 2006.

32. Koc, B.; Ko, H.-P. Piezoelectric actuator and lens driving device. US Patent 7,764,449, 29 May 2008.

33. Koc, B.; Ko, H.; Jeong, H.S. A miniature Piezoelectric Motor for Phone Camera Auto Focusing Application. In Proceedings of the 11th International Conference on New Actuators, Bremen, Germany, 9–11 June 2008; pp. 184–187.

34. Koc, B.; Tressler, J.F.; Uchino, K. A Miniature Piezoelectric Rotary Motor Using Two Orthogonal Bending Modes of a Hollow Cylinder. In Proceedings of the 7th International Conference on New Actuators, Bremen, Germany, 19–21 June 2000; pp. 242–245.

35. Sashida, T. Motor device utilizing ultrasonic oscillation. US Patent 4,562,374, 16 May 1984.

36. We Evolve Our Ultrasonic Motor. Available online: http://www.shinsei-motor.com/English/index.html (accessed on 24 November 2015).

37. Takano, T.; Tomikawa, Y.; Ogasawara, T.; Sugawara, S.; Konno, M. Ultrasonic Motors Using Piezoelectric Ceramic Multimode Vibrators. IEEE Trans. UFFC 1990, 37, 224–228.

38. Tamai, J.; Okumura, I.; Tsukimoto, T.; Fujimoto, K. Vibration wave driving apparatus and a vibration member, and manufacturing method of the apparatus and the member. US Patent 5,949,178, 22 April 1996.

39. Focus and Zoom Drive Mechanisms Using Ultrasonic Vibration. Available online: http://www.canon.com/technology/ canon_tech/ explanation/cp_tech.html#usm (accessed on 28 January 2016).

40. Wischnewskiy, W. Piezoelectric drive excited by longitudinal and flexural waves. US Patent 6,806,620, 31 July 2000.

41. Ueha, S.; Kurosawa, M.; Masuda, M. Ultrasonic Driven Type Motor. US Patent 4,885,499, 10 February 1989.

42. Marth, H. Untersuchungen zur Dimensieonierung und zum Betriebsverhalten von Vibrationslinearmotoren und deren Einsatz in Positioniereinrichtungen für Magnetköpfe. Ph.D. Thesis, TU, Dresden, Germany, 1980. (In German).

43. Mori, K.; Ogiso, T. Actuator. US Patent 4,613,782, 18 March 1985.

44. Kurosawa, M.K.; Kodaira, O.; Tsuchitoi, Y.; Higuchi, T. Transducer for High Speed and Large Thrust Ultrasonic Linear Motor Using Two Sandwich-Type Vibrators. IEEE Trans. UFFC 1999, 45, 1188–1195.

45. Hills, C. Piezo-Electric Motor. US Patent 6,373,170, 10 March 2000.

46. Iula, A.; Pappalardo, M. A High-Power Traveling Wave Ultrasonic Motor. IEEE Trans. UFFC 2006, 53, 1344–1351.

47. Morita, T.; Kurosawa, M.K.; Higuchi, T. Cylindrical Micro Ultrasonic Motor Utilizing Bulk Lead Zirconate Titanate (PZT). Jpn. J. Appl. Phys. 1999, 38, 3347–3350.

48. SQUIGGLE Micro Motor Technology. Available online: http://www.newscaletech.com (accessed on 20 November 2015).

49. Kurosawa, M.; Nakamura, K.; Okamoto, T.; Ueha, S. An Ultrasonic Motors Using Bending Vibrations of a Short Cylinder. IEEE Trans. UFFC 1989, 36, 517–521.

50. Ichiro, O. Electro-mechanical energy conversion element and a vibration wave actuator using the electro-mechanical energy conversion element. US Patent 5,814,919, 23 April 1996.

51. Koc, B.; Thelen, M. Piezoelectric Rotary Motor with Independently Driven Orthogonal Electrodes of a Piezoelectric Plate. In Proceedings of the 12th International Conference on New Actuators, Bremen, Germany, 18–20 June 2012; pp. 258–261.

52. Pohl, D.W. Dynamic Piezoelectric Transducer Devices. Rev. Sci. Instrum 1987, 58, 54–57.

53. Higuchi, T.; Watanabe, M. Apparatus for effecting fine movement by impact force produced by piezoelectric or electrostrictive element. US Patent 4,894,579, 23 May 1987.

54. Miyano, M.; Kawasaki, T.; Kuwana, M.; Miyazawa, M.; Ueyama, M.; Tasaka, Y. Lens Moving Apparatus. US Patent 5,587,846, 24 December 1996.

55. Yoshida, R.; Okamoto, Y.; Higuchi, T.; Hamamatsu, A. Development of Smooth Impact Drive Mechanism (SIDM). J. JSPE 1999, 65, 111–115. (In Japanese).

56. Niedermann, P.; Emch, R.; Descouts, P. Simple Piezoelectric translation Device. Rev. Sci. Instrum. 1988, 59, 368–369.

57. Howald, L.; Rudin, H.; Güntherodt, H.J. Piezoelectric inertial stepping motor with spherical rotor. Rev. Sci. Instrum.1992, 63, 3909–3912.

58. Aleksandrovich, B.V.; Viktorovic, B.A.; Evich, K.V.V. Inertia Step Motor. RU Patent 2,465,712, 27 October 2008.

59. Thomas, P.; Desailly, R. Optical adjustment mounts with piezoelectric inertia drive. Patent WO2008087469A2, 18 January 2007.

60. Kortschack, A.; Rass, C. Method for controlling a multi-actuator drive. Patent WO2013128032A1, 4 March 2013.

61. Huebner, R. Piezoelectric rotary drive for a shaft. US Patent 9,106,158B2, 1 August 2012.

62. Takagi, T. Vibration Driven Motor. US Patent 5,917,268, 29 June 1999.

63. Kanbara, T. Electro-Mechanical Transducer lens drive mechanism for a vibration compensating system. US Patent 5,768,016, 26 July 1995.

64. Karrai, K.; Heines, M. Inertial rotation device. US Patent 6,940,210B2, 4 November 2003.

65. Yamagata, Y.; Higuchi, T.; Saeki, H.; Ishimaru, H. Ultrahigh vacuum precise positioning device utilizing rapid deformations of Piezoelectric Elements. J. Vac. Sci. Technol. A 1990, 8, 4098–4100.

66. Meyer, C.; Sqalli, O.; Lorenz, H.; Karrai, K. Slip-stick step-scanner for scanning probe microscopy. Rev. Sci. Instrum.2005, 76, 063706.

67. Bansevičius, B.; Blechertas, V. Ultrasonic motors for mass-consumer products. Ultragarsas 2006, 4, 50–52.

68. Koc, B. Piezoelectric Motor, Operates by Exciting Multiple Harmonics of a Square Plate. In Proceedings of the 12th International Conference on New Actuators, Bremen, Germany, 14–16 June 2010.

69. Morita, T.; Yoshida, R.; Hosaka, H. Resonant-Type Smooth Impact Drive Mechanism Actuator Operating at Lower Input Voltages. Jpn. J. Appl. Phys. 2013, 52, 07HE05.

70. Tuncdemir, S.; Ural, S.O.; Koc, B.; Uchino, K. Design of Translation Rotary Ultrasonic Motor with Slanted Piezoelectric Ceramics. Jpn. J. Appl. Phys. 2011, 50.

71. Brisbane, A.D. Position Control Device. US Patent 3,377,489, 9 April 1965.

72. Galutva, G.V.; Ryazantsev, A.I.; Presnyakov, G.S.; Modestov, J.K. Device for Precision Displacement of a Solid Body. US Patent 3,684,904, 15 April 1970.

73. May, W.G. Piezoelectric Electromechanical Translation Apparatus. US Patent 3,902,084, 30 May 1974.

74. Marth, H.; Gloess, R. Verstelleantrieb aus Bimorphelementen. DE Patent 4,408,618A1, 15 March 1994.

75. Johanson, S.; Bexell, M.; Lithell, B.O. Fine Walking Actuator. US Patent 6,798,117, 10 July 2002.

76. Müller, K.D.; Marth, H.; Pertsch, P.; Gloess, R.; Zhao, X. Piezo-Based, Long-Travel Actuators for Special Environmental Conditions. In Proceedings of the 10th International Conference on New Actuators, Bremen, Germany, 14–16 June 2006; pp. 149–153.

77. Eriksson, T.; Holmgren, B. Particle accelerators having electromechanical motors and methods of operating and manufacturing the same. US Patent 8,653,762, 23 December 2010.

78. Maeno, T.; Takayuki, T.; Miyaki, A. Finite-Element Analysis of the Rotor/Stator Contact in a Ring-Type Ultrasonic Motor. IEEE Trans. UFFC 1992, 39, 668–674.

79. Hagedorn, P.; Wallaschek, J. Traveling wave ultrasonic motors, part I: Working principle and mathematical modelling of the stator. J. Sound Vib. 1992, 155, 31–46.

80. Hagood, N.W.; McFarland, A.J. Modeling of a piezoelectric rotary ultrasonic motor. IEEE Trans. UFFC 1995, 42, 210–224.

81. Honda, T.; Kato, K. Fundamental properties of friction and wear on friction drive by traveling wave type ultrasonic.Jpn. J. Tribol. 1993, 47, 1073–1084.

82. Flynn, A.M. Piezoelectric Ultrasonic Motors. Ph.D. Thesis, MIT, Cambridge, MA, USA, December 1997.

83. Pirrotta, S.; Sinatra, R.; Meschini, A. Evaluation of the Effect of Preload Force on Resonance Frequencyies for a Traveling Wave Ultrasonic Motor. IEEE Trans. UFFC 2006, 53, 746–753.

84. Sattel, T.; Hagedorn, P.; Schmidt, J. The Contact Problem in Ultrasonic Traveling–Wave Motors. J. Appl. Mech. 2010, 77, 031014.

85. Mashimo, T.; Terashima, K. Dynamic analysis of an ultrasonic motor using point contact model. Sens. Actuators A Phys. 2015, 233, 15–21.

86. Uzunovic, T.; Golubovic, E.; Sabanovic, A. Piezo LEGS Driving Principle Based on Coordinate Transformation.IEEE/ASME Trans. Mechatron. 2015, 20, 1395–1405.

87. Mashimo, T.; Terashima, K. Experimental Verification of Elliptical Motion Model in Traveling Wave Ultrasonic Motors. IEEE/ASME Trans. Mechatron. 2015, 20, 2699–2707.

88. Amano, T.; Ishii, T.; Nakamura, K.; Ueha, S. An ultrasonic actuator with multi degree of freedom using bending and longitudinal vibrations of a single stator. In Proceedings of the IEEE Ultrasonic Symposium, Sendai, Japan, 5–8 October 1998; Volume 1, pp. 667–670.

89. Mashimo, T.; Toyama, S.; Ishida, H. Design and Implementation of Spherical Ultrasonic Motor. IEEE Trans. UFFC2009, 56, 2614–2521.

90. Masayuki, M.; Toyotoshi, K.; Minoru, K.; Masayuki, M.; Yoshihiro, T. Lens moving apparatus. US Patent 5,587,846, 14 July 1995.

91. Cho, J.-W.; Lim, J.-H.; Park, S.-H. Vibrating element, fabrication method thereof, and ultrasonic motor having the same. US Patent 8,237,330, 10 February 2009.

92. Song, C.; Gehlbach, P.L.; Kang, J.U. Active tremor cancellation by a "Smart" handheld vitreoretinal microsurgical tool using swept source optical coherence tomography. Opt. Express 2012, 20, 23414–23421.

93. Seifabadi, R.; Song, S.E.; Krieger, A.; Cho, N.B.; Tokuda, J.; Fichtinger, G.; Iordachita, I. Robotic system for MRI-guided prostate biopsy: feasibility of tele operated needle insertion and ex vivo phantom study. Int. J. CARS 2012, 7, 181–190.

94. Su, H.; Weijian, S.; Cole, G.A.; Harrington, K.; Fischer, G.S. Haptic System Design for MRI-Guided Needle Based Prostate Brachytherapy. In Proceedings of the IEEE Haptics Symposium, Waltham, MA, USA, 25–26 March 2010; pp. 483–488.

95. Yang, S.; MacLachlan, A.R.; Riviere, N.C. Manipulator Design and Operation of a Six-Degree-of-Freedom Handheld Tremor-Canceling Microsurgical Instrument. IEEE/ASME Trans. Mechatron. 2015, 20, 761–772.

96. Priester, A.M.; Natarajan, S.; Culjat, M.O. Robotic Ultrasound Systems in Medicine. IEEE UFFC 2013, 60, 507–523.

3

SHAPE MEMORY ACTUATORS FOR MEDICAL REHABILITATION AND NEUROSCIENCE

Simone Pittaccio and Stefano Viscuso

[1] National Research Council of Italy, Institute for Energetics and Interphases, Lecco,, Italy

1. INTRODUCTION

Actuators based on shape memory alloys (SMA) proved to be particularly advantageous with respect to other actuation technologies when they are embedded in applications requiring strict compliance to a set of compatibility (e.g. mechanical, biological, weight, ...) and environmental constraints. Most noteworthy are the uses in miniaturised components, lightweight systems, sensing-actuating systems (e.g. actuators interacting with changing environmental variables), low-noise or low-impact appliances (e.g. actuators with reduced interaction with the environment) and self-sensing controllable devices. With particular regard to these preferred applications, SMA can also play a role in solving specific actuation problems in the fields of Medical Rehabilitation and Neuroscience. The main characteristics expected of SMA actuators for these fields are light weight and portability; self-adjustment to evolving needs; applicability of actuation in shielded environments (with bioimaging and diagnostic devices: magnetic resonance imaging (MRI, fMRI), magnetoencephalography (MEG), and the like).

Taking the move from these key ideas, this Chapter will discuss some innovative uses and some implemented examples of biomedical devices based on the Shape Memory Effect (SME), explaining their advantages and limitations. In particular, it will address the following main topics: analysis of the relevant literature and background; the derivation of technical constraints and required material properties given the expected biomechanical and or clinical cases; the mathematical design of sample actuators and their implementations; ideas for their clinical application and control. It is important to notice that these same concepts can be easily transferred to other important industrial engineering fields.

1.1. Biomedical Background

Neuromuscular Rehabilitation is the science and practice of supporting the recovery of lost or impaired motor function, particularly due to damage of the Central Nervous System (CNS). The established approach to treatment is physiotherapeutic, pharmacological and surgical, the choice of a specific line of action depending on the exact diagnosis, the general state of the patient, the severity of the damage, the time distance from the acute event, the evolution of sequelae, the presence of co-morbidities, etc.

In the general case, a patient undergoing neuromuscular rehabilitation was struck by a primary event (e.g. a stroke, a traumatic brain injury, an infection, hypoxia, a tumour,...) producing severe neuronal loss. According to the affected area of the CNS different scenarios may result, including a state of paresis (essentially, tetra- or hemi-plegia), and possibly other types of impairment, such as cognitive, sensory and so on. The recovery process from these initial insults often does not proceed independently of a series of interconnected secondary phenomena, which, if left uncontrolled, can undermine all hopes of an optimal final outcome. These phenomena include muscular contracture, spasticity, CNS remodelling and learned non-use.

With no ambition to be exhaustive, it can be pointed out here that the main mechanisms by which the human body, and in particular the neuromuscular system, react to the primary event are mediated by two functional disruptions of the physiological status, namely, *immobility* and *disuse*. Immobility is the absence of movement, especially of a limb or body joint; disuse is the lack of muscular activation and consequent mechanical loading of the skeleto-muscular structure. It is evident that both conditions are tightly connected to and often caused by the state of paresis. *Contracture* (shortening of the muscles) can occur due to prolonged (days or weeks) permanence of the body segment in a given position. This can be the case e.g. for a flexed elbow when patients sitting on wheelchairs keep their hands in their laps; or for an ankle joint that is maintained in an overly extended posture by gravity during continuous rest in bed. Muscular contracture produces, as a major consequence, a forceful and rigid ill-positioning of the adjacent joint, reduction of the available articular range of motion, and malformation. It can lead, in time, through reduced mobility, to structured calcifications of joints and retraction. Alongside contractures, the development of *spasticity* is one of the most common and severe secondary events in the sub-acute phase of paresis. Spasticity is an abnormal and speed-dependent stiffening of the muscles due to hyperactive stretch reflex. This exaggerated response to muscle lengthening exacerbates the impossibility of movement and subjects the patient to discomfort and pain as soon as the limb is mobilised. In turn, contrasting immobility becomes an even harder challenge. The picture resulting from the instauration of contracture and spasticity is given the name of *spastic paresis*, which is a painful, disfiguring and severely disabling condition, and constitutes a serious obstacle to patients' recovery [1-2]. In fact, the increased rigidity of muscles and joints further hinders any voluntary or involuntary motion or use of the limb, thus creating a vicious circle towards greater disruption of neuromuscular functionality.

Besides producing peripheral modifications, the prolonged immobility and disuse of the limb appears to affect negatively the plastic reorganisation and remapping of brain structures, by which live neurons take over the role of lost ones. The manner of such remapping is bound to have a paramount influence on ultimate functional healing, and this is one more motivation for trying and avoiding those conditions in the first place. Their removal could shun the worsening of muscle stiffness and hypertone, while a sustained proprioceptive (sensorial) stimulation through movement may support appropriate cortical reorganisation. However, a complete picture of brain remodelling after a CNS insult and the effect of rehabilitation in modulating such a reorganisation is still not available. Studies of Neuroscience investigating in particular rehabilitation processes are certainly of great interest for physicians, neurologists, researchers and physiotherapists, as they can shed some light on those phenomena.

Recent trends in neuromuscular rehabilitation rely on pharmacological treatments against mild and severe spasticity, such as denervating neurotoxins (BOTOX) or myorelaxation agents (Baclofen) [3]; in most severe cases of contracture the approach can resort to surgical interventions such as tendon lengthening or transfer. The most widely used technique for all severities and types of spastic paresis is however physical therapy, which consists in static repositioning by means of splints and casting, controlled and gradual mobilisation of the limbs, muscular stretching and active exercising. It has been widely acknowledged that active work-out is a particularly important part of the rehabilitation program [4-5]. However, if autonomous control by the patient is absent or insufficient, passive mobilisation is utilised to try and maintain the viscoelastic properties of muscles and periarticular tissues. Furthermore, it is reasonable to think that minimising immobility and disuse, even by means of passive motion, can be advantageous for the prevention of the other adverse sequelae of CNS damage or to contrast their worsening.

Robotic tools are becoming very popular, as they make it possible to extend the duration of therapeutic sessions without the need for an individual involvement of human therapists, and reduce costs. In fact, though time-extensive physiotherapy is known to influence positively the rehabilitation outcome, organisational issues may limit the practical application of this principle, especially for passive mobilisation, in which case the physiotherapist has to move each affected joint one by one. In most situations, robotic tools provide repetitive and repeatable passive mobilisation of the whole limb [6], or joint by joint [7]. Moreover, several rehabilitation robots can assist active work-out, by completing patient's efforts to move [8] with the aim of improving muscle recruitment, by perturbing limb trajectories to enhance active control, or by contrasting movement [9] to strengthen muscles. Some drawbacks can be identified concerning the current use of robots in rehabilitation. First, as many of these machines are large, massive and costly, in a typical clinical environment the number of robots is limited and their use at home is ruled out. Second, most part of them require minimal skills by the patients, e.g. trunk control, equilibrium or cognitive abilities: that excludes very early or severe subjects, although they could benefit from extended sessions of passive mobilisation as well.

Robotic devices have been used for the mobilisation of body parts also in the field of Neuroscience [10-15]. Machines to manage the motion of the effectors (muscles) can in fact be used to support the measurement and visualisation of the interconnections between the activities of peripheral segments and the brain structures involved in their control. In this type of applications tight compatibility with Bioimaging and Biosignal technologies is mandatory [13].

1.2. Overview Of Shape Memory Alloys

Robotic devices for Rehabilitation, like all robotic devices, are based on one or more actuators to transfer motion and displace a load. The load, in the case of Rehabilitation, will often consist of body limbs and joints, comprehensive of their mechanical properties, inertia and gravity. As was reported in Section 1.1, there is still a want of appropriate robots, meeting the needs of frailer patients and organisational requirements of the clinical structure. Much investigation in this respect is devoted to the manufacture of light-weight, portable and/or wearable devices. To the ears of the materials scientist or technologist these properties and the capability for actuation are typical features of a special class of metallic alloys, namely Shape Memory Alloys (SMA).

SMA are a compositionally heterogeneous class of materials, comprising a number of intermetallic compounds made up of transition and post-transition metals of different groups. It is beyond the scope of this Chapter to relate about all the different available compositions and their properties, which can be found in reference texts and scientific literature [16-19]. The most relevant compound of this class, employed in most technological applications to date is nearly-equiatomic NiTi, which is also supplied commercially in many semi-finished forms. The general name of NiTi (or TiNi) or Nitinol refers to several grades of the alloy found in the quasi-stoichiometric range from 49%at Ni to 51%at Ni. There are also some important ternary compositions based on NiTi, such as NiTiCu, NiTiHf, NiTiPt, NiTiNb, NiTiCr, with optimised characteristics for dedicated uses.

The Shape Memory Effect (SME) is an athermal reversible martensitic transformation producing macroscopic strain recovery upon heating above a certain characteristic temperature, generally referred to as A_f in the specialised literature. In the case of binary NiTi, the stable low-temperature phase (below the M_f temperature) is a B19' martensite, while the high-temperature structure is a B2 parent phase. $M_f$and A_f are separated by a temperature difference of around 20°C in the solution-treated state. The effect of cold working is to strengthen and embrittle the material, and suppress the transformation. A controlled work-hardening however produces beneficial effects on the mechanical properties. In this state the material forward and backward transformations occur across spread-out temperature ranges and the hysteresis is also increased. Ageing treatments are necessary to optimise mechanical and functional properties and to adjust the characteristic temperatures. Typically, NiTi is aged at 350-650°C and water quenched, and this process also shape-sets the material in the shape "to be remembered". Treatment temperature and duration have to be honed to the application and size of the specimen. The mechanical properties, as described by a stress-strain tensile loading curve, may vary but, in the most representative cases, show an initial linear-elastic range, followed by a long flat plateau and a final increase. Most of the macroscopic shape change happens along the plateau region (up to 6% long) and corresponds to the microstructural phenomenon of martensite detwinning. There are two phenomenological varieties of the unloading curve, according to the temperature of the test. If the test is carried out below M_f, unloading occurs along a line and strain decreases only minimally, so no macroscopic shape recovery is often observable. Full recovery is obtained by successive heating above the A_f temperature. This is called the SME proper. However, if the test is carried out above A_f the unloading curve is different and shows a long recovery plateau, at lower stresses than the loading one, and a final linear decrease towards zero-strains. The curve, in this case, is hysteretic and the shape recovery is attained without any need to heat the material. When the test is at room temperature T_r,andA_f lies below T_r, SME is given the special name of Pseudoelastic Effect (PE). In all cases where the test temperature falls between M_f and A_f the behaviour is intermediate. Cold-work, ageing and precipitation of second phases can have very significant effects on all aspects of the mechanical curves, and in particular, on the height and separation of the plateaux, and cycling stability. So has the precise test temperature: first of all, in the case of SME loading occurs through the deformation of preformed martensite, resulting in lower stress values of the plateau and lower linear elastic modulus than in the case of PE, where austenite is initially loaded until stress starts inducing the formation of detwinned martensite; furthermore, through the Clausius-Clapeyron effect, stress is proportional to temperature and thus plateau stress levels are increased by an increase in test temperature, in particular in the case of PE. Finally, the compositions in the range 49%at-50.5%at Ni (Ti-rich) tend to show SME proper, while the Ni-rich ones (50.6%at-51%at Ni) have a pseudoelastic behaviour at room temperature. As a general trend, the higher the Ni content, the lower are the transformation temperatures.

1.3. State Of The Art Of Sma In Rehabilitation and Neuroscience

Many efforts were made in the last 20 years for developing SMA-rehabilitation tools. Figure 1 shows the evolution of the number of papers, patents and conference contributions over this period. It is evident that in the last five years several groups focused on studying the matter, issuing some two thirds of the overall production. All papers dealing with implanted devices were not included in the search, for a number of reasons: because the typical fields of use of implantable devices are hardly connected to Neuromuscular Rehabilitation; because the design of parts to be utilised inside the human body is based on assumptions and limitations rather different from those distinctive of external actuators; because the domain of implantable devices in NiTi is well covered by extensive and comprehensive reviews and is now an established application. A choice was made to include only papers addressing directly the issue of Neuromuscular Rehabilitation or, indirectly, suggesting manners of applying static or dynamic external forces to reposition or move body parts. Particular attention was paid to applications including SMA *actuators*, i.e. devices exploiting the SME to transform heat into mechanical work.

1.3.1. Repositioning

Repositioning is the set of procedures aiming at contrasting ill-postures and malformations caused by contractures and spasticity. Static orthoses are often used to impart these treatments and may be as rigid as castings or partially compliant. Many authors imagined the use of SMA in orthotics as passive components that apply static corrective forces [20-30]. In these applications SMA are mostly employed for their pseudoelastic properties, even though some designs are also based on SME. A series of papers [20-25] showed applications of pseudoelastic NiTi for spastic limb repositioning (elbow, ankle). The authors suggest that pseudoelasticity can be an interesting solution to the disuse and immobility problems during the orthotic repositioning therapies, in that it safeguards residual motor capabilities (voluntary or reflex) and decreases contact pressure by yielding under muscular jerks. They also showed possible advantages of this type of devices for correcting equinus gait [25]. In [26], pseudoelastic NiTi was utilised to remodel deformed auricles. Though not strictly correlated to the main subject of Neuromuscular Rehabilitation, this paper presents an approach in which the anatomic and biomechanical constraints are integrated in the design procedure. In [27], PE is employed to try and correct flat foot malformation and provide increased stability.

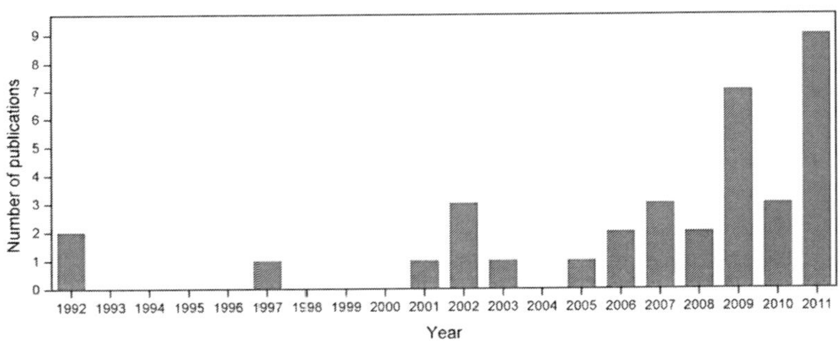

Figure 1. Histogram of the number of publications about SMA-based rehabilitation devices in the last 20 years.

On the other hand, [28] used the Shape Memory Effect to stretch gradually spastic wrist and fingers of paretic patients. The authors of this particular report also carried out measurements of joint resistance to movement, and designed the NiTi elements on the basis of these results. Thanks to the biomechanical design, the joints could be positioned at the neutral angle when the SMA was activated. The main characteristic of SMA exploited for this application was its ability to change shape, which eased putting on the device on malformed limbs. In an international patent [29], SMA are cited as a possible means to produce adjustable degrees of knee joint distraction and assuage contact overload between femur and tibia in unicompartmental osteoarthritis. Another international patent [30] discloses the use of SMA elements as possible actuators in a dynamically adjustable shoe to adapt to congenital or acquired deformities of the foot.

1.3.2. Functional Exercise

Passive physiotherapy consists in repetitions of movements, muscular stretching and proprioceptive stimulation imparted by the hands of a therapist or by robots. Some authors [31-41] investigated the possibility of employing SMA to make portable or wearable devices to act as functional exercising robots.

Some papers focused on finger movement, in particular the review [31] reported an example of rehabilitation glove fully actuated by SMA wires showing the two-way Shape Memory Effect. Another attempt to move fingers is reported in [32,33]. This paper describes the characterisation of bending wire actuators on a dummy finger and a temperature-controlled switch, and clearly evidences the issue of the force-speed trade-off. In [34] another device for passive mobilisation of the flaccid fingers is presented. This paper dealt with biomechanical constraints and design issues connected to the use of SMA springs as actuators, also proposing several modelling equations. Interestingly, both [33,34] employed the composition NiTiCu, which is known to have a narrower thermal hysteresis and lower detwinning stress than binary NiTi.

Other papers [35-41] concentrated on the lower limb. In [35], a concept for a paediatric boot is described. In the intention of the authors, a control system would activate two NiTi wires producing slow movement of the ankle and provide the possibility for home passive mobilisation. Two different implementations of a wearable ankle mobiliser were published by another group [36-41]. In the first paper [36] a description of the design procedure is presented, including mechanical and power dimensioning of SMA linear actuators based upon biomechanical considerations. Pre-clinical trials are also reported in the text. The following publications [40-41] dealt with an amagnetic rotary actuator and its use for the construction of a biosignal-controlled ankle exerciser that could be used in two different modes: fully passive and haptic-assistive, according to the evolving state of the patient during recovery. Trials on healthy subjects are described in [41].

A different use of SMA is proposed in [42], where a NiTi-wire was utilised to produce fingertip tactile stimulation with the purpose of providing a means for sensory feedback in haptic rehabilitation. Several implementations are presented and also qualitative tests on healthy volunteers are reported.

1.3.3. Muscle Toning

Some robotic tools provide resisting forces that contrast active exercising by the patients. This could help increase muscle strength and control abilities. A few publications propose SMA-based devices for manifestly therapeutic uses and for generic muscle strength training. In the international patent [43], a set of wearable pseudoelastic garments are proposed to favour muscle toning during daily life activities. Though this application was not described as a therapeutic tool, it shows similarities with techniques that are used in Neuromuscular Rehabilitiation. Two papers by another group [32,33] suggest the use of SMA actuators for applying resisting forces to active movement. Unfortunately, the authors did not discuss either the level of force required or the consequent dimensioning of the actuator.

1.3.4. Assistive Robotics

Other proposed applications of SMA in Rehabilitation comprise a number of studies that present active tools to support [44-47] or take over [48-51] impaired motor functions.

An M.Sc. thesis and a following paper by the same authors [44,45] report about the design and control of a downscaled bio-inspired mechanical model of an ankle joint wearing a SMA-actuated orthosis to support the walking capabilities of plegic subjects. The thesis discusses the biomechanical constraints and both publications clearly explain the dimensioning of the device. The authors also discuss the relationship between the size and number of SMA elements and the actuation properties. Another group issued a paper [46] and a patent [47] describing the use of SMA actuators to manufacture soft wearable orthoses for the knee joint. The garments were intended to provide sufficient torque to support the impaired gait of neurological patients, but the authors eventually found that the cycling speed was somewhat slower than appropriate.

An international patent [48] describes an orthotic device for restoring the grasping function in paretic subjects. The SMA actuation is intended to substitute entirely the lost function by allowing index and middle finger tips to reach jointly the tip of the thumb. A more complex multi-joint device with the same general purpose was developed in a Ph.D. thesis [49], where motors (possibly SMA-based ones) are however not mounted on the patient's arm, but lie at a distance, being connected to the limb by Bowden cables. This manuscript also includes biomechanical measurements and a control system. Another patent [50] imagines orthotic devices activated by

SMA elements aiming at accurately simulating the mobile capabilities of human limbs. An unusual application is provided by [51], where SMA wires are employed as actuators to regain lost facial mobility and the ability to smile. The activation of the wires, aligned along the main affected muscle groups, is triggered by bioelectric signals acquired on the contralateral healthy side of the face.

1.3.5. Neuroscience

The scope of limb-moving actuators in the field of Neuroscience is mainly to provide proprioceptive, kinaesthetic or tactile stimulation in order to investigate neural responses to those stimuli. Increasing interest has being expressed during the last few years [10-15] to the effect of finding specialised tools that can be compatible with the instrumentation used to conduct Neuroscience experiments, e.g. magnetic resonance imaging (MRI, fMRI), magnetoencephalography (MEG), electroencephalography (EEG), near-infrared spettroscopy (NIRS).

SMA actuators have found little application in this field as yet: only two examples of SMA-based devices for use in Neuroscience were identified in the literature [37-41]. These publications describe a linear and a rotary actuator designed to comply with electromagnetic noise constraints, and be compatible in particular with MEG and fMRI. Most other devices were implemented using alternative technologies such as pneumatic and piezoelectric. The main reported advantage of SMA actuators is again portability and wearability.

1.3.6. Final Remarks

Most authors recognised as a major advantage of using SMA in orthotics the light weight and compactness of the actuators. These factors, in fact, suggest that these materials can be used to build portable and/or wearable devices, and oriented the concepts of many designs towards compact robots, light static or dynamic splints, or even soft wearable garments, often for use in the home environment. Going from concept to actual implementation of the devices, the most challenging issue in this field appears to be the dimensioning of the actuators to address at the same time long-strokes and fairly-high-force requirements. In particular, human joints, bones and muscles move or deform across broad ranges of angular motion and several centimetres in length, and possess non-negligible masses. Understanding this is a first fundamental step of an engineering approach to the problem of actuation for Rehabilitation. Some papers did not find an optimal solution to the design question, because there is a serious trade-off between size and power variables to be overcome. In particular, high forces require large cross-sections of SMA: big diameters or many units working in parallel, or both. Any one of these choices, however, brings along some drawbacks: large diameters generally imply larger currents (for Joule's heating) and longer heating and especially cooling times; thin actuators are faster, but building large bundles or arrays of those require cross-insulation of the parallel fibres, larger volumes or surfaces to store them on board and the right strategy of electric connection (in series to minimise the current but requiring high voltages, or *vice versa* for the parallel configuration). It must be kept in mind, as an extra constraint, that patient's safety issues would tend to favour solutions with both low currents and voltages. Furthermore, when precise timing of the movements, matching some physiological rhythm (such as gait), or speed control are parts of the desired functionality of the device, actuation frequency is also an important factor, imposing tighter limitations on the acceptable surface-volume ratio and thermal exchange properties (viz. heating and cooling times) of the SMA actuators. The most successful approaches appeared to be characterised by a precise preliminary analysis of the biomechanics of the joint to be moved, and an optimal and simultaneous modelling solution of the overall functioning and sizing of the SMA actuator: this method made it possible to balance advantages and drawbacks against all standing trade-offs.

In terms of control systems, some designs implemented fine closed-loop architectures to set, maintain or change speed and position of the effector. These systems attracted the attention mostly of designers of dynamic tools, such as those to support the gait or the motion of the fingers. Simpler approaches were used to obtain automatic switching on and off of the actuators, based on some measurable variable (position, temperature, time,...). A few papers employed biosignals (particularly muscular bioelectric activity measured from the skin surface) to trigger actuation. We find this idea particularly interesting, dealing with Neuromuscular Rehabilitation, because it may bring the residual abilities of the patients into the design picture.

Actuators for Neuroscience share a number of characteristics with those described for the field of Rehabilitation. Besides the constraints depending on biomechanics and SMA properties, special requirements must be fulfilled in order for these devices to be able to be used in electromagnetically shielded environments. In particular, if the Joule's Effect is chosen as a means to heat SMA elements and produce phase transformation, care will have to be taken in controlling any unwanted stray magnetic fields generated by the current flow.

2. USES OF SMA ACTUATORS IN MEDICAL REHABILITATION

This section will provide guidelines for designing SMA actuators aiming at moving human joints. General statements and suggestions will be accompanied by the discussion of a case study taken from the authors' experience, i.e. an ankle mobiliser conceived as a lightweight and portable tool for providing neurologic patients (stroke, traumatic brain injury, paresis) with motor support and sensorial stimulation not only during the very early phase of rehabilitation when they are still unable to exercise, but also during the subsequent phase of motor relearning, thanks to a specially-designed biofeedback-based control system.

2.1. Biomechanical, Bioengineering and Clinical Specifications

The initial step of a biomechanical design is identifying the anatomical compartment, the pathology of interest and aim of the medical device. This is however just the basic set of information to tackle the problem and provides a very rough description of the actual boundary conditions affecting the following steps. Further human-related design constraints that ought to be considered in developing rehabilitation devices span from the mechanical properties of the joint, to the typical characteristics of target patients, and as far as considering the psychological impact.

2.1.1. Tissues and Joints

Any biomedical device, either implantable or external, should take into account the peculiar characteristics of living tissues. Focusing on the musculoskeletal system, a great variety of tissues can be described: cortical bone, cancellous bone, muscular fibres, tendon, ligament, cartilage, synovial fluid, fibrous tissues... All these tissues show non-linear, viscoelastic and hysteretic properties. Some of them, such as the muscular tissues, also display active mechanical behaviour. Moreover, tissue characteristics may vary significantly with age, sex, ethnicity, fitness, pathology or malformation. The properties of biological tissues work together to produce the observed features of joints and limbs, which therefore also display non-linear, viscoelastic and patient-dependent behaviour.

Human joints may differ significantly from one another in terms of dimensions, degrees of freedom, neutral position, range of motion, associated muscular power and weight of the adjacent limb segments, neighbouring structures, etc. A "general purpose" actuator is hard to imagine, but dedicated devices can be tailored to target specific joints and pathologies. Furthermore, once the application has been selected, a statistical analysis of the target population could be carried out and possibly *ad-hoc* assessment of joint characteristics, biometric parameters, classic associated malformations or impairments and comorbidities, so that an appropriate average set of biomechanical parameters is acquired for a "typical user".

Modelling of the joint behaviour is often an essential step in the design process. The desired application can direct modelling choices as to whether a detailed or simplified description of the joint should be included. For example, it may be a purpose of the new device to activate only one degree of freedom (d.o.f.) of a di- or tri-axial joint; in this case, the extra d.o.f's can be excluded from the joint model, but it must be realised that the neglected mobility can produce unwanted stresses or sliding, or pain when the actuator is in use, and it is up to the designer to decide if not controlling those events is acceptable, or else they ought to be controlled by some other means (e.g by fixing and preventing movements along the extra d.o.f.'s). Joint characteristics cannot be considered as unchangeable data, but have to be connected to the current design case: for instance, in judging what a suitable range of motion (ROM) for the actuator should be, the designer must consider the joint type, but also the application (what is the required movement for the rehabilitation exercise?), the pathology (is it likely to change the ROM? Does the actuator have to work to restore the physiological ROM?), the relationships with posture (e.g. certain positions of adjacent joints can affect the available ROM of the target one), the influence of other d.o.f.'s, etc. In relation to this, it can also be decided if joint mechanical properties can be linearised within the ROM of interest.

2.1.2. Expected Loads

Considering the force or torque generated by the actuator, it is important to keep in mind that joints have passive and active torque-angle characteristics that can be activated by the patient voluntarily or unconsciously.

A passive response is elicited from the muscles and periarticular structures when the joint is moved from its neutral angle, i.e. a resting position where forces producing joint flexion are balanced by those producing joint extension (including or not including gravity, according to limb position in space). As mentioned above, this response is generally nonlinear, viscoelastic and hysteretic, even in physiologic conditions. It can be evaluated by means of a motor-dynamometer-encoder system, in such a manner that the joint can be mobilised passively at controlled speed across a set range of motion, and resisting torque and position can be acquired simultaneously throughout the movement. By changing movement speed the viscoelastic characteristics can be evidenced, while conducting tests both in the flexion and extension direction (or inversion and eversion; or adduction and abduction; etc.) can reveal the hysteresis. The obtained torque-angle curves can be used as load curves in the design of the actuator. The absolute values of passive joint torques strongly depend not only on pathology (viz. contracture, retraction, fibrosis, etc.), but also on age, sex and muscular trophism, and of course change from joint to joint in the body. For this reason we can only give here some examples to clarify the order of magnitude of these torques. Maximum physiological passive torques can be as low as 50-60Ncm for finger joints [52], 100-300Ncm for the elbow [53] and up to 500-1000Ncm for the ankle joint [53-54], also depending on the angle of the knee. In healthy conditions, joint passive stiffness is not prevalently affected by movement speed. The same may not be true in some pathological cases. Muscle contracture and joint retraction can cause the reported values of joint passive stiffness to increase several times and can even make joints completely rigid to any practical purpose.

Apart from a passive response, joints can resist imposed movement also through the activation of the associated muscle groups. Muscular power can be delivered voluntarily (active motion) or by reflex activation. Involuntary active motion can also be produced by neurological conditions such as cloni, dystonia, chorea, etc. During limb and joint passive mobilisation muscle stretching can occur: a dedicated physiologic reflex produces instantaneous muscular contraction when the muscle fibres are elongated over a certain limit. This so called *stretch reflex* is much exasperated in spastic syndromes and may produce an involuntary, uncontrolled increase of resisting loads. Spasticity strongly depends on the speed of stretching; preconditioning the limb by a few cycles of slow manual stretching can help reduce momentarily the intensity of stretch reflex. Very severe spasticity, however, may correspond to the absolute impossibility of mobilising a joint.

Furthermore it must be remembered that the limbs adjacent to a joint possess a mass, and are therefore subjected to gravity and inertial forces connected to accelerations produced even in different parts of the body. One important case is gait. Where and when any or all of these components of the resisting or facilitating loads for the actuation must be taken into account has to be decided for each new application separately, in particular considering the typical position held by the target patient while the actuator is functioning, whether the patient will be prone to uncontrollable movements (jerks, dystonia, convulsions, etc.), or, on the contrary, whether some physiological movements will be typically impeded or suppressed (paresis, coma, castings, etc).

2.1.3. Interfaces

The forces and strokes generated by the actuator must be suitably transmitted to the patient's limb, so connective or fastening elements have to be carefully considered in relation to local anatomical constraints, such as the size of the limbs or body parts and the available surrounding space. In particular, when designing a wearable device, additional dimensional and weight constraints must be taken into account. Generally, the device should be attached to the limb segments that articulate at the target joint, in order to promote their reciprocal movement. Designing such connections should take into account the dimensions of limb segments (that strongly depend on age, sex and pathology) and the available space (determined primarily by joint position and orientation relative to the rest of the body, sex and malformation, etc.). For example, a device to move the shoulder should be designed taking into account the sex of the patient, especially if a frontal strap is required; in the case of a knee device, on the other hand, the lateral aspect of the joint could provide more free space for attachment than the medial one, so that interference with the other leg can be avoided. Another important factor linked to the transmission of forces and movement is the device-skin interface. This must be carefully designed so that no excessive pressure, friction or relative motion occurs during activation of the actuator. Some of these problem can arise due to misalignment of the actuator rotation axis with respect to the corresponding anatomic joint axis. In case of wearable applications, also the weight and stability of the device on the limb should be taken into account.

2.1.4. Other Issues

Safety is a mandatory constraint when designing medical devices in general. For the particular case of rehabilitation actuators based on SMA, most risks could arise from electric, mechanical or thermal hazards. IEC 60601-1 is a widely accepted standard for medical electric equipment, and many countries tend to require compliance with IEC 60601-1 before commercialization of an electrical medical device is permitted [55]. Even if the medical device is intended for use only in the scientific laboratory, reference to this set of prescriptions may suggest precautions and guide designers and engineers in the development of safe devices. In connection with this latter standard, international standard ISO 14971 specifies a process for identifying the hazards associated with the use of medical devices and carrying out a suitable risk analysis [56]. Consequently, good practice requires to try and minimise all the identified risks. Moreover, when the medical device is approaching clinical validation, another norm to be considered is ISO 14155, which addresses the issues of designing, conducting, recording and reporting of clinical investigations carried out on human subjects [57]. All of these guidelines are valid also for SMA-based devices and may help consider further implications of new designs for the human body.

Dealing with patients, there are still other questions to be considered. Even if the device is functional and efficient and the actuator fulfils all the technical requirements, they will not be employed in the clinical setting if they are not acceptable to the patient. At the basis of *acceptability* there is a set of physical and psychological factors. Of course, the medical device should not produce nuisance or pain during short or prolonged use. Those problems, in the particular case of rehabilitation appliances and actuators, may depend on badly-designed interfaces with the skin or on movements imparted too fast. In fact, as already stated, contracted and spastic muscles are very responsive to stretch, especially when it is applied rapidly. During mobilisation, the increased muscle tone due to spastic reflexes may result in much discomfort for the patient. The use of SMA actuators in general has the advantage of providing low movement rates; however, severe spasticity may be hard to treat even with slow active devices. Other acceptability issues are often connected to psychological aspects. In particular, wearable devices should not embarrass the patient with noisy, bulky or cumbersome structures. For children, shape and colour can also be important features. SMA actuators are likely to be silent and miniaturised, and thus suitable in this respect, but all other components may not (i.e. power supply). Other impediments for patients employing rehabilitation devices actively or for assistive therapies may arise from their possible co-existing cognitive impairment: this consideration supports the use of very clear instructions and user interfaces, especially using non-verbal pictorial communication, lights, vibrations or sounds.

2.2. Actuator Design

2.2.1. Strokes and Loads

Rehabilitation devices in most cases are intended to generate a mutual rotation of limb segments around an anatomical joint. By employing rotary actuators, it is possible to provide directly angular strokes and torques. On the other hand, linear actuators can be coupled with suitable lever arms and achieve the same results indirectly. The first step in designing an SMA-based rehabilitation actuator is to define, by careful analysis of the biomechanical constraints and clinical requirements, the appropriate neutral position, range of motion and resisting loads. The neutral position and range of motion will help determine a correct linear or angular stroke for the actuator; the resisting loads will set the linear force (and lever arm) or torque requisites. Of course, the two quantities will influence each other, for example different choices of lever arm may affect the linear stroke.

Once the required stroke and load have been determined, the geometrical parameters of SMA element can be worked out. Different types of actuators need different formulas for determining the parameters: for example, wire actuators have only two parameters, length and cross-section that directly depend on stroke and load respectively, while springs (helical, torsion or flat) parameters comprise coil dimensions, wire diameter and number of turns, which all depend on both stroke and load. Moreover material properties influence actuator parameters, especially in springs where anisotropy may play an important role [34]. Attention should be paid in particular to the critical stresses needed to detwin martensite both in the heated (above A_f) and cooled (below M_f) states. This consideration can guide the choices about material composition and thermomechanical treatment, because those factor have great impact on the height of the plateaux and the mechanical hysteresis. In most actuation applications, in fact, it must be guaranteed that the load to be displaced is low enough that stress in the material is lower than the critical stress for the heated state; on the other hand, sufficient loading should be provided to detwin martensite and ensure actuator deformation and position reset during cooling for most cyclic applications. Whatever the SMA geometry, it can be stated in general terms that cross-section depends mainly on the maximum desired work output, through the mean level of stress acting on it. The question of choosing the level of stress is quite complex, and also involves the expected fatigue life of the actuator and the strains recovered upon activation. For example, it is reported that commercial NiTi-wire actuators can remain stable for more than 10^5 cycles at 200MPa and 4% of strain in traction [58], but other optimal compromises can be found, provided that stress does not exceeds 400MPa and strains 5%. The use of parallel actuators can help reduce required work from each actuator, i.e. given a stress level, cross-section can be reduced. Dimensioning cross-section should also take into account cooling rates, as will be discussed in the next section.

2.2.2. Power

Heating and cooling are key points concerning SMA actuation. The general equation that drives these processes is

$$P_{in}(t) = c_p m \frac{dT}{dt} + m\Delta H \left|\frac{d\xi(T)}{dt}\right| + \frac{dW(\theta(t),t)}{dt} + P_{diss}(t). \tag{1}$$

Considering the heating (actuation) phase, $P_{in}(t)$ is the heating power, $c_p m \frac{dT}{dt}$ is the power needed to increase SMA element temperature (c_p is the constant-pressure specific heat and m is the mass), $m\Delta H \left|\frac{d\xi(T)}{dt}\right|$ is the power involved in phase transformation (ξ is the fraction of austenite and increases in time as transformation goes on, ØHis the transformation enthalpy), $W(\theta(t),t)$ is the mechanical output (useful work, inertia and frictional losses) which varies along the movement trajectory $\theta(t)$, and$P_{diss}(t)$ is the power dissipated by thermal convection, thermal conduction and (less important in this case) thermal irradiation. Considering the cooling phase, and imagining that a bias force acts to restore the starting actuator configuration (by deforming the SMA element), $P_{in}(t)$ is again a heating power that could be used to help control the cooling rate, $c_p m \frac{dT}{dt}$ is the power to be dissipated in order to decrease SMA element temperature, $m\Delta H \left|\frac{d\xi(T)}{dt}\right|$ is the power involved in phase transformation (enthalpy in this case is negative), $W(\theta(t),t)$ accounts for the mechanical energy stored in detwinned martensite plus inertial and loss terms, and $P_{diss}(t)$ is the power dissipated. The contribution of $P_{diss}(t)$ is present both during the heating and cooling phases of an actuation cycle, but it may differ significantly if any active cooling system is included. Equation 1 gives a simplified description of the process, but permits to focus on some key points. First, SMA mass is an important factor when designing power consumption. Second, as all the diverse forms of dispersed energy depend on the amount of exposed surface, surface/mass ratio is the driving contribution to determine cooling time. Therefore, thin actuators (maximising the surface/mass ratio) should be preferred to bulky ones if cyclic, especially high-frequency actuation is desired. Having in mind which use is intended for the device provides a guidance for the choice of heating source and consequently gives some hints for the actuator design. The best strategy for providing controllable, cyclic actuation is heating by Joule's effect, but, especially for non-cyclic and one-shot activation, also direct thermal transfer may be suitable. Focusing on Joule's effect, the energy provided depends on the injected current and the electric resistance of the SMA element according to Equation 2:

$$P_{in}(t) = P_{Joule}(t) = I(t)^2 \cdot R_{SMA}(\xi) \qquad (2)$$

where the electric resistance RSMA(*F*)RSMA(F) can be expressed as

$$R_{SMA}(\xi) = \rho(\xi) \cdot \frac{L(t)}{S} \qquad (3)$$

with $\rho(\xi)$ representing the SMA resistivity, S the cross-sectional area of the SMA element, L(t) the length over which the current flows, that obviously changes as actuation goes on. Given a length of SMA wire (which depends directly on the stroke required from the actuator), the thinner the wire, the higher is the electrical resistance. Moreover, mass-dependent contributions to Equation 1 are also minimised, and heating efficiency increased.

SMA actuators are subjected also to the Clausius-Clapeyron relation, that can be expressed as:

$$\theta^* = \theta + \frac{\sigma}{C} \tag{4}$$

where $\boldsymbol{\theta}$ is one of the transformation temperatures, which is effectively increased to $\boldsymbol{\theta}^*$ when a stress σ is applied to the material. This is a very important issue when purchasing shape memory materials or dimensioning SMA actuators powering: the transformation temperatures as provided by DSC analysis will be shifted when the material is loaded. Typical values of C for NiTi range from 5 to 9 MPa/°C, i.e. an actuator designed to bear 200MPa would have transformation temperatures increased by as much as 40°C. However, actuation temperatures cannot be increased indefinitely: in fact, above a certain temperature (M_d), part of the deformation cannot be recovered and material undergoes plastic slip.

2.2.3. Choice Of The Sma Element

SMA actuators can be provided in a number of shapes, which may differ significantly from one another in terms of available stroke and force. The first distinction is based on the deformation mode imposed onto the material. Wires are subjected mainly to traction, which produces a uniform loading of the material cross-section: this fact makes it possible to produce high force actuators, but, on the other hand, the available stroke is limited by the fact that linear deformation can be as high as 8% only, 4% in most cyclic operating modes. Other simple SMA actuators are linear springs (tension or compression springs), whose principal mode of deformation is torsion. Springs of this kind are helical coils of SMA wire arranged around a central axis: this allows for large elongations along the central axis depending on spring ratio and the number of turns. On the contrary, force is limited by the fact that torsional loading is not uniform in the material cross-section. Some authors proposed also bending actuators, made of SMA ribbons [27] or SMA wires [33]. For the same purpose, also torsion springs and flat springs can be employed, which share the same principal mode of deformation, i.e. flexion. The advantage of this actuation is that angular stroke is directly available, but the non-uniform cross-section loading limits the available torque.

The problem of moving human joints generally involves large angular strokes coupled with fairly high levels of force or torque. The simple configurations just described most of the time provide solutions only for very long wires or thick cross-section springs, ribbons or bars, which are often deemed impractical for a number of reasons. Long wires have to be housed in a suitable manner, to limit length into a compact three-dimensional structure. A way to do so is coiling the SMA wire along a series of pulleys, whose diameter should be sufficiently large with respect to the wire diameter, to avoid strain concentrations. On the other hand, large cross-section actuators are impractical because they need high electric power to reach the transformation temperature. Moreover, cooling down bulky wires or ribbons may require too long a time for most cyclic applications. A possible solution for increasing force output in springs and ribbons is arranging many actuators in parallel or in bundles. Provided space around the joint is sufficient, attention should be posed to electrical insulation and electrical connections, in order to provide suitable solution to tension-current requirements for actuation, keeping in mind also the safety issues.

2.3. Actuator Control

The problem of controlling actuation is very important for many applications in Rehabilitation, as patient's needs and responses may vary during the evolution of therapy or even during the same session of exercise. Control strategies apply mainly to assistive robots, but may also be employed in passive mobilisation devices. As the focus of this Chapter is not on control systems, only a brief overview of possible applications to SMA rehabilitation devices will be discussed.

2.3.1. Control Of Passive Mobilisation Cycles

Two major strategies exist to control cyclic heating and cooling of SMA actuators: in *open-loop control*parameters are predetermined by a set routine, whereas *closed-loop control* employs a feedback signal of any nature to adjust heating and cooling parameters (timing, current or voltage, active cooling systems, etc). The main drawback of open-loop control is that any perturbation to the trajectory would not be compensated, as the current profile is predetermined. This could happen, for example, if patient exerts any unexpected active contraction or in the case of unpredicted changes in environmental conditions. However, there are applications in which open-loop control is still feasible, for example repetitive passive motion of flaccid limbs. Closed-loop controls need monitoring of a feedback variable, which could be the SMA temperature, the joint angle, actuation speed, SMA force or any other measurable physical quantity. Apart from rare examples of feedback loops based on monitoring of SMA resistance [59], closed-loop controls rely on dedicated sensors that are better included from the initial steps of the device design.

Another distinction can be made on different types of control of movement trajectory. In many applications an ON/OFF actuation is suitable: the focus is only on the initial and final positions, while the detailed trajectory is controlled only mechanically and the speed only in average terms. In this case, open-loop control may be practical, and heating can be achieved with a very simple current profile, e.g. a step or a ramp. Experimental tests are a viable solution to adjust the heating and cooling parameters. In particular, attention should be paid to the experimental conditions, including the orientation of the actuator in the gravitational field, as thermal convection is strongly dependent on that.

On the other hand, when a precise trajectory of movement over time is required for the application, alternatively open-loop or closed-loop control strategies can be applied. In the open-loop approach,Equation 1 should be solved by imposing the desired mechanical output $W(\theta(t),t)$ and by calculating the time-varying current profile to be injected. However, the most reliable way to provide trajectory control is closed-loop control, in which the current input is continuously adjusted to match the prescribed position, thus counterbalancing perturbations.

2.3.2. Assistive Control

As rehabilitation implies that patient's capabilities evolve during therapy, the characteristics of therapeutic tools could be devised to change accordingly. A way to implement this is to trigger SMA actuators only when the patient makes some effort to move the paretic limb: thus, a measure of the "effort" should be chosen to implement such a biofeedback-based closed-loop control. In some cases, mechanical variables could be employed: for example, the SMA actuator completes movement only if the patient generates at least a certain level of voluntary muscular force or flexes the joint over a minimal extent. On the other hand, possible biosignals for closed-loop control can be the surface electromyographic signals from the muscles that control the movement to be rehabilitated. The level of the biofeedback variable above which actuators are switched on may depend on the patient, pathology and progression of therapy, and therefore should be adjusted to match patient's capabilities with rehabilitation goals.

Besides the biofeedback signal, another key point for assistive rehabilitation is how to encourage the patient to participate actively in the rehabilitative session. Visual or verbal or acoustic feedback are the most common way to prompt, instruct and reward the patients during the exercise, but when selecting the communication format attention should be paid to their possible co-existing cognitive impairment.

2.3.3. Gait Triggering

A very peculiar example of precise trajectory control is gait, during which any wearable active device should be able to activate according to the walking rhythm of the patient. A closed-loop control is desirable, and can be implemented with mechanical sensors (which detect angles at different joints or contact forces when the foot contacts the ground) and/or electromyographic monitoring of the leg muscles. Gait is a quite demanding application for SMA in terms of frequency of activation, as even slow walk takes place at over 1Hz. This influences dramatically actuator design, i.e. diameter and number of SMA elements. For gait applications, movement trajectory must be controlled fully within the step and from step to step: in other words, both heating and cooling phases should have adjustable parameters so that actuation can adapt to changing stride or external perturbations. The possibility of modulating input current profiles has already been discussed and may be sufficient for the control of heating; cooling, on the other hand, shall not be left to natural convection in this case, but an active system for accelerating and tuning heat transfer will often have to be adopted.

2.4. Case Study: Passive/Assistive Sma Device For Ankle Dorsiflexion

2.4.1. Design Aims and Clinical Constraints

Ankle is one of the body joints that often suffer from the sequelae of stroke, i.e. immobility, disuse, contracture and spasticity. After stroke, a rehabilitation program should be carried out as soon as the patient's conditions are sufficiently good, but in the acute phase the patient is unable to exercise actively. In these circumstances, moving the limb passively can help maintaining joint flexibility and normal muscle tone, while also contrasting contracture, spasticity and sensory deprivation. A rehabilitation device for promoting repetitive dorsiflexion of the ankle in the first few weeks after the acute event could help improve clinical outcomes. The fact of limiting mobilisation to just one degree of freedom (sagittal plane of the joint) is appropriate because dorsi-plantarflexion is the principal motion involved in gait. It is regarded as a matter of importance that the device be able to manage patient's changing conditions during recovery. As needs evolve during patient's improvements, this therapeutic device should be able to guide and sustain gradual recovery by providing commensurate aid. This includes exploiting even initial attempts at voluntary motion and turn those into effective workout. To this end appropriate control will have to be included.

The device is intended for patients in the first weeks after a stroke, i.e. we should expect adult subjects having a flaccid ankle (no active control of the muscles, mild or no hypertone) and possibly increased ankle stiffness. Figure 2 shows the passive viscoelastic ankle characteristic of a healthy subject at rest (no spasticity, no contracture, no voluntary movement) and a typical chronic flaccid patient with mild contracture. The ankle joint can move in the sagittal plane in the approximate range -30°/+20° (negative being towards plantarflexion) even though the range of -15° to +10° is the functional range of utmost interest in gait. These ranges become reduced with evolving contracture and spasticity, and in particular the positive degrees are progressively lost. In the range of interest (-15°/+10°) the characteristics of the presented healthy and paretic joints differ mostly for dorsiflexed angles (greater than around +2°, in this case), where increased stiffness in the paretic ankle can be observed. The typical acute flaccid patient will have intermediate characteristics between the healthy one and the chronic.

Given this information, the biomechanical constraints on the design can be summarised as follows:

1) the resisting forces which have to be considered in actuator dimensioning are the viscoelastic passive characteristic of the joint and the foot weight. No active contribution from the muscles are expected.

2) Choosing a reduced range for the device, biased towards dorsiflexion, has the advantage that the calf muscles are kept under a relative stretch (physiological neutral angle is around -25°); furthermore it is appropriate to avoid forcing the joint excessively towards the extreme dorsiflexion position (conservative approach to limit insurgence of spastic reflexes). A target ROM for mobilisation was therefore was chosen as -5°/+10°. Of course, the actual working range will depend on the passive characteristic of the specific ankle, but is not expected to be much different from the target one.

3) For a target population represented by this type of subjects it is reasonable to assume that passive resisting torque does not exceed 400Ncm in the range of interest.

Considering the general conditions of the patients in the first weeks after a stroke, it would be wise to design a device that can be employed in the bed. However, the knee joint should be sustained in a flexed position, so that the bi-articular *gastrocnemius* muscle is not pre-stretched and full ankle range of motion is available. As a way of compromise, an angle of 10° for the leg rest can be assumed, with the foot positioned lower than the knee. It should be taken into account that, in such a configuration, the contribution of foot weight to the resisting force is limited. Figure 3 shows the total resisting torque, comprising different foot weights, and the same viscoelastic resistance from a typical patient. The expected maximal resisting torque can be approximately calculated in 400Ncm. As the curve hysteresis is not large for this joint (cf. Figure 2), the same values used for dorsiflexion can be also utilised when considering the movement towards plantarflexion and the loading level associated to martensite detwinning.

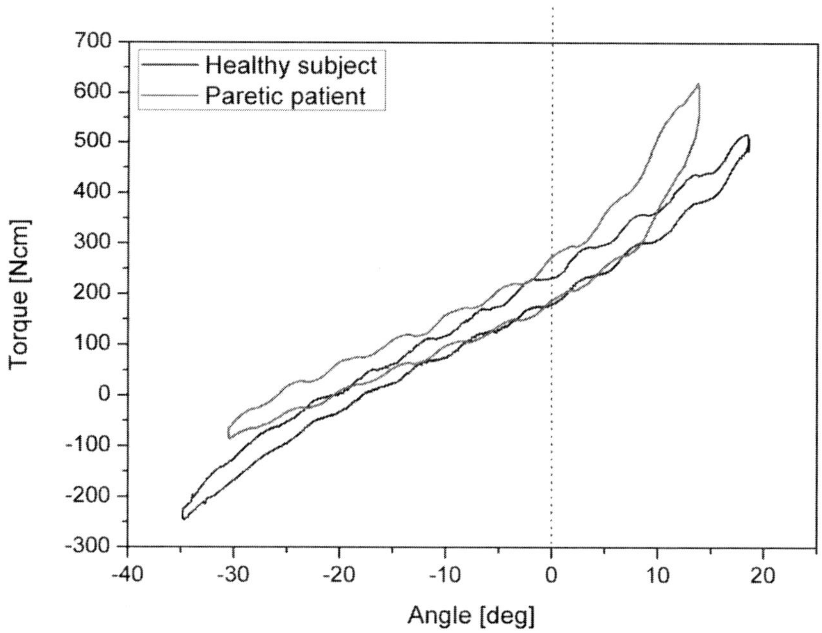

Figure 2. Measured passive characteristic of the ankle in a healthy subject vs. a paretic patient with mild contracture and ROM limitation. The main difference appears for positive dorsiflexion angles

As the device is not intended for walking, there are no direct constraints on cycle duration, provided that a suitable number of cycles can be delivered to the patient's ankle in a therapeutic session (at least 100 cycles/hour), and movement speed is not too elevated, lest spastic responses are aroused. The actuator should provide repeatable movement and maintain the same characteristics for a number of sessions, as it may be impractical to adjust or change the SMA actuator too often. For this reason, maximal strain should be limited to 3% and stress to 300MPa.

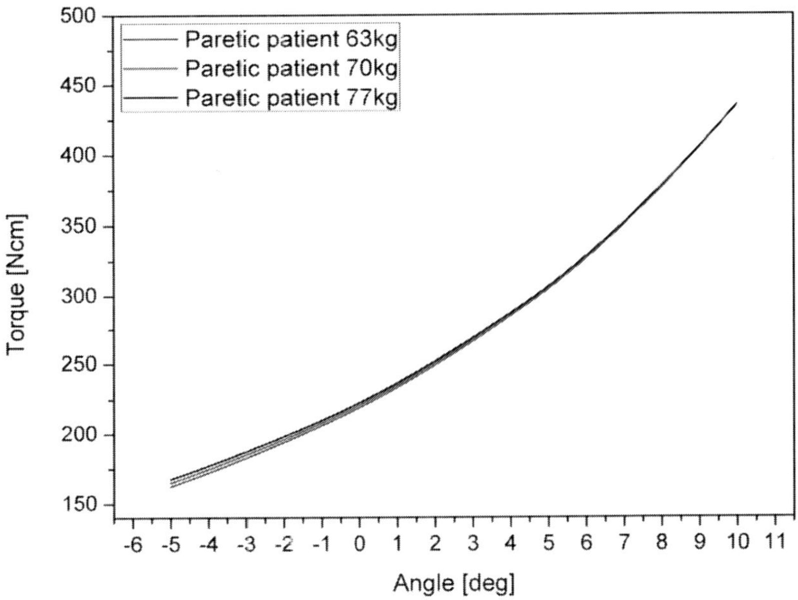

Figure 3. Influence of foot weight on the total resisting torque at the ankle. Differences in foot weight (proportional to the body weight) account for little change due to the lying position of the limb.

2.4.2. General Concept Of The Device

Figure 4 shows a simple embodiment of the concept. Two thermoplastic shells lined with soft foam are modelled on a prototype human lower limb of average size. These shells are hinged together and strapped by Velcro® bands respectively on the frontal aspect of the shin and the foot, in such a manner that the ankle and the hinge are perfectly aligned along the same axis (this minimises unwanted friction and sliding between the orthosis and the skin). The choice of planar hinges was also made to control (fixate) ad/abduction and in/eversion, although this solution makes the device unsuitable for patients that have already developed severe malformations out of the sagittal plane (unusual in the acute phase). With this structural configuration, two linear actuators are fixed on the front of the leg shell and transfer actuation pull to the foot shell through inextensible threads.

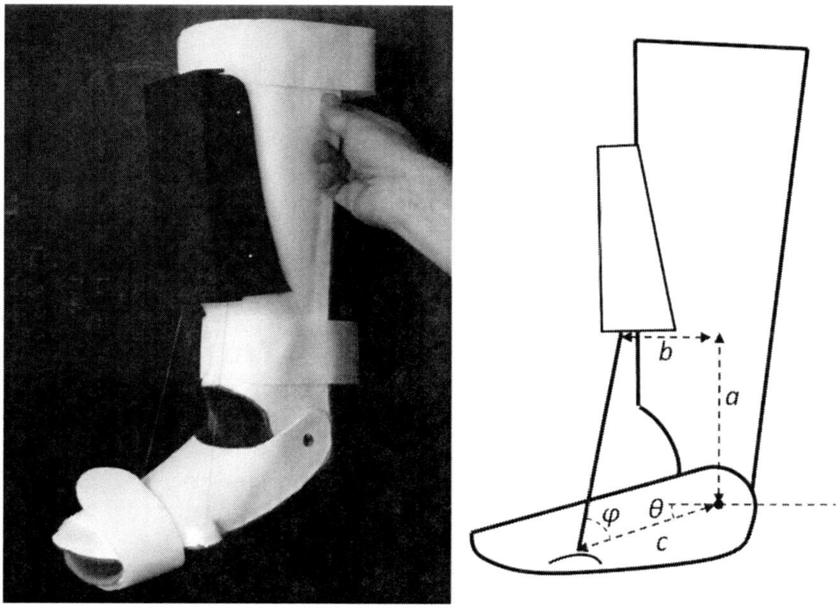

Figure 4. Implementation of an exerciser for the ankle joint with linear SMA actuators. The device is designed to provide cyclic joint flexion in the rehabilitation of neurological patients in the acute phase. On the left, schematic representation of the main design dimensions to calculate the lever arm.

The linear stroke and force output required of the SMA actuators depend, through the lever arm, on the fixation points for the actuators (on the shin), and the distal ends of the inextensible threads (foot shell). Actuators produce dorsiflexion, while plantarflexion and actuators position reset is left to gravity and viscoelastic resistances.

Power supply is to be provided by a dc-generator, which is not on board the wearable device, as the intended use is not for walking. This choice helps limiting weight and keep the device stable on the patient's leg during use.

An open-loop control strategy is adopted for the use as passive exercise device, whereas a simple closed-loop control for assistive rehabilitation employs the electromyographic activity from *tibialis anterior* as control signal, picked up by surface electrodes. As for the power supply, it was decided that all the components needed for control are not mounted on board the wearable device, even though integration would be possible by developing *ad-hoc* electronics.

2.4.3. Actuator Design and Characterisation

With the selected set of geometrical parameters (Figure 4: a=16cm, b=8.5cm, c=15cm), a linear stroke of 2.5cm is required to produce an angular movement across the range -5°/+10°. Considering a linear deformation of 3%, the required amount of wire for linear actuation would be 83cm. However, in our case it would be impractical to have a free-standing wire of such a length as a linear actuator. For this reason, the NiTi wire should be confined into a more compact actuator. In dimensioning the length of wire now we must take into account also the local deformations that inevitably will be imposed on the SMA wire when coiling it. The actuator was designed as an insulated cartridge wherein the necessary reference length of NiTi wire is led back-and-forth between two arrays of ten pulleys. An end of the wire is connected to an inextensible thread that transmits the force to the foot, while the other one is fixed to the housing. A pseudoelastic spring is connected to the moving end of the NiTi wire in order to keep it just taut, and its pull is negligible during actuation.

Considering the force requirement, each actuator should bear half of the resisting torque T_r showed inFigure 3, and the amount of force varies at varying ankle angles θ (defined between the foot direction and the horizontal, Figure 4):

$$F(\theta) = \frac{T_r(\theta)}{c \cdot \sin\varphi} \qquad (5)$$

where ϕ is the angle between the inextensible thread and the foot and depends on θ according to:

$$\varphi(\theta) = \frac{\pi}{2} - \theta - \arccos\left[\frac{c \cdot \sin\theta + a}{\sqrt{(c \cdot cos\theta - b)^2 + (c \cdot sin\theta + a)^2}}\right] \qquad (6)$$

With a maximum torque of 200Ncm, Equations 5 and 6 give a maximum load of 13N on each actuator. By selecting a 250μm-diameter commercial NiTi wire, stabilised for actuation, the maximum stress on the wire cross-section is 266MPa. In order to complete plantarflexion, full martensite detwinning should be achieved. It is certainly possible to find stabilised wires for which martensite detwinning occurs with less than 100MPa, corresponding to around 5N pulling on a 250μm-diameter wire. With the expected resisting torque in Figure 3 and the geometrical parameters, force requirement for martensite detwinning is always satisfied in the range -5°/+10° and thus cyclic actuation will take place in that range. Considering a diameter of the pulleys of 12mm, the maximum amount of localised strain is 2%. In order to respect the 3% limit for the maximal deformation, only 1% of strain is available for actuation and can be employed for calculating the length of wire to be coiled within each actuator, that is to say, 250cm.

Having limited strains to 3% and stress to 300MPa, a cycling life of 30k-100k cycles is expected [58].

2.4.4. Power Dimensioning

In order to evaluate the parameters for Joule's effect heating, various experimental tests were carried out on samples of NiTi wire. First, the transformation temperatures were investigated by means of differential scanning calorimetry (on DSC 220 SSC/5200 - Seiko Instruments, Tokyo, Japan), showingA_f = 351K and M_f = 274K. However, calorimetry gives information about the material with no loads applied, whereas Equation 4 suggests that transformation temperature depends on the load applied. For this reasons, tensile tests on the material at different temperatures were conducted using an MTS 2/M thermo-mechanical test machine (MTS Systems, Eden Prairie, MN, USA) equipped with a 2kN load cell. The material was deformed up to an engineering strain of 5%, at 365K, 380K, 390K and 400K. The pseudoelastic plateau values varied as a function of temperature with a ratio of 8.402 MPa/K, which is exactly the constant C in Equation 4. With the estimated stress level of 266MPa, full transformation and strain recovery can be achieved at 383K.

Activation tests on the wire were conducted injecting different currents (0.65A, 0.7A, 0.75A, 0.8A) in the wire at a constant strain of 3% for a set period of 13s, in order to evaluate what current value is most appropriate to provide the working load of 13N, minimizing current expenditure and heating time. As expected, the higher the current the faster is actuation. Patient safety considerations should be taken into account, as well. A compromise solution can be accepted with 0.7A, which allows for reaching 13N within 6s. With the selected current, voltage is limited to 35V, considering an average resistance of 49.5 Ω (45-54Ω during transformation from fully austenitic to fully martensitic). Notice that the rehabilitation device mounts two actuators, which need a 0.7A current each. The electrical configuration of the two actuators could be a series or a parallel circuit: in the first case, the dc-generator should be able to provide 0.7A at 70V, in the latter case 1.4A at 35V. According to IEC 60601-1 specifications, dc tension preferably does not exceed 60V, i.e. in our case the two linear actuators are better connected in parallel.

Heating parameters can be extracted by testing a free-standing NiTi wire, and by applying 35V to the ends of the cartridge actuator suspending a 13N weight. It was demonstrated that nominal linear stroke is reached in around 6s (mean dorsiflexion speed 2.5°/s). On the other hand, cooling times depend strictly on the geometrical arrangement of the wire in space. A compact actuator may have a considerably slower cooling and position reset time, as tests on the actuator confirms. Basically, if the vertical free-standing wire cools down in approximately 10s, cooling and actuator position reset takes place in 30s. The full cycle thus lasts 36s, which makes it possible to deliver 100 cycles/hour to the patient's ankle, as required by the design specifications.

2.4.5. Control Strategy

An open-loop control is adopted for using the device as a passive exerciser. A computer routine was written in LabView (National Instruments, Austin, TX, USA) to control an electronic relay (NI9481 - National Instruments, Austin, TX, USA) that closes the electric circuit between the dc-generator and the actuators.

For the assistive therapy, a closed-loop architecture is preferred, with surface electromyographic signal (sEMG) from *tibialis anterior* as control variable representing the patient's attempt to move his ankle. Three Ag/AgCl electrodes (positive, negative and reference) were used to pick up the signal. Analogical waveforms were acquired, pre-amplified and band-pass filtered (18-478Hz) before being digitalised (sampling at 1000Hz) using an NI9205 (National Instruments, Austin TX, USA) connected to an ordinary laptop computer running a LabView routine. Pre-amplifying and filtering stages were assembled using conventional 8-pin PDIP components, and included also a feed-back loop towards the body (akin to the driven right leg stage used in electrocardiographers). Further signal manipulation is built into the control software, including rectification and low-pass filtering (3Hz), so that the obtained waveform is sufficiently smooth that it can be employed as a measure of instantaneous muscular activation.

Through the user interface of the control routine two patient-specific sEMG threshold values can be selected. The lower one is set to the minimal required level of exercise (which can lie even lower than the muscular motor threshold, in some cases); the upper one is set to an appropriate activation representing the ultimate (or a higher) therapeutic goal, i.e. an effective motion.

At the start of the assistive exercise session, a visual cue is presented to the patient to dorsiflex the ankle. Then the system is set on hold waiting for the sEMG from *tibialis anterior* to cross one of the threshold values. If the lower threshold is reached, then the system waits a few milliseconds for the upper threshold also to be reached. If this latter event does not occur before the time-out, then power is supplied to the actuators and the motor task is completed for the patient. If, on the contrary, the upper threshold is reached, then a visual feedback is provided to the patient that the higher goal was hit, while the device does not intervene to support the movement.

2.4.6. Experimental Tests On The Device

The assembled device was tested with a static load of 40N fixed on the foot shell at a distance of 10cm from the hinge axis, while measuring the angular displacement by means of electrogoniometer SIM-HES-EG 042 (Signo Motus, Messina, Italy). The device was tested for 4000 cycles, with the actuators powered with a 6s step of 1.4A (0.7A per cartridge) and 30s allowed for cooling. The results of these tests (Figure 5) demonstrated that stroke is quite stable across cycling. Plantarflexion occurs more rapidly after cycling because of the instauration of two-way shape memory. Following this result, by employing cycled wire or wire stabilised for two-way shape memory, cycle duration could be reduced from 36s to 30s without changing the stroke, if desired (reaching 120 cicles/hour).

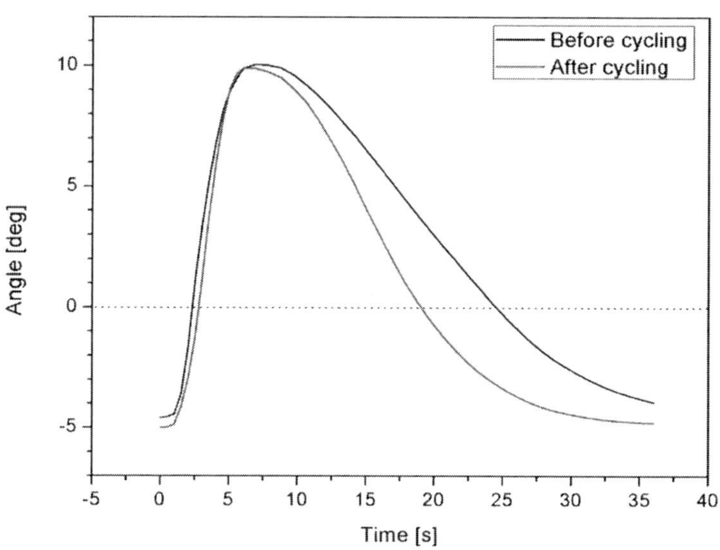

Figure 5. Angular stroke of the SMA-based device for ankle dorsiflexion: first activation and after 4000 cycles against a resisting torque of 400Ncm.

The closed loop control strategy was tested on three healthy volunteers (28.17±6.08 years old). Ag/AgCl electrodes were placed on the belly of the *tibialis anterior* muscle of the dominant side, the corresponding distal muscle-tendon junction, and the internal malleolus (driven electrode). Subjects were asked to perform a maximal isometric contraction at the ankle neutral position, then to sustain the minimum voluntary activation they could manage. Subsequently, values were set for the lower (110% of minimum individual contraction) and upper (60% of individual maximum isometric contraction) thresholds. Then, the control routine was launched and subjects were asked to follow on-screen instructions (graphic and written) trying to respond with just a supra-minimal contraction when cued to dorsiflex the ankle.

The measured joint angle and sEMG time courses for a representative subject are shown in Figure 6. When the lower threshold was crossed, the system triggered the powering of the orthosis, which completed the movement of dorsiflexion (assisted active session). It can be appreciated how passive mobilisation can be triggered by a very subtle muscular contraction, which may correspond only to a very slight movement (~1°). The movement produced by the orthosis as a consequence of a minimal contraction brings along some degree of reflex sEMG activity: this may also be thought of as an interesting result to the effect of rehabilitative exercise.

3. SMA ACTUATORS FOR NEUROSCIENCE

The development of SMA devices for use in Neuroscience is a very challenging task as diagnostic equipment utilised in this field of research is generally very sensitive to electromagnetic noise. Whereas some techniques (e.g. electroencephalography or near-infrared spectroscopy) are less affected by environmental conditions, magnetic resonance imaging (MRI, fMRI) and magnetoencephalography (MEG) acquisitions can be carried out only in highly shielded Faraday cages, to abate electromagnetic noise. Therefore, materials and devices must comply with a number of requirements, in order to be allowed into the acquisition room. In the next paragraph, these two diagnostic techniques will be discussed separately, as their peculiar characteristics demand different compatibility constraints.

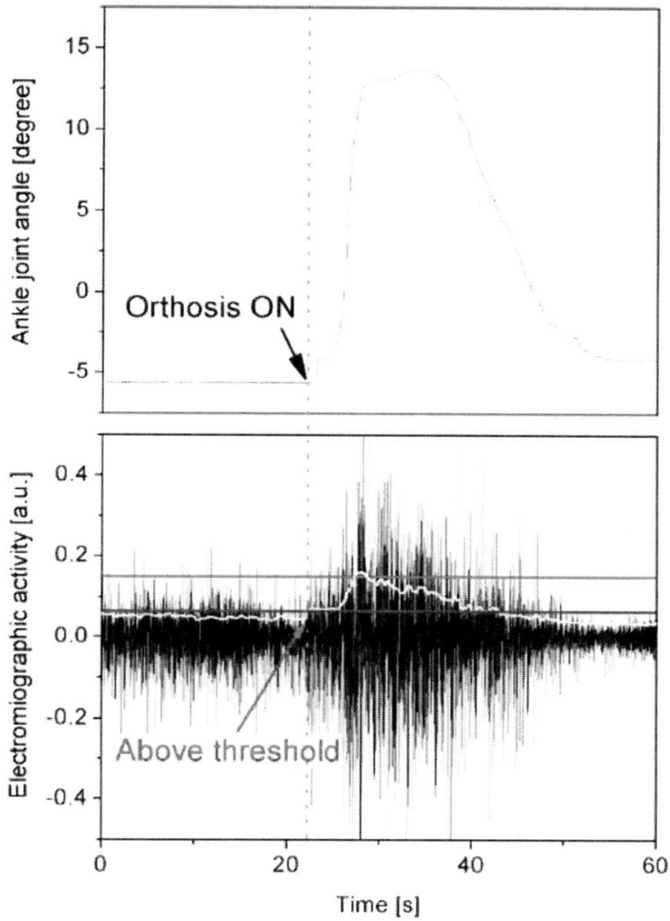

Figure 6. Top: angle timecourse measured during an EMG-triggered activation of the device for a healthy volunteer. Bottom: EMG recording during the trial. When the subject's muscular activity crosses the lower threshold, actuators are triggered to complete the movement of dorsiflexion.

3.1.Electromagnetic Constraints

3.1.1. Functional Magnetic Resonance Imaging (Fmri)

fMRI is a magnetic resonance procedure utilised in Neuroscience for measuring brain activity by detecting associated haemodynamic responses. MRI equipment uses a strong static magnetic field to polarise the nuclear spins of atoms in living tissues: the higher the magnetic field, the higher the polarisation of the spins, thus, the higher signal-to-noise ratio. Commercial scanners can generate polarisation fields up to 3T, but efforts are being made to increase the available magnetic field far beyond [13]. In addition to this static magnetic field, the MR scanner uses rapidly-varying magnetic field gradients for spatial encoding during the imaging sequence. The use of strong polarisation magnetic fields entails safety precautions, as high forces can be exerted especially on objects made of ferromagnetic materials, which may become dangerous projectiles for the patient, medical personnel, and the instruments. Besides safety issues, ferromagnetic materials generally affect image quality, as they alter the homogeneity of the static magnetic field. Another potential source of artefacts is the presence of objects made of conductive materials inside the acquisition room. In fact, the magnetic field gradients can induce electrical fields and currents flowing through the conductive materials (eddy-currents): these currents, in turn, induce magnetic fields which inevitably interact with the MR fields and affect the quality of the images. In much the same way as eddy-currents, it can be inferred that any current flowing inside conductive materials could generate image artefacts.

Given these strict constraints, a possible solution for moving the limbs of a patient during fMRI investigations by an actuator is to keep the actuator and control system outside the scanning room and transfer the mechanical energy to the subject via pneumatic, hydraulic, or mechanical (pulleys, ropes, etc.) means. These systems are usually complex and suffer from long transmission lines accompanied by dissipative and delay effects. It would be desirable to place actuators directly inside the acquisition room, but this limits the possibilities of using many categories of actuators. Of course, electromagnetic actuators are, in general, not compatible with the MRI environment, but also pneumatic or hydraulic motors could hardly get into the acquisition room, because of their bulky metallic components. Generally speaking, in order to be utilised in an MRI acquisition room, actuators should not include ferromagnetic and conductive materials. However, it is reported that small parts made of MR-incompatible materials do not compromise safety or generate image artefacts as long as they are sufficiently small and appropriately positioned relative to the imaged area [13]. This clears the way to the use of SMA actuators in this field, provided that some specific design rules are respected and that they are applied to body segments sufficiently distant from the head, which is the area of principal interest in Neuroscience.

3.1.2. Magnetoencephalography (Meg)

MEG is a technique for investigating neuronal activity in the living human brain by recording magnetic fields produced by the electrical currents flowing in cortical neurons. These weak magnetic fields (ranging 10^{-14}T-10^{-12}T), can be detected by employing arrays of SQUID (Superconducting Quantum Interference Device) gradiometers, which convert the magnetic flux threading a pick-up coil into voltage. Since the magnetic fields to be measured are extremely small as compared to the Earth's magnetic field (10^{-5}-10^{-4}T), MEG measures are carried out in a shielded room that minimises interference. Of course, any electromagnetic noise should be avoided inside the acquisition room: there is no safety issue for the patient connected with this constraint, but if not respected the measure would be impossible because SQUID channels would saturate rapidly and would remain unusable as long as the pick-up coils are in the superconductive state. Unlike the case of MRI, theoretically any material could enter an MEG acquisition room. In practice, objects made of conductive materials should remain still inside the shielded room, as any movement would generate an artefact. These constraints are very demanding for most actuation technologies, but design solutions are possible for SMA actuators.

3.2. Compatible Design Guidelines

Some design indications could be useful for devices intended for either fMRI or MEG studies. In particular, the SMA element should have a shape or should be arranged in a way that limits to a great extent the magnetic fields induced by eddy-currents and by the current used for Joule's heating. For example, the coil of a spring is not particularly indicated, as the intensity of the magnetic field is proportional to the number of turns. Moreover, the SMA actuator should be supplied with a dc-current rather than other types of time-varying currents (pulse-width modulation, sinusoidal current...). In fact, this could help reduce the magnetic field induced by currents flowing into the SMA elements. Power generators and control systems should be positioned outside the acquisition room and wires passed through. As these cables could act as antennas that radiate electromagnetic noise, some countermeasures should be adopted: shielded or twisted cables can limit pick-up of stray frequencies, while using proper filters when connecting the wires inside and outside the acquisition room can help rejecting the time-varying components of the power signal. Coming to the control strategy, an open-loop may be preferred (for its simplicity), as both fMRI and MEG data analysis generally need a precise windowing of signals that matches different phases of movement, in order to extract features of interest with statistical significance. So, closed-loop architectures could be used e.g. to control precise abidance to set movement speed evolution, but might not be strictly necessary if the experimental protocol only requires repeatable ranges of motion in a given time. In fact, within the very protected environments of the shielded rooms, often no major disturbances are expected to intervene and perturb the movement.

Additional recommendations apply only to devices for use in the MRI acquisition room: in particular, ferromagnetic materials should be avoided for safety reasons, even if small parts could be tolerated in some cases. On the other hand, as we said before, during MEG acquisition conductive materials should not move inside the shielded room. Translating this information into design specifications, a first recommendation would be that the moving parts should not mount conductive materials, including the SMA actuators. However, the principle of SMA actuation is that metal moves! The geometry of the SMA element should be conceived so that macroscopic changes of space occupancy are avoided, thus limiting the amount of movement. For example, in SMA helical springs the linear length can vary considerably during actuation, spanning in many cases more than 300% the unloaded length. On the other hand, traction wires, having a limited elongation, change their space occupancy less dramatically. Interestingly, it should be noticed that the slow actuation rates typical of SMA could be an advantage for limiting artefacts influencing MEG measurements. In fact, it is reasonable that artefacts have mostly the same time frequency components as the movement that generated them, i.e. typically less than 1Hz for SMA actuation. Cortical oscillations range 1-100Hz, but when investigating sensorimotor representation (as expected in the case of Neuroscience for Neuromuscular Rehabilitation), the range of interest reduces to 8-100Hz, giving the possibility to devise suitable signal filtering stages.

3.3. Case Study: Amagnetic Rotary Actuator For Ankle Dorsiflexion During Meg And Fmri Acquisitions

3.3.1. Concept

A possible device for ankle dorsiflexion could be devised as a leg part and a foot part connected by planar hinges parallel to the ankle joint, in a similar way to the device presented in section 2.4.2. Two SMA rotary actuators mounted externally with respect to the hinges promote the dorsiflexion of the foot part with respect to the leg part, while plantarflexion and position reset is left to viscoelastic resistance and foot weight (Figure 7). It should be noticed that the requirements for the ranges of movement are different in this application, as the purpose is to produce a clearly detectable movement without muscular stretching. This can be done maintaining joint angle negative or slightly positive (i.e. extended). The objective of the study was to test healthy subject, so it can be estimated that maximum resisting torque will be in the range 200-250Ncm (see Figure 2).

A rotary SMA actuator can be generally described as a structure in which the SMA wire connects parts that can rotate relative to one another about a central axis: when the elongated SMA wire is heated above A_f and recovers its deformation, the linear stroke ΔL is converted into a rotation $\Delta\theta$ of the moving parts. However, it is likely that a conspicuous length of wire is needed to produce a suitable amount of rotary stroke $\Delta\theta$ for ankle dorsiflexion. As we already said, in such cases coiling the wire could improve conveniently overall compactness of the actuator. Unfortunately, this design decision contrasts with the aim of fabricating an amagnetic actuator, as the electric current flowing in coils of wire produces, according to Ampère's Law, an overall magnetic field along the central axis of the solenoid, which is proportional to the number of turns. A very simple solution can be found exploiting the same physical laws of electromagnetic fields. In fact, the magnetic field vector is directed according to the right hand rule with respect to the direction in which the current flows that induced it. Thus, theoretically, given two identical and concentric solenoids traversed by the same electric current in opposite directions, the magnetic fields generated by the coils will mutually annihilate. Moreover, no ferromagnetic materials should be employed in the implementation of all other components of such actuator, as well as the assembled device.

3.3.2. Actuator Implementation and Mechanical Characterisation

An implementation of the invented amagnetic actuator is shown in Figure 8. Two discs of approximately 10cm in diameter are connected through a central shaft allowing reciprocal rotation, leaving a clearance of 2cm between the two discs. One of the two discs mounts, along its circumference, 6 pins bearing 3 pulleys each, to create a spiralling sequence, along which the SMA wire can be coiled suitably. These pulleys have a double triangular groove, which makes the double coiling as compact as possible. The two ends of the SMA wire are fixed to one disc, while the mid-point of the wire connects to the other disc. The two halves of the wire are would along the pulleys thus creating two concentric coils very close to one another. By providing sufficient current flowing between the two ends of the SMA wire, shape recovery occurs generating a linear stroke ΔL that is converted into a reciprocal rotation $\Delta\theta$ of the two discs. Notice that the torque generated is given by twice the cross-section of the wire, while the induced magnetic field is self-compensated to a large extent, as the two coils are traversed by the same electric current in opposite directions [40].

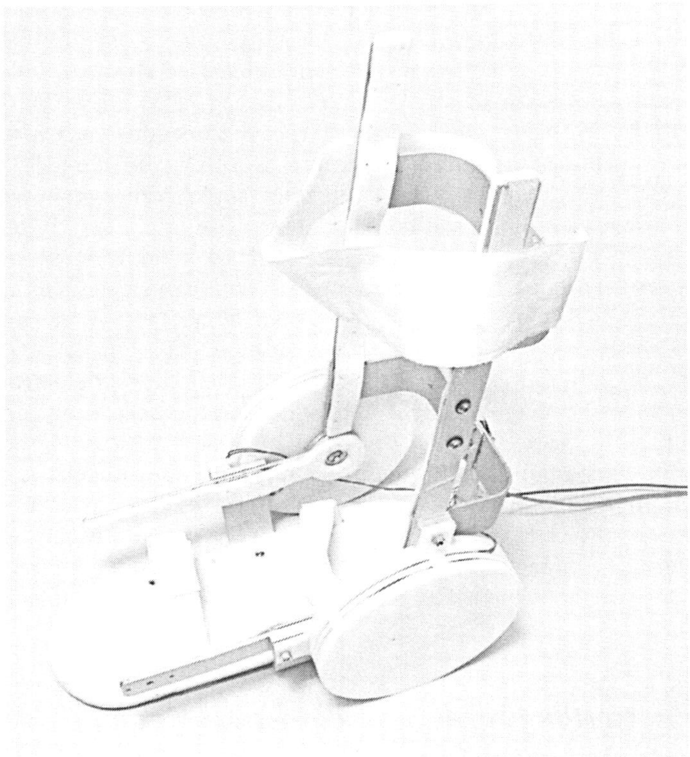

Figure 7. Implementation of an amagnetic device for ankle dorsiflexion with rotary actuators

As already described for the linear actuator in paragraph 2.4.3, wire length depends on the expected angular stroke and the working strain level, while torque output is connected to the wire diameter and mean stress on the cross-section. Moreover, the geometrical parameters of the assembled actuator affect both wire length and torque output. The rotary actuator was designed to provide an angular stroke up to 40° against resisting torques in the range 120Ncm-250Ncm. By limiting the linear strain to 3.8%, the required length of NiTi wire can be calculated in 219cm. The selected NiTi wire is 250μm in diameter, corresponding to stresses of 130-270MPa in the above-mentioned range of torques. Taking into account the localized strain on the wire resting on the pulleys, the total maximal strain reaches 4.42%, which in combination with the chosen stress level should guarantee suitable fatigue life (several thousands of cycles). Commercial stabilised NiTi wire was utilised, displaying large deformability (4.5%) at room temperature for stress levels as low as 150MPa [58]. The power dimensioning procedure was similar to the one described in paragraph 2.4.3 and led to utilizing 0.7A at 30V for each actuator. Control is achieved with an open-loop strategy, by which acquisition windows can be synchronised to the ankle movement. Power supply and control appliances are left outside the acquisition room, and shielded cables are let into the shielded room through suitable access vents.

Figure 8. Amagnetic SMA rotary actuator, consisting of two coils of wire traversed by the same current in opposite directions. This arrangement produces self-annihilation of induced magnetic fields.

In the present implementation, ferromagnetic materials were avoided. However, it was not possible to exclude completely metallic components, such as the bearing balls, the shafts and the pins. After careful consideration, it was decided that the rotating discs and the frame of the device could be made of an aluminium alloy, in order to minimise structural sections. This choice did not affect the acquisition of biosignals, as will be discussed in the next paragraphs.

Technical tests were carried out on the assembled device, to assess its characteristics. Increasing loads were attached to the foot part of the orthosis 13cm from the axis of rotation, while the device was held aloft by a static support. The weights used in this test produced resisting torques in the range 28-250Ncm. A direct current injection at 30V for 7s was applied to the actuators, connected in parallel; then 30s were allowed for position reset through natural cooling and the action of the weights. The resulting angular upward and downward strokes were measured by means of electrogoniometer SIM-HES-EG 042 (Signo Motus, Messina, Italy). Figures 9 shows the results. At lower values of resisting torques (i.e. in the range 28Ncm-120Ncm), angular stroke is not stabilised and steadily increases from 24° to 36°. For torques above 120Ncm, angular stroke is quite stable at 36°. Curves steadily shift towards negative angles (zero being the horizontal position, and negative in the direction of plantarflexion) with increasing torques. This increase in stroke is caused by the incomplete detwinning of martensite at lower torques: it is worth noticing that anyway, even at the lower values of measured angular stroke, suitable angular strokes are obtained.

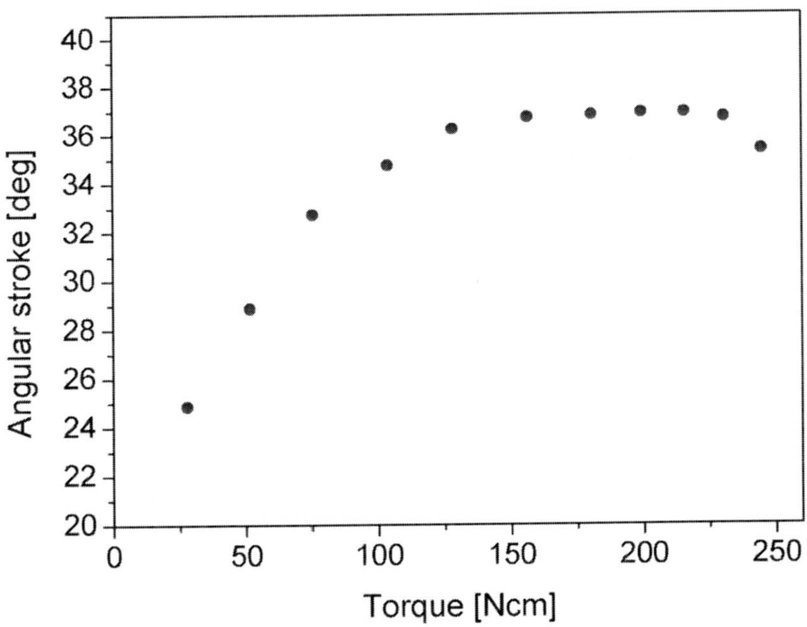

Figure 9. Angular stroke of the SMA-based exerciser for different resisting static torques. The dotted line is just a guide for the eye.

3.3.3. Tests In A Meg Shielded Room

In order to assess compatibility with the MEG technology, neural activity from a healthy subject's brain during repeated passive mobilisation of the ankle by the device was carried out with an MEG system composed of 153 SQUID channels covering the whole surface of the scalp. The electric power generator and the programming computer were kept out of the shielded room, and cables were passed through suitable vents in the shims. The NiTi wire was activated by a current pulse (ramp to 0.7A in 1s, then flat for 9s).

These tests revealed no significant noise on SQUID channels. Figure 10 (left) shows the signal recorded by one representative channel during MEG testing of the healthy volunteer. At t=0s, the actuators on board the orthosis were switched on. There is no artefact at that moment, indicating that the level of noise in the acquisition room has not changed. Furthermore, the right graph in the same Figure demonstrates that the frequency components in the actuator-OFF and actuator-ON states are the same. The variation in intensity of the signal power spectral density (PSD) can be totally accounted for by a change in the cortical reactivity of the subject under testing: this variation is an important part of the measured quantities of interest.

Incidentally, the use of conductive materials in the construction of the device did not affect the acquisition. This is probably due to the slow movement provided by the actuators, which generates low-frequency artefacts that can be suitably eliminated during routine filtering and windowing of the MEG data.

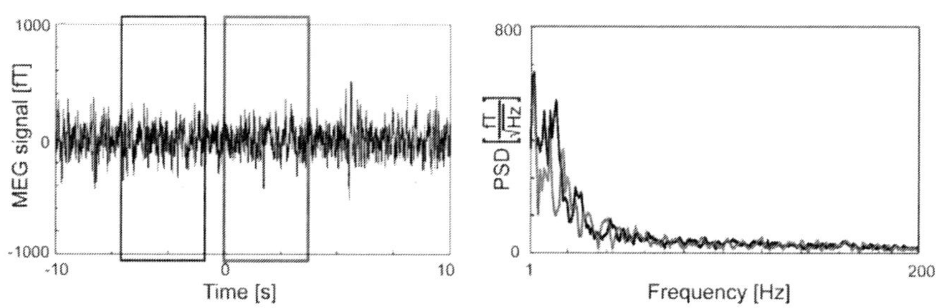

Figure 10. Left: MEG signal recorded by a representative SQUID channel from a healthy subject during passive mobilisation of the ankle by the amagnetic exerciser. Right: spectral components of the recorded MEG signal, before and after switching on the actuators at t=0s (black and red line, respectively).

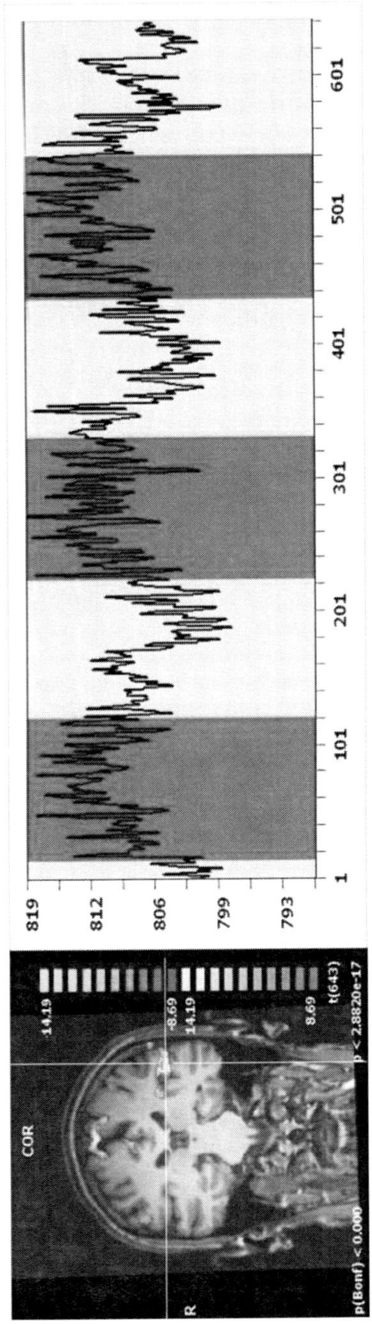

Figure 11. Left: fMRI image of metabolic activity in a healthy brain during passive mobilisation of the ankle. Right: time-course of the BOLD signal (arbitrary units) in the area of the brain highlighted in the left figure.

3.3.4. Tests In A Mri Acquisition Room

Magnetic resonance images were collected in a Philips 3T Achieva X-Series Magnetic Resonance scanner. fMRI signals were treated to extract the Blood Oxygen Level Dependent (BOLD) image, which depicts changes in neural metabolism related to neural activity. fMRI images (Figure 11) were not affected by any artefact. The BOLD signal was clearly collected from all parts of the brain and in some areas seemed to be temporally dependent on the movements of the ankle. The use of conductive materials in the implementation of the device did not affect the acquisition. The distance between the gantry (or head coil) and the ankle probably helped limit any influence on image encoding.

4. CONCLUSIONS

This Chapter showed some innovative applications of Shape Memory Alloys requiring deep understanding of the interaction of the material characteristics with the complex constraints imposed by the human body. In particular, it was explained how the design plan should be laid considering the many aspects connected to the state of the target patient, and the technical requirements should be chosen to meet very well identified needs. The field of Medical Rehabilitation is an interesting domain for exploiting the functional properties of NiTi-based alloys in making new lightweight and portable actuators. The Neuroscience applications introduced in this Chapter, on the other hand, albeit representing a niche sector per se, both make the most of the typical design techniques employed for rehabilitation devices, and provide a development ground for interesting industrial actuators with amagnetic characteristics. It is hoped that the SMA-based design strategies presented here will be of inspiration to engineers interested in utilising shape memory actuation for biomedical, robotic, aerospace or automotive applications with tight and mandatory external constraints where compactness, light weight or wearability are desired features of the device.

ACKNOWLEDGEMENT

The authors would like to thank the staff of ITAB, University of Chieti, where the tests employing MEG and fMRI were conducted: in particular Gian Luca Romani, Filippo Zappasodi, Vittorio Pizzella and Gabriella Tamburro for the MEG measurements, Cosimo Del Gratta, Antonio Ferretti and Mauro Gianni Perrucci for the fMRI trials. The financial support from Regione Lombardia (Mind in Italy Project) is most gratefully acknowledged.

REFERENCES

1. Gracies J-M2005Pathophysiology of spastic paresis I: paresis and soft tissue changes, Muscle Nerve 31535551

2. Gracies J-M2005Pathophysiology of spastic paresis II: emergence of muscle overactivity, Muscle Nerve 31552571

3. O. Simon, A. P. Yelnik, 2010Managing spasticity with drugs. Eur J Phys Rehabil Med. 46401410

4. European Stroke Organisation (ESO) Executive Committee, ESO Writing Committee2008Guidelines for management of ischaemic stroke and transient ischaemic attack, Cerebrovasc. Dis. 25457507

5. P. W. Duncan, R. Zorowitz, B. Bates, J. Y. Choi, J. J. Glasberg, G. D. Graham, R. C. Katz, K. Lamberty, D. Reker, 2005Management of Adult Stroke Rehabilitation Care: a clinical practice guideline, Stroke. 36: e100e143.

6. S. Hesse, 2008Treadmill training with partial body weight support after stroke: a review. NeuroRehabilitation. 235565

7. H. I. Krebs, S. Rossi, S. Kim, P. K. Artemiadis, D. Williams, E. Castelli, P. Cappa, 2011Pediatric anklebot. IEEE Int. Conf. Rehabil. Robot 15

8. P. Staubli, T. Nef, V. Klamroth-Marganska, R. Riener, 2009Effects of intensive arm training with the rehabilitation robot ARMin II in chronic stroke patients: four single-cases. J Neuroeng Rehabil. 17: 6:46.

9. S. E. Fasoli, H. I. Krebs, J. Stein, W. R. Frontera, N. Hogan, 2003Effects of robotic therapy on motor impairment and recovery in chronic stroke. Arch. Phys. Med. Rehabil. 84477482

10. F. Alary, C. Simões, V. Jousmäki, N. Forss, R. Hari, 2002Cortical activation associated with passive movements of the human index finger: an MEG study. Neuroimage. 15691696

11. T. W. Boonstra, H. E. Clairbois, A. Daffertshofer, J. Verbunt, B. W. van Dijk, P. J. Beek, 2005MEG-compatible force sensor. J Neurosci Methods. 144193196

12. Suminski AJ, Zimbelman JL, Scheidt RA2007Design and validation of a MR-compatible pneumatic manipulandum. J Neurosci Methods. 163255266

13. N. V. Tsekos, A. Khanicheh, E. Christoforou, C. Mavroidis, 2007Magnetic resonance-compatible robotic and mechatronics systems for image-guided interventions and rehabilitation: a review study. Annu Rev Biomed Eng. 9351387

14. H. Onishi, M. Oyama, T. Soma, M. Kubo, H. Kirimoto, H. Murakami, S. Kameyama, 2010Neuromagnetic activation of primary and secondary somatosensory cortex following tactile-on and tactile-off stimulation. Clin Neurophysiol. 121588593

15. C. Hollnagel, M. Brügger, H. Vallery, P. Wolf, V. Dietz, S. Kollias, R. Riener, 2011Brain activity during stepping: a novel MRI-compatible device. J Neurosci Methods. 201124130

16. H. Funakubo, 1987Shape memory alloys. New York: Gordon and Breach Science Publishers. 275p.

17. K. Otsuka, C. M. Wayman, 1998Shape Memory Materials, Cambridge: Cambridge University Press. 287 p.

18. K. Otsuka, X. Ren, 2005Physical metallurgy of Ti-Ni-based shape memory alloys, Progr. in Mater. Science. 50511678

19. Lagoudas DC2008Shape Memory Alloys: Modeling and Engineering Applications. Berlin: Springer. 435 p.

20. S. Viscuso, S. Pittaccio, M. Caimmi, G. Gasperini, S. Pirovano, S. Besseghini, F. Molteni, (200, alloy. Pseudoelastic, for. devices, elbow. spastic, relaxation, In: Lim CT, Goh JC, editors. ICBME 2008th International Conference on Biomedical Engineering, IFBME Proceedings 23. Berlin: Springer. 15841587

21. S. Pittaccio, S. Viscuso, 2009The use of muscle "creep" as opposed to relaxation in stretching braces: a pseudoelastic device. In: Dössel O, Schlegel WC, editors. World Congress on Medical Physics and Biomedical Engineering 2009, IFMBE Proceedings 25/9. Berlin: Springer. 178181

22. S. Pittaccio, S. Viscuso, 2009Customizable neuro-mechanical model of a hemiplegic elbow interacting with a pseudoelastic dynamic orthosis. In: Dössel O, Schlegel WC, editors. World Congress on Medical Physics and Biomedical Engineering 2009, IFMBE Proceedings 25/9. Berlin: Springer. 182185

23. S. Viscuso, S. Pittaccio, M. Caimmi, G. Gasperini, S. Pirovano, E. Villa, S. Besseghini, F. Molteni, 2009Pseudoelastic Nitinol-Based Device for Relaxation of Spastic Elbow in Stroke Patients. J Mater Eng Perform 18805813

24. Patent Number:WO/2011Joint For Articulations With Pseudo-Elastic Elements.

25. S. Pittaccio, S. Viscuso, E. Beretta, A. C. Turconi, S. Strazzer, 2010Pilot studies suggesting new applications of NiTi in dynamic orthoses for the ankle joint. Prosthet Orthot Int 34305318

26. A. Hanafusa, T. Isomura, Y. Sekiguchi, H. Takahashi, T. Dohi, 2002Orthosis Design System for Malformed Ears Based on Spline Approximation. In: Dohi T, Kikinis R, editors. Medical Image Computing and Computer-Assisted Intervention- MICCAI 2002. Berlin: Springer. 227234

27. Y. Shimizu, Kobayashi, 2011The Development of Foot Orthotics using Shape Memory Alloy for Preventing Falls. In: Werner B, editor. 2011 IEEE 11th International Conference on Bioinformatics and Bioengineering. Washington: Conference Publishing Services. 117120

28. M. Takami, K. Fukui, S. Saitou, I. Sugiyama, K. Terayama, 1992Application of a shape memory alloy to hand splinting. Prosthet Orthot Int. 165763

29. Patent Number: US2008A1. Orthotic Apparatus And Method Of Operation.

30. Patent Number: US2011A1. Dynamically Adjustable Orthotic Device.

31. Machado LG, Savi MA2003Medical applications of shape memory alloys. Braz J Med Biol Res. 36683691

32. Lai YJ, Yeh LJ, Chiu MC2010An Assessment of the Body Joint Bending Actuator using a Shape Memory Alloy. J Appl Sci, 1019731977

33. Lai YJ, Yeh LJ, Chiu MC2012An experimental investigation on shape memory alloy dynamic splint for a finger joint application. Sens Actuator A-Phys 173210218

34. M. Torri, S. Viscuso, S. Pittaccio, A. Nespoli, S. Besseghini, 2006Biomechanical design of a shape memory alloy spring for the activation of a flaccid hand rehabilitation device. In: Venugopalan R, Wu MH, editors. Medical Device Materials III: Proceedings of the Materials

& Processes for Medical Devices Conference. Materials Park: ASM International. 237242

35. M. B. Van Kuren, S. Gillette, P. Mejia, T. Stoever, A. Walker, 2005Design considerations for a wearable pediatric rehabilitative boot. In: ICORR 2005, 9th International Conference on Rehabilitation Robotics. 400403Available: http://ieeexplore.ieee.org/stamp/stamp.jsp?tp=&arnumber=1501128.Accessed 2012 Mar 28.

36. S. Pittaccio, S. Viscuso, M. Rossini, L. Magoni, S. Pirovano, E. Villa, S. Besseghini, F. Molteni, 2009SHADE: A shape-memory-activated device promoting ankle dorsiflexion. J Mater Eng Perform 18824830

37. S. Viscuso, S. Pittaccio, F. Tecchio, F. Zappasodi, M. Rossini, L. Magoni, S. Pirovano, S. Besseghini, F. Molteni, 2009Assessment of the peripheral performance and cortical effects of SHADE, an active device promoting ankle dorsiflexion. In: Lim CT, Goh JC, editors. ICBME 2008- 13th International Conference on Biomedical Engineering, IFBME Proceedings 23. Berlin: Springer. 961965

38. S. Viscuso, S. Pittaccio, F. Zappasodi, F. Tecchio, 2009A magnetically quasi-transparent tool for ankle passive mobilization in investigations on cortical involvement using MEG. In: Dössel O, Schlegel WC, editors. World Congress on Medical Physics and Biomedical Engineering 2009, IFMBE Proceedings 25/7. Berlin: Springer. 410413

39. S. Pittaccio, F. Zappasodi, S. Viscuso, F. Mastrolilli, M. Ercolani, F. Passarelli, F. Molteni, S. Besseghini, P. M. Rossini, F. Tecchio, 2011Primary sensory and motor cortex activities during voluntary and passive ankle mobilization by the SHADE orthosis. Hum Brain Mapp 326070

40. Patent Number:WO/2011Rotational Actuator.

41. S. Pittaccio, S. Viscuso, 2011An EMG-Controlled SMA Device for the Rehabilitation of the Ankle Joint in Post-Acute Stroke. J Mater Eng Perform 20666670

42. G. V. Merrett, C. D. Metcalf, D. Zheng, S. Cunningham, S. Barrow, S. H. Demain, 2011Design and Qualitative Evaluation of Tactile Devices for Stroke Rehabilitation. Presented at IET Assisted Living 2011, London, UK. Available: http://eprints.soton.ac.uk/271802/.Accessed 2012 Mar 28.

43. Patent Number: US2001A1. Exercise Kit And Method Of Using Same.

44. Esfahani ET2007Developing an Active Ankle Foot Orthosis Based On Shape Memory Alloys. M.Sc. thesis. Available: http://etd.ohiolink.edu/

send-pdf.cgi/Tarkesh%20Esfahani%20Ehsan. pdf?toledo1197561946. Accessed 2012 Mar 28.

45. Esfahani ET, Elahinia MH2007Stable Walking Pattern for an SMA-Actuated Biped IEEE/ASME Trans Mechatron 12534541

46. L. Stirling, C. H. Yu, J. Miller, E. Hawkes, R. Wood, E. Goldfield, R. Nagpal, 2011Applicability of Shape Memory Alloy Wire for an Active, Soft Orthotic. J Mater Eng Perform 20658662

47. Patent Number: WO2011A2. Actively Controlled Orthotic Device.

48. Patent Number: US 5103807.Shape Memory Alloy Orthotic Device.

49. Abolfathi PP2007Development of an Instrumented and Powered Exoskeleton for the Rehabilitation of the Hand. Ph.D. thesis. Available:http://ses.library.usyd.edu.au/bitstream/2123/3690/1/01pp-abolfathi-2008 -thesis.pdf.Accessed 2012 Mar 28.

50. Patent Number: US 6379393 B1. Prosthetic, orthotic, and other rehabilitative robotic assistive devices actuated by smart materials.

51. D. Jayatilake, A. Gruebler, K. Suzuki, 2010Robot Assisted Smile Recovery. In: Kordic V, editor. Cutting Edge Robotics 2010. Rijeka: InTech. 333350

52. A. Esteki, J. M. Mansour, 1996An experimentally based nonlinear viscoelastic model of joint passive moment. J Biomech. 29443450

53. Given JD, Dewald JPA, Rymer WZ1995Joint dependent passive stiffness in paretic and contralateral limbs of spastic patients with hemiparetic stroke. J Neurol Neurosurg Psychiatry 59271279

54. Lin CCK, Ju MS, Chen SM, Pan BW2008A Specialized Robot for Ankle Rehabilitation and Evaluation. J Med Biol Eng 287986

55. IEC 60601-1. Medical electrical equipment- Part 1: General requirements for basic safety and essential performance.

56. ISO 14971. Medical devices- Application of risk management to medical devices.

57. ISO 14155.Clinical investigation of medical devices for human subjects- Good clinical practice.

58. L. Fumagalli, F. Butera, A. Coda, 2009SmartFlex® NiTi Wires for Shape Memory Actuators. J Mater Eng Perform 18691695

59. Liu SH, Huang TS, Yen JY2010Tracking Control of Shape-Memory-Alloy Actuators Based on Self-Sensing Feedback and Inverse Hysteresis Compensation. Sensors 10112127

4

A TECHNOLOGY FOR SOFT AND WEARABLE GENERATORS

Thomas G. McKay[1], Benjamin M. O'Brien[1] and Iain A. Anderson[1, 2]

[1] The Biomimetics Laboratory of the Auckland Bioengineering Institute, The University of Auckland, Level 6, Auckland, New Zealand

[2] Department of Engineering Science, School of Engineering, The University of Auckland, Level 3, Auckland, New Zealand

1. INTRODUCTION

Never again needing to recharge the batteries of portable electronic devices is an exciting prospect.The uptake of portable electronics is increasing and recharging or replacing batteries is not only an inconvenience, but also contributes to an environmental hazard.Furthermore, with technologies such as GPS, pressure sensors, and cameras becoming smaller and cheaper, smart clothing such as the Adidas miCoach range are now a reality [1].For seamless integration of smart devices into clothing, the inconvenience of battery replacement or recharging needs to be eliminated.

Conveniently there is an alternative source of energy in the exact location where portable and wearable devices operate: biomechanical energy from human movements.Interestingly, the device often credited as being the first ever power generator was a biomechanical energy harvester.17th century engineer Otto von Guericke produced a sulfur globe which was charged through the triboelectric effect[1] - when it was rubbed with a dry hand [2].However, four centuries later the ability to harvest appreciable amounts of energy from biomechanical motions remains a research challenge.In this chapter we will discuss recent progress towards harvesting biomechanical energy using a soft and wearable electro-active polymer technology.

To provide sufficient power for portable devices a biomechanical energy harvester needs to supply in the order of a few Watts.For instance, an insulin pump consumes approximately 5 Watts [3] and the Nexus One smartphone (HTC Corporation, Taiwan) consumes 25 milli-Watts in standby, 330 milli-Watts in idle, and 750 milli-Watts during a phone call [4].So to eliminate the need to recharge batteries, a similar or greater quantity of power needs to be sourced.

Starner and Paradiso provide a good review of the energy available for scavenging from human movement and identify walking as a rich source of energy [5].They highlight that 13Watts of power is available from the heel-strike of a 68Kg person if the sole of their shoe is compressed by 1cm when walking at 2 steps per second.Furthermore, a typical running shoe midsole dissipates a relatively large amount of energy as heat. Shorten's analysis of the energetics of a shoe midsole worn by a 76Kg runner suggests that a running shoe will dissipate between 2 and 10 Joules per step[6].A well designed energy harvesting shoe could instead turn this energy into electricity, without altering the comfort or energy expenditure of the person wearing the shoe.

The most prevalent energy harvesting technology, electromagnetic generators, have been used to harvest energy from human gait, but additional mechanisms are required to condition the mechanical energy [5, 7].Donelan et al. developed a knee-brace generator for harvesting energy from human gait.Their generator contained auxiliary components including a gear train, bearings, and a separate input shaft to convert the mechanical energy to suit their electromagnetic generator.Their system cost on average 59W of metabolic power to carry without harvesting energy, whereas an additional 5W of metabolic power was required to produce 4.8W of electrical power [7].Although their harvesting mechanism was extremely efficient, they could have achieved larger efficiency gains if their system's mass was reduced and the device did not alter gait patterns.This begs the question: why are auxiliary mechanical components required for an electromagnetic biomechanical energy harvester?

The system by Donelan et al. was relatively heavy because it required additional componentry for it to work efficiently: electromagnetic generators produce more energy during a single rotation or stroke as velocity increases and are poorly suited for the low induced velocities associated with walking unless augmented mechanically.Thus the viability of harvesting energy from human motions could be improved by utilizing a different energy harvesting technology.

We propose the key characteristics of an ideal technology for harvesting biomechanical energy below:

1) Efficient operation at low biomechanical "walking" speeds.

2) Ability to couple directly to large walking motions.

3) A high energy density.

4) Good mechanical impedance matching to muscle.

5) Cheap to produce and maintain.

The first two criteria eliminate the need for additional mechanisms to condition the mechanical energy when it is transferred to the generator. The first three criteria therefore provide low mass/bulk. By having good impedance matching to muscle, the generator will be comfortable to wear, reducing its effects on the person wearing it. For the mass market, it is essential that the generator is low cost because consumers are unlikely to pay a premium for generators that produce power of approximately 1 Watt.

The research efforts of the authors of this chapter have focused on an energy harvesting technology called dielectric elastomer generators (DEG) which have been identified as a highly promising technology for biomechanical energy harvesting [8-11].Furthermore, DEG fit the five listed criteria particularly well, so this chapter will focus on our recent developments towards portable and wearable DEG. For the interested reader, Anderson et al. provide a good, broad review of dielectric elastomer transducers [12].

2. THE DIELECTRIC ELASTOMER GENERATOR

DEG are a class of variable capacitance generatorsthat are fabricated from a rubbery dielectric material sandwiched between stretchable electrodes. They have excellent impedance matching to natural muscle [13], can be fabricated from a wide range of low cost materials (commonly acrylic or silicone membranes sandwiched between carbon-based electrodes), have demonstrated extremely high energy densities, can undergo strains in excess of 100% [14], and have the ability to work over a wide frequency range without sacrificing efficiency [15].

Mechanical energy can be converted to electrical energy by cyclically deforming a DEG and placing charges on its flexible electrodes in the deformed state.DEG are typically produced from incompressible polymers so that an area stretch results in a decreased thickness.Relaxation of the charged, deformed DEG forces the opposite charges apart and packs like charges closer together and this transfers the mechanical energy to the electrical charges.The energy flows during an ideal DEG cycle are highlighted by the red arrows in Figure 1.The system receives 1 unit each of mechanical and electrical energy and converts the unit of mechanical energy into electrical, so that it outputs two units of electrical energy.

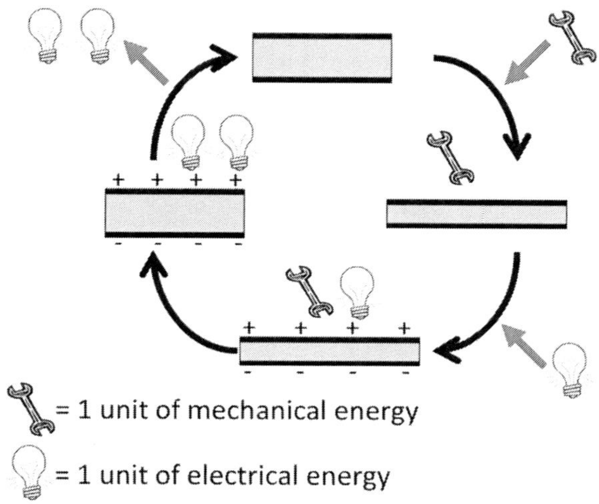

Figure 1. Schematic of the DEG states during a generation cycle.The grey area represents the dielectric and the black area represents the electrodes.From top moving clockwise, mechanical energy is input to the system deforming the DEG; an electrical energy input then charges the stretched DEG; the mechanical energy is then transferred to the charges by separating opposite and compressing like charges together.The electrical energy is then extracted and the cycle repeats (From [**16**]).

The major advantage that DEG hold for biomechanical energy harvesting is their ability to directly harvest low frequency motions without any gear mechanisms.To illustrate why this is true we will briefly describe their fundamental energy harvesting mechanism.As illustrated in Figure 1, DEG convert mechanical energy to electrical when the deformation of a stretched charged DEG is relaxed.During this relaxation period the thickness of the dielectric increases and the electrode area decreases, both resulting in a reduction of the capacitance.If charge is trapped on the generator during this relaxation phase, the voltage on the DEG will increase and there will be an increase in energy given by equation 1 where *Cd* and *Vd* are the capacitance and voltage of the DEG in its deformed state, and *Cr*and *Vr*are the relaxed DEG's capacitance and voltage, respectively.

$$E = 0.5C_r V_r^2 - 0.5C_d V_d^2 \qquad (1)$$

Since we are considering the case where the charge on the DEG is fixed during the relaxation period, and that the charge on a capacitor can be calculated from its voltage and capacitance (*Q=CV*), we can relate the voltage of the deformed DEG to the voltage of the generator in its relaxed state using equation 2. We substitute equation 2 into equation 1 to get equation 3.

$$V_R = \frac{C_d}{C_r} V_d \tag{2}$$

$$E = 0.5\left(\frac{C_d^2}{C_r} - C_d\right) V_d^2 \tag{3}$$

These equations emphasise that along with driving voltage, *Cr* and *Cd* are the key parameters that influence the energy output of a DEG. The capacitance is dependent on the geometry of the DEG and the driving voltage is controlled by the generator's associated electronics. Neither of these parameters are dependent on the velocity at which the generator is deformed, so the fundamental mechanism of DEG is not dependent on driving velocity.

Although we have highlighted that DEG have a highly suitable mechanism and characteristics for harvesting biomechanical energy, Figure 1 highlights that DEG need an electrical circuit that will control the flow of charge onto and off of the generator at appropriate stages of the energy harvesting cycle. Such circuitry can add weight and is typically composed of stiff and bulky parts.Furthermore the requirement for the electronics to replace charge delivered to the load has traditionally reduced the portability of DEG because they have been either tethered to the grid or used batteries that need periodic replacement [10].This chapter will focus now on recent developments that eliminate the need for a secondary power source and reduce external DEG circuitry mass and stiffness.

3. PORTABLE DIELECTRIC ELASTOMER GENERATOR ELECTRONICS

A passive DEG circuit for controlling the charge state of the DEG appropriately as it is mechanically cycled called the Self-Priming Circuit (SPC) [17, 18] has been developed to eliminate the need for a secondary power source and thus improve DEG portability.The circuit works as a charge pump that provides energy in a higher charge form than the energy supplied to it.The self-priming circuit is configured so that it harvests energy from a DEG and then supplies that energy in a higher charge form to a load or to the DEG when it requires priming.This effectively boosts the charge of the generated energy before it is used, thus a secondary power source is not necessary because the generated energy is used to replace circuit charge losses and charge delivered to the load.A schematic of the simplest form of an SPC is given in Figure 2, showing that an SPC is simply a capacitor bank that has diodes to convert the topology of the circuit between a low capacitance when it is charged (Figure 2c, capacitors in series) and a high capacitance when it discharges (Figure 2d, capacitors in parallel).This toggling of state provides the SPC with a higher output capacitance which converts the energy to a higher charge form. This can be explained using equation 4, which provides two expressions for the energy (E) stored on a capacitor. If the energy is conserved when the capacitance (C) of the SPC increases, then the charge (Q) must increase too, and since the SPC is an adiabatic process the increased charge is accompanied by a decreased voltage (V).Thus the SPC outputs energy in a higher charge, lower voltage form than the input.

$$E = \frac{Q^2}{2C} = \frac{CV^2}{2} \tag{4}$$

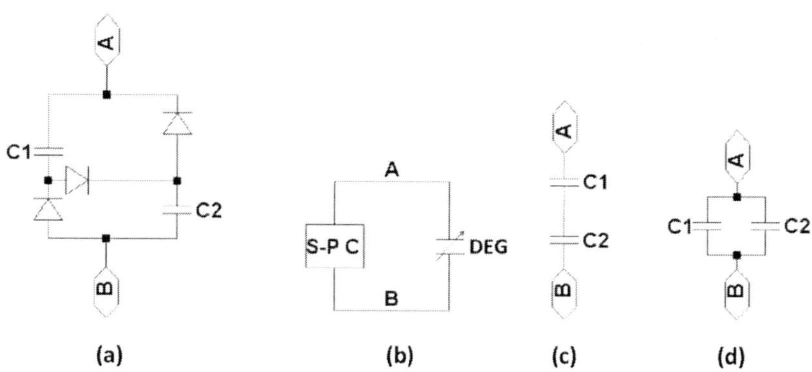

Figure 2. Schematic of a self priming circuit (a) which connects to a DEG in parallel (b), the diodes control current flow so that the SPC capacitor bank takes on the form given in (c) when energy is transferred to it from the DEG, when energy is transferred off the SPC the diodes configure it to take on the higher charge form given in (d).

The SPC has additional benefits for portable DEG.First, the SPC is passive and requires no active switching or control, so it does not require any power to drive it.Second, the SPC accumulates charge from cycle to cycle, this means that the system voltage climbs (see Figure 3).The ability to boost its own voltage is highly desirable because as demonstrated in equation 3, generated energy climbs with priming voltage. This high voltage hastraditionally been supplied by a high voltage power supply or converter, so the SPC eliminates the need for these typically high cost components.

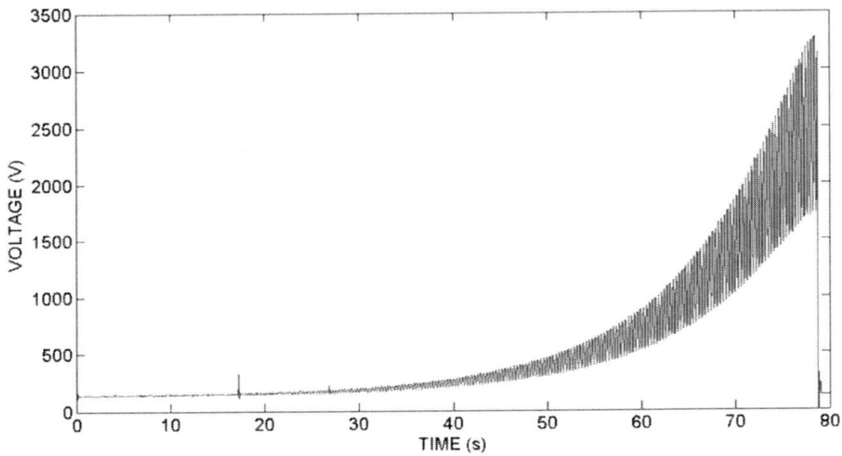

Figure 3. The output voltage waveform of a self-priming generator mechanically deformed at 3 Hz.The voltage climbs from cycle to cycle because the generated energy accumulates in the form of extra charge stored on the generator and priming circuit (From [**19**]).

Although the SPC is low cost, low complexity, and autonomous it still adds considerable mass and stiffness to the DEG system. The SPC consists of diodes and capacitors. The function of the capacitors is to store priming charges and the diodes control the transfer of charge to and from the DEG, so that an appropriate generation cycle is achieved.We will now discuss how these functions have been integrated onto the DEG membrane to produce a generator that can be fabricated entirely from soft elastomers.

4. DIELECTRIC ELASTOMER GENERATORS WITH INTEGRATED SOFT ELECTRONICS

The DEG membrane is essentially a soft capacitor and one can take advantage of this, using the DEG to provide the energy storage function of the SPC.This means the generator can be fabricated with its external circuitry consisting solely of a few diodes [16, 20].One configuration that has been used to integrate the SPC storage function into the DEG membrane is given in Figure 4 where DEG1 is electrically configured as an SPC [16].Because the generator's elastomer membranes are integrated into the self-priming circuit, this system has been referred to as the integrated self-priming circuit [21].The generator was configured into two DEG membranes which were deformed so that as one membrane was stretched the other was relaxed.When the voltage of DEG1A and DEG1B connected electrically in parallel exceeded the voltage of DEG2, energy was transferred from the SPC to DEG 2 through the path shown in Figure 4c; when the voltage of DEG2 exceeded that of DEG1A and DEG1B connected in series, energy was transferred onto the SPC from DEG2 through the path shown in Figure 4d.

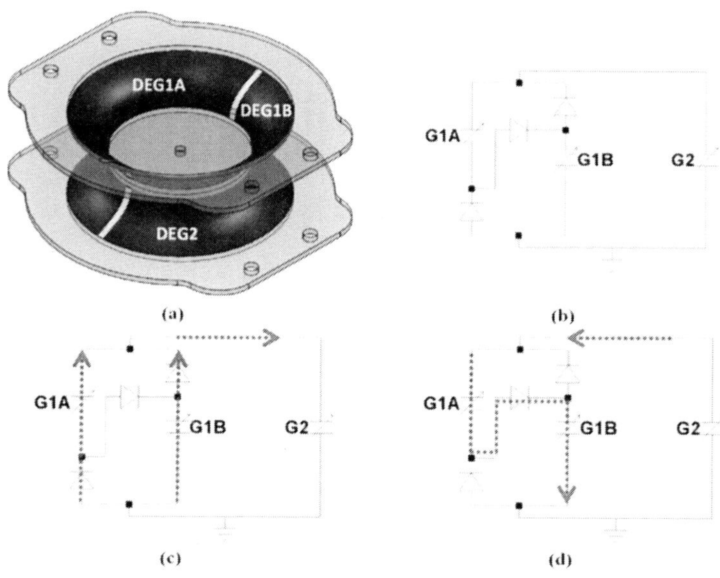

Figure 4. A segmented generator which consists of two membranes that are antagonistically deformed when the inner hub displaces up and down (a).The top membrane DEG1 can be electrically interconnected with diodes to form a self-priming circuit (b).The paths along which current flows off (c) and onto (d) the SPC are also illustrated.

For a small scale portable and potentially wearable generator the diodes of the SPC can represent a significant mass.For instance, the prototype of the generator shown in Figure 4 consisted of DEG membranes with a combined mass of 0.35 grams, whereas the diodes weighed 0.63 grams.If we could remove the diodes, 64% of the total mass (ignoring the mass of the frame) could be eliminated.

The function of the diodes is to control the charge transfer between the DEG and SPC.The diodes simply allow current to flow along one path when one diaphragm is stretched, and along an alternative path when that diaphragm is relaxed.This means that the diodes can be replaced by switches that are toggled at the appropriate time.Since this timing is dependent on the material stretch state, the diodes can be replaced by stretch-sensitive switches coupled to the DEG.

Stretch sensitive electronics called Dielectric Elastomer Switches (DES) have been used to integrate the functionality of the diodes into the DEG membrane.DES consist of piezoresistive electrodes fabricated directly onto a highly stretchable dielectric elastomer membrane.They exhibit very large changes in resistance with stretch.O'Brien et al. first presented the concept [22] and characterized M-shaped DES as illustrated in Figure 5[23].These DES had a resistance of several MΩ in their rest state, which increased to several GΩ when they were stretched to approximately 1.4 times their original length (seeFigure 5b) [23].

The generator illustrated in Figure 4 was redesigned by placing DES onto the membrane to replace the hard diodes as illustrated in Figures 6 and 7.The SPC capacitors were fabricated onto diaphragm 1.When diaphragm 1 was relaxed the switches Q1 and Q2 were also relaxed and therefore conducted, configuring the SPC capacitors into a parallel topology (high charge form).When diaphragm 1 was stretched so too were Q1 and Q2, so they no longer conducted, but diaphragm 2 was simultaneously relaxed, causing Q3 to conduct, connecting the SPC capacitors in series.This generator can be fabricated entirely from soft elastomers, so we refer to it as the soft generator.

The output voltage of a prototype soft generator is given in Figure 8.When compared to the output of the self-priming circuit given in Figure 3, the soft generator's voltage climbs very rapidly.But the most profound advantages are the ability to produce DEG that maintain the advantages of high flexibility, softness, low volume, low cost, low component count, and low mass at the system level.

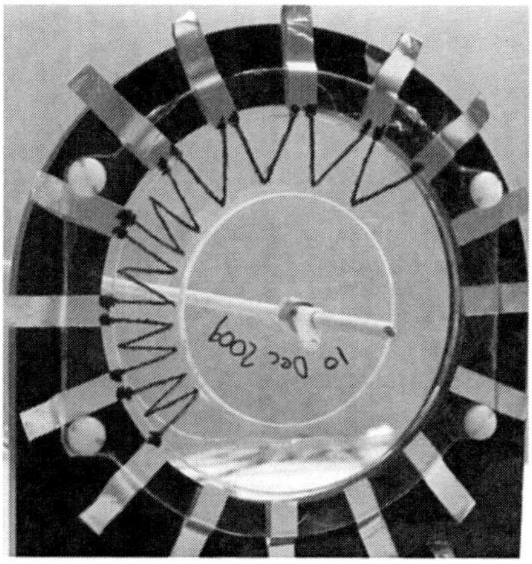

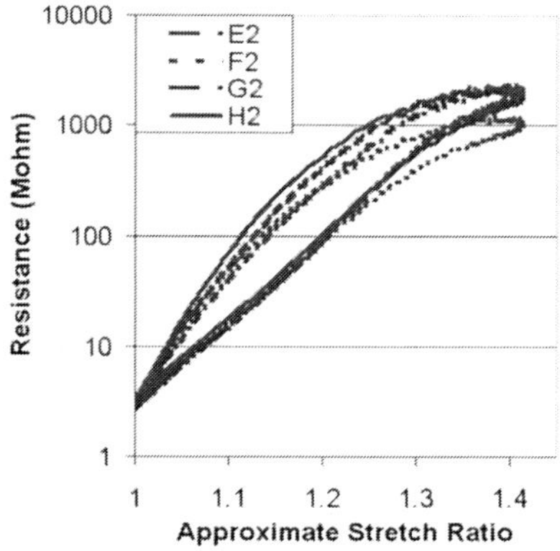

(a) (b)

Figure 5. Carbon powder-based DES applied to a dielectric elastomer diaphragm (black "M" and "V" shaped tracks) for characterisation by O'Brien et al. (a), and a plot of measured DES resistance versus approximate radial stretch ratio showing that the resistance climbed several orders of magnitude when the DES were stretched (b).Images taken from O'Brien [**23**].

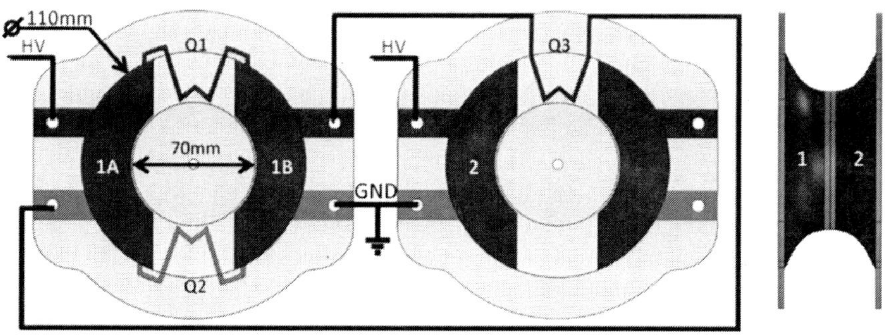

(a) (b)

Figure 6. A schematic of the soft generator.The switches Q1-Q3 control the charge flow within the self-priming circuit in a similar manner to the diodes in the integrated self-priming circuit.The two diaphragms in (a) are connected together to form the antagonistic pair shown in (b).(From [**24**]).

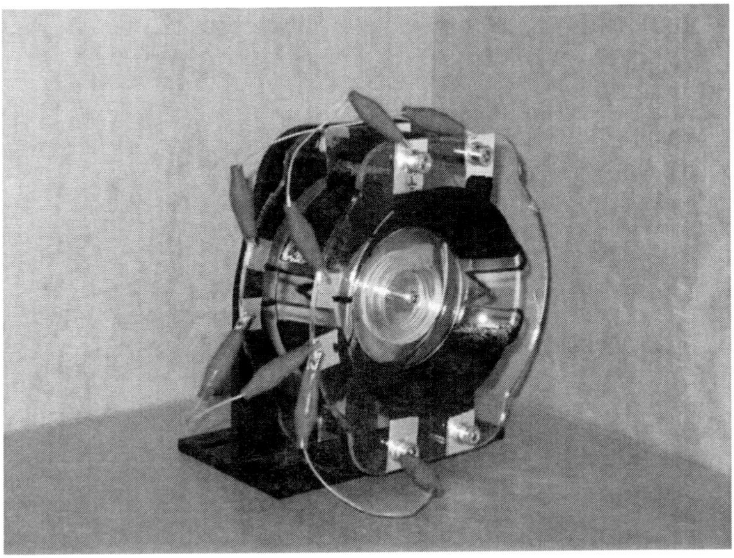

Figure 7. A photograph of a soft generator prototype.The large black regions are the generator electrodes and the thin "M-shaped" electrodes are the piezo-resistive switches.

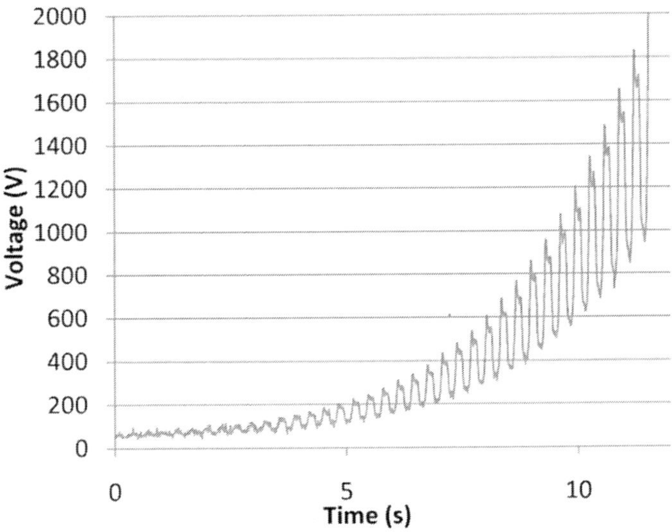

Figure 8. The output voltage waveform of a soft generator initially primed to 10 Volts then mechanically deformed at 3 Hz showing that the generator rapidly boosts its operating voltage through accumulation of generated energy (From [**24**]).

The total mass of biomechanical energy harvesters has a great influence on the metabolic cost of their use and can be easily compared in a quantitative manner.In Figure 9 we compare the energy density of each DEG system described in this chapter with the mass of their associated electronics included in the calculations (see equation 5).

$$EnergyDensity = \frac{EnergyGenerated}{Mass_{dielectric} + Mass_{ExternalCircuit}} \tag{5}$$

The energy and energy densities produced by similarly sized DEG membranes (~0.3 grams) mechanically cycled at a rate of 3 Hz and operating at 2 kV are compared in Figure 9.The soft generator energy density was superior to both the integrated and external SPC generators because their respective external circuit masses are approximately 0 grams, 0.6 grams, and 13.4 grams.The soft generator's energy density of 9.5 mJ/g is highly competitive with the predicted practical maxima of electromagnetics (4 mJ/g) and piezoelectrics (17.7 mJ/g) at the ~1cm3 scale [25], demonstrating the utility of DEG for small-scale energy harvesters.

The recent developments discussed in this chapter have provided progress towards wearable, soft power generators becoming a reality, but there are still issues that need to be addressed:

1) The switching technology used in the soft generator, DES, are in their infancy.Material and process developments are required to create more reliable DES with resistances that can be tuned to their application.

2) The prototypes discussed in this chapter are 1 to 2 orders of magnitude smaller than those required to produce sufficient power to drive the more power hungry portable electronics such as smart phones, which begs the question:How do we expect them to scale?Figure 9 illustrates that the integrated SPC generator produced more energy than the soft generator, so as the generator is scaled up and the diode mass becomes insignificant the integrated generator may outperform the soft generator in terms of energy density.When scaled up to produce an order of a Watt, are the soft generator's advantages sufficient to provide a worthwhile alternative to more complex DEG control strategies that will provide superior energy production [26-29].

3) The soft and integrated generators have an antagonistic pair of membranes.What is the best configuration for harvesting energy from a given biomechanical source?

4) DEG provide the best energy production when operated at high voltages (typically kV range).Compact energy-efficient circuitry is required to reduce this voltage down to consumer electronics levels

(~5 Volts) or perhaps the high voltage energy can be used to drive another high voltage device such as a dielectric elastomer actuator.

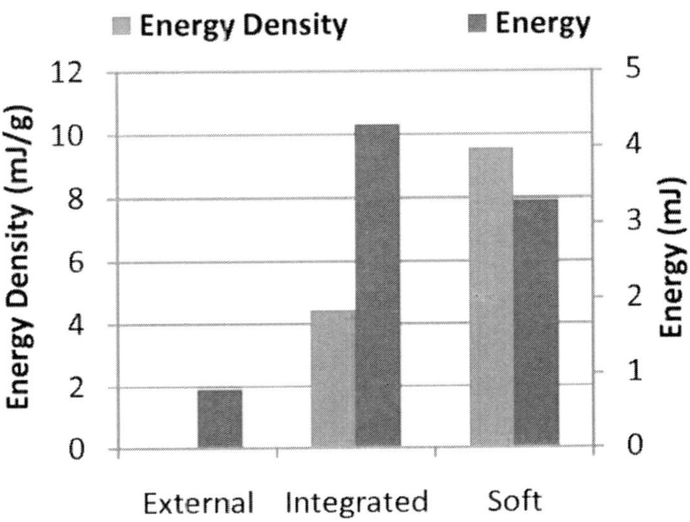

Figure 9. The energy and energy density produced by the external SPC, integrated SPC, and soft generators.

With these challenges comes an exciting future: the emergence of smart, wearable soft devices.The DEG membranes are multifunctional; they can be operated as actuators or generators and simultaneously sense strain [12].Furthermore DES have been used to fabricate rubber-based NAND-gates and memory elements, the two primitives required to build a computer [30-32].Perhaps the future will not only include soft generators to power portable electronic devices, but it may also include a new breed of smart, multifunctional self-powered portable soft devices.These possibilities open the door for smart devices to be integrated directly into clothing.So soft wearable generators will not only revolutionize the use of today's portable devices, but they will be an integral part of a distributed body-area network of sensors and smart devices that will improve future prosthetic devices, sports monitors, and video gaming interfaces.

ACKNOWLEDGEMENT

This work was supported by a University of Auckland Bioengineering Institute FRDF postdoctoral fellowship and a Rutherford Foundation postdoctoral fellowship from the Royal Society of New Zealand. The authors acknowledge the support of the Auckland Bioengineering Institute, their colleagues in the Biomimetics lab, Kristal Murray, Michelle O'Brien, and Suzie Pilkington.

NOTES

[1] - The triboelectric effect refers to the transfer of electrons between two materials when they come into contact.Some materials are more susceptible to donating electrons when the contact is separated and some are more likely to hold on to electrons so the materials remain charged when separated.

REFERENCES

1. Adidas miCoach.2012cited 2012]; Available from: www.micoach.com/.

2. Schiffer, M.B. and C.L. Bell2003Draw the Lightening Down: Benjamin Franklin and Electrical Technology in the Age of Enlightenment. Berkeley: University of California Press, Ltd.

3. D. Jia, J. Liu, 2009Human power-based energy harvesting strategies for mobile electronic devicesFrontiers of Energy and Power Engineering in China2746

4. A. Carroll, G. Heiser, 2010An Analysis of Power Consumption in a Smartphone. USENIX ATC'10.

5. T. Starner, J. A. Paradiso, 2004Human Generated Power for Mobile Electronics in Low Power Electronics Design C. Piguet Editor CRC Press. 4541

6. M. R. Shorten, 1993The energetics of running and running shoes.Journal of BiomechanicsSupplement 1(0): 4151

7. J. M. Donelan, Q. Li, V. Naing, J. A. Hoffer, D. J. Weber, A. D. Kuo, 2008Biomechanical Energy Harvesting: Generating Electricity During Walking with Minimal User EffortScience807810

8. C. Jean-Mistral, S. Basrour, 2010Scavenging energy from human motion with tubular dielectric polymerElectroactive Polymer Actuators and Devices (EAPAD) 2010 7642: 764209764212

9. C. Jean-Mistral, S. Basrour, J. J. Chaillout, 2008Dielectric polymer: scavenging energy from human motionProc. SPIE. 6927: 692716

10. R. Pelrine, R. D. Kornbluh, J. Eckerle, P. Jeuck, S. Oh, Q. Pei, S. Stanford, 2001Dielectric elastomers: generator mode fundamentals and applications. Proc. SPIE. 4329: 148

11. H. Prahlad, R. Kornbluh, R. Pelrine, S. Stanford, J. Eckerle, S. Oh, 2005Polymer Power: Dielectric Elastomers and Their Applications in Distributed Actuation and Power Generation in Proc. ISSS: Bangalore, India. 100

12. I. A. Anderson, T. A. Gisby, T. G. Mc Kay, B. M. O'Brien, E. P. Calius, 2012Multi-functional dielectric elastomer artificial muscles for soft and smart machinesJournal of Applied Physics041101041120

13. Y. Bar-Cohen, 2004Electroactive Polymer Actuators as Artificial Muscles: Reality, Potential, and Challenges SPIE:New York.

14. R. Pelrine, R. Kornbluh, P. Qibing, J. Joseph, 2000High-speed electrically actuated elastomers with strain greater than 100%. Science. 287(5454): 836839

15. R. Pelrine, H. Prahlad, 2008Generator Mode: Devices and Applications in Dielectric Elastomers as Electromechanical Transducers F. Carpi, et al. Editors Elsevier: Oxford, UK. 146155

16. T. G. Mc Kay, B. M. O'Brien, E. P. Calius, I. A. Anderson, 2011Realizing the Potential of Dielectric Elastomer Generators. Proc. SPIE 7976: 79767910

17. T. Mc Kay, B. O'Brien, E. Calius, I. Anderson, 2010Self-priming dielectric elastomer generators. Smart Materials and Structures. 19(5).

18. R. E. Pelrine, R. D. Kornbluh, J. S. Eckerle, H. Prahlad, 2008Circuits for electroactive polymer generators SRI International.

19. T. G. Mc Kay, B. O'Brien, E. Calius, I. Anderson, 2010Self-priming dielectric elastomer generators. Smart Mater. Struct. 19(055025): 055025

20. T. Mc Kay, B. O'Brien, E. Calius, I. Anderson, 2010An integrated, self-priming dielectric elastomer generator. Applied Physics Letters. 97(6).

21. T. Mc Kay, B. O'Brien, E. Calius, I. Anderson, 2010An integrated, self-priming dielectric elastomer generator. Applied Physics Letters. 97(062911): 062911

22. B. M. O'Brien, E. P. Calius, T. Inamura, S. Q. Xie, I. A. Anderson, 2010Dielectric elastomer switches for smart artificial muscles. Appl. Phys. A.: 15

23. B. O'Brien, Fabrication. Simulation, of. Control, Actuator. Biomimetic, Auckland. Arrays, Institute. Bioengineering, 2010The University of Auckland.

24. T. G. Mc Kay, B. M. O'Brien, E. P. Calius, I. A. Anderson, 2011Soft generators using dielectric elastomers. Applied Physics Letters. 98(14): 142903142903

25. S. Roundy, P. K. Wright, J. M. Rabaey, 2004Energy Scavenging For Wireless Sensor NetworksBoston, MA: Kluwer Academic Publishers. 212.

26. J. Due, S. Munk-Nielsen, R. O. Nielsen, 2010Energy harvesting with Di-Electro Active Polymers. IET Conf. Pub. 2010(CP563): WE244WE244

27. C. Graf, J. Maas, D. Schapeler, 2010Energy harvesting cycles based on electro active polymers. Proc. SPIE 7642: 764217764212

28. R. Kaltseis, C. Keplinger, R. Baumgartner, M. Kaltenbrunner, T. Li, P. Machler, R. Schwdiauer, Z. Suo, S. Bauer, 2011Method for measuring energy generation and efficiency of dielectric elastomer generators. Appl. Phys. Lett. 99(16): 162904

29. S. J. A. Koh, X. Zhao, Z. Suo, 2009Maximal energy that can be converted by a dielectric elastomer generator. Appl. Phys. Lett. 94(26): 262902

30. B. M. O'Brien, I. Anderson, 2011An Artificial Muscle Ring Oscillator. Mechatronics, IEEE/ASME Transactions on. 99p. 1-4.

31. B. M. O'Brien, E. P. Calius, T. Inamura, S. Q. Xie, I. A. Anderson, 2010Dielectric Elastomer Switches for Smart Artificial Muscles. Applied Physics A: Materials Science & Processing. 100: 385389

32. B. M. O'Brien, T. G. Mc Kay, S. Q. Xie, E. P. Calius, I. A. Anderson, 2011Dielectric Elastomer memory. Proc. SPIE. 7976: 797621797627

5

STATUS AND PERSPECTIVES OF MULTIFERROIC MAGNETOELECTRIC COMPOSITE MATERIALS AND APPLICATIONS

Haribabu Palneedi [1,2,†], Venkateswarlu Annapureddy [2,†], Shashank Priya [3] and Jungho Ryu [2,*]

[1] Department of Materials Science and Engineering, Korea Advanced Institute of Science and Technology (KAIST), Daejeon 34141, Korea

[2] Functional Ceramics Group, Korea Institute of Materials Science (KIMS), Changwon 51508, Korea

[3] Bio-inspired Materials and Devices Laboratory (BMDL), Center for Energy Harvesting Materials and Systems (CEHMS), Virginia Tech, Blacksburg, VA 24061, USA

ABSTRACT

Multiferroic magnetoelectric (ME) composites are attractive materials for various electrically and magnetically cross-coupled devices. Many studies have been conducted on fundamental understanding, fabrication processes, and applications of ME composite material systems in the last four decades which has brought the technology closer to realization in practical devices. In this article, we present a review of ME composite materials and some notable potential applications based upon their properties. A brief summary is presented on the parameters that influence the performance of ME composites, their coupling structures, fabrications processes, characterization techniques, and perspectives on direct (magnetic to electric) and converse (electric to magnetic) ME devices. Overall, the research on ME composite systems has brought us closer to their deployment.

KEYWORDS

multiferroic; magnetoelectric; ferromagnetic; ferroelectric; magnetostrictive; piezoelectric; composites; thin film; thick film; core/shell; Janus-type structure

1. INTRODUCTION

The multifunctional properties of multiferroics enable the design of novel electronic devices for various sensing, transduction and memory applications. Muliferroic materials are characterized by two or more ferroic orders, such as ferroelectric, ferromagnetic, or ferroelastic, and the interactions between these order parameters, as shown in Figure 1. In multiferroic magnetoelectric (ME) materials, coupling occurs between the magnetic and electric subsystems. This enables the control of dielectric polarization P by a magnetic field H (direct ME (DME) effect: $\Delta P = \alpha_H \Delta H$) and the manipulation of magnetization M by an electric field E (converse ME (CME) effect: $\mu_0 \Delta M = \alpha_E \Delta E$), with μ_0 denoting the vacuum permeability. The ME response is quantified in terms of the ME coupling coefficient (α_H or α_E), which represents the coupling efficiency between the electric and magnetic fields. This is considered as the figure of merit for the strength of ME coupling [1]. This article discusses the present status and recent progress in the development of ME materials and their applications in devices.

Single phase magnetoelectrics are chemically homogenous and isotropic compounds. They exhibit intrinsic ME coupling but require the co-existence of magnetic moments and electric dipoles with long-range ordering. From a fundamental point of view, the coupling between the magnetic and polar sublattices in single phase ME compounds is fascinating. However, due to the mutual exclusion of ferromagnetism and ferroelectricity, only few monolithic ME materials exhibiting non-zero coupling at room temperature have been found so far. Most of the single phase materials possess either low permittivity or low permeability at room temperature and thus exhibit weak ME coupling which hinders their applications. For example, $BiFeO_3$, the archetype of single phase ME compounds, displays good ferroelectricity but weak ferromagnetic properties above room temperature [2].

ME composites, the focus of this review, consist of physically separated magnetic and electric order phases. These composites show coupling with orders of magnitude larger than those found in single phase materials at room temperature [3]. Current understanding suggests that ME coupling in composites occurs extrinsically in three different ways mediated through

(i) strain, (ii) charge carrier, and (iii) spin exchange. Among these mechanisms, the strain-mediated ME coupling has been widely studied, while the investigations on the other two mechanisms are still in early stages [4]. In this paper, the ME composites based on the strain-mediated coupling will be discussed in detail, and their applications in potential devices will be highlighted.

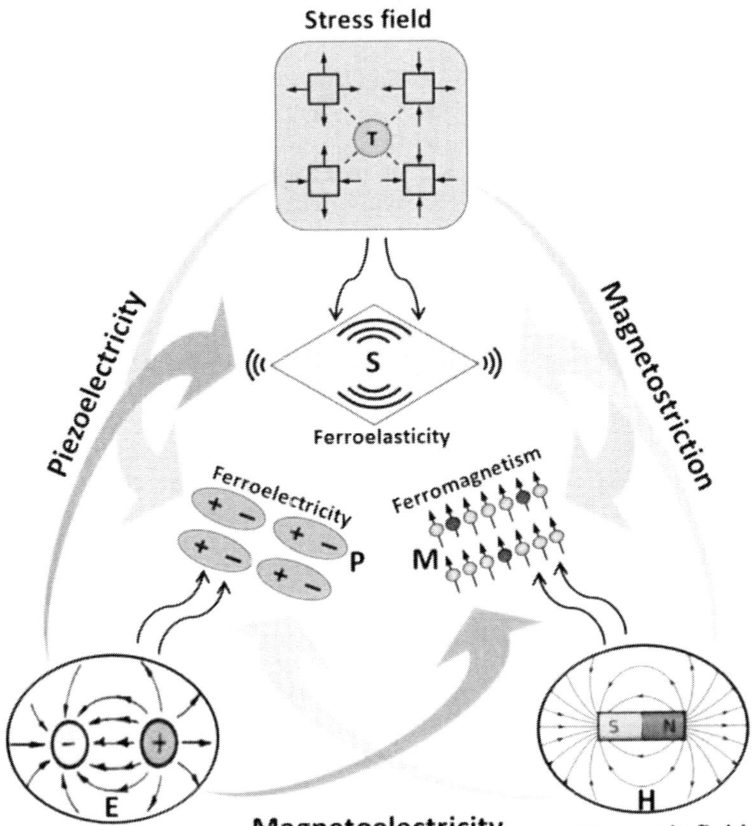

Figure 1. Schematic illustrating magnetic-elastic-electric couplings in multiferroic materials. Here, M is magnetization, S is mechanical strain, and P is dielectric polarization.

The strain-mediated ME effect in composites is a product tensor property and results from the elastic coupling between the piezoelectric and magnetostrictive components as illustrated in Figure 2 [5,6]. In DME coupling, the applied magnetic field generates strain in the magnetic layer via the magnetostriction effect, and this strain is transferred to the piezoelectric layer resulting in an electric displacement or a dielectric polarization through the piezoelectric effect. In CME coupling, an external electric field induces strain in the ferroelectric layer due to the inverse piezoelectric effect, and the strain transferred to the magnetic layer produces a magnetization change or domain reorientation by the piezomagnetic effect.

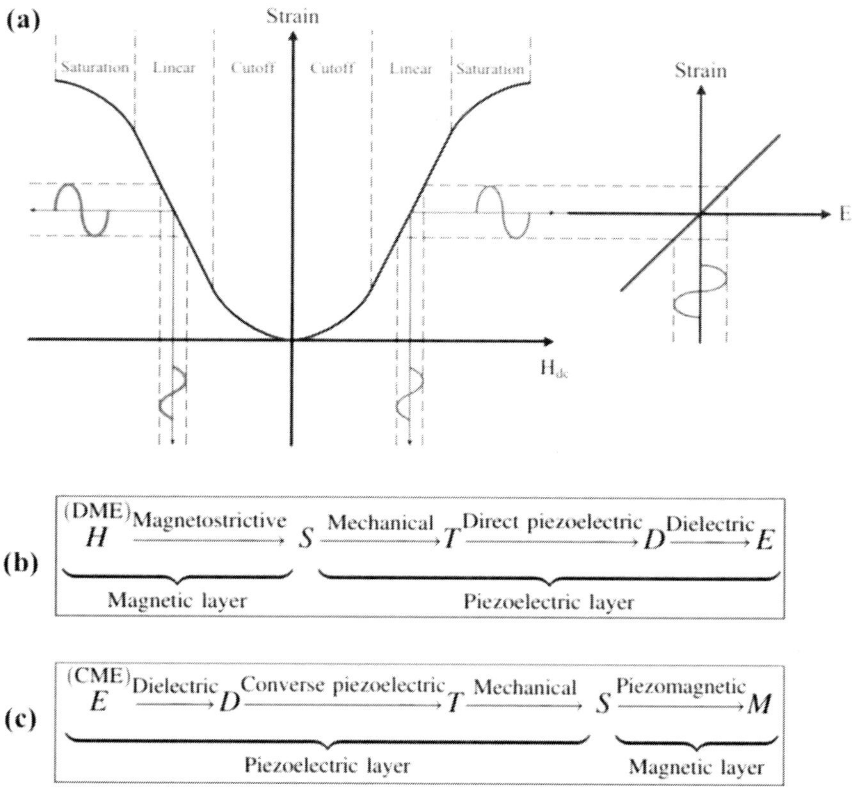

Figure 2. Schematic representation of the (**a**) ME effect utilizing the product property (Adapted from [6]). (**b**) DME effect and (**c**) CME effect in composites [7]. Here, H is magnetic field, S is mechanical strain, T is mechanical stress, D is electric displacement, E is electric field, and M is magnetization.

2. FACTORS AFFECTING ME COUPLING IN COMPOSITES

In recent years, ME composites have been extensively studied due to the ease of fabrication and design flexibility. ME composites have been developed with a diverse set of materials, properties, and microstructures using a variety of processing techniques. The parameters and variables that should be taken into account for developing high performance composites with strong ME coupling are summarized in this section.

2.1. Connectivity and Interface Bonding

The phase connectivity and interfacial bonding of the piezoelectric and magnetostrictive constituents have a strong influence on elastic coupling in the composite and its ME response. Good mechanical bonding at the interface between the composite phases, characterized by the interface coupling factor (k_c), facilitates efficient strain transfer. Though several connectivity schemes have been proposed for designing two-phase composites [8], ME composites have been commonly prepared with 0-3, 1-3, and 2-2 connectivity, as shown in Figure 3a, d, and g, respectively. Here, the numbers represent the connectivity of the magnetic and piezoelectric phases, respectively. In the 0-3 particle-matrix composites, magnetic particles are embedded in the piezoelectric matrix. 1-3 cylinder-matrix composites are formed by embedding magnetic fibers/rods/tubes/wires in the piezoelectric matrix. The particles (in the 0-3 composite) and fibers (in the 1-3 composite) can be either randomly dispersed or periodically aligned [9]. A 2-2 laminate composite consists of alternating magnetic and piezoelectric layers. Such laminates can be prepared in different shapes and geometries, including discs, squares, rectangles, and rings, with different dimensions. They can be arranged as unimorphs and bimorphs as well as bilayered and multilayered structures. Further, for all of the above composites, the volume fraction and dimensions of the constituents can be altered to tailor the properties of the composite. The 2-2 composites preserve the physical characteristics of individual phases, and they are comparatively simpler to fabricate. These composites can be poled to a higher degree since the piezoelectric and the low resistivity magnetic phases are separated. Results of both theoretical and experimental studies have shown that the 2-2 layered composites exhibit higher ME responses compared to the 0-3 and 1-3 composites. There have been other composite structures reported in the literature. Park *et al.* [10] reported the development of a 3-2 structured composite consisting of a $(Ni_{0.6}Cu_{0.2}Zn_{0.2})Fe_2O_4$ [NCZF] phase with 2D connectivity dispersed in a $0.8Pb(Zr_{0.52}Ti_{0.48})O_3–0.2Pb(Zn_{1/3}Nb_{2/3})O_3$ [PZNT] matrix. Recently, Gillette *et al.* [11] fabricated a quasi-one-dimensional ME composite by inserting a magnetostrictive wire (FeNi/FeGa/FeCoV) into a PZT tube where the tube-wire interface bonding was made with silver paste. Comparing the results from these prior studies, it can be stated that 2-2 composite structure has inherent advantages in terms of fabrication and performance.

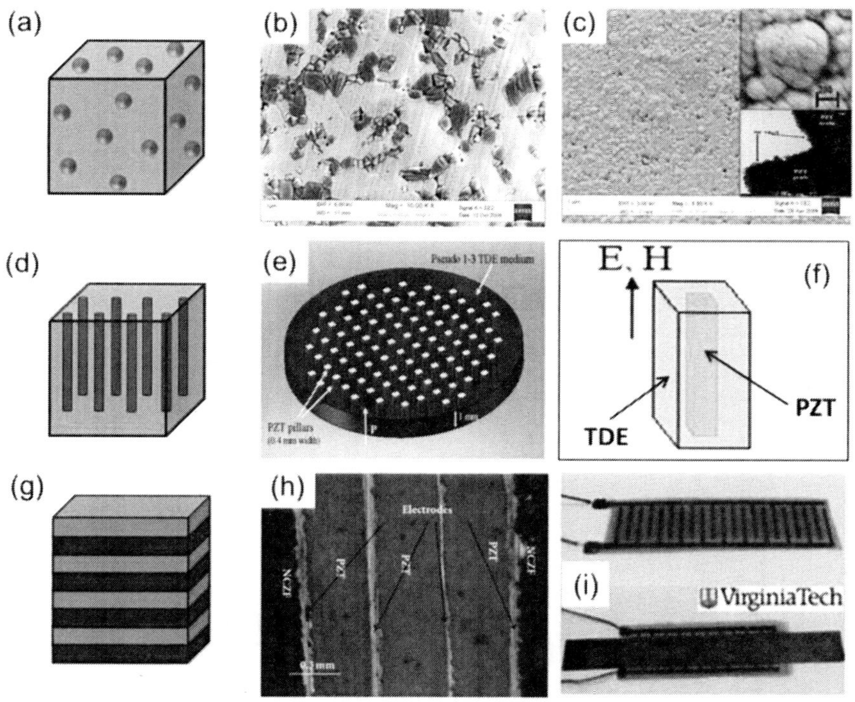

Figure 3. Development of bulk ME composites with different phase connectivity: (**a**)–(**c**) 0-3 connectivity [13,18]; (**d**)–(**f**) 1-3 connectivity [20,21]; and (**g**)–(**i**) 2-2 connectivity [22,24].

2.2. Materials and Their Properties

The selection of suitable materials is a primary step in fabricating a composite with good ME response. Various piezoelectric and magnetostrictive materials commonly used for synthesizing ME composites are listed in Table 1. Enhanced piezoelectric and ferroelectric properties can be achieved in ceramics through (a) composition selection (ideally near morphotropic phase boundary (MPB) or polymorphic phase transition (PPT)) and modification (by doping); and (b) microstructure design (via domain engineering and texturing) [12]. Among piezoelectric materials, PZT-based ceramics have been widely employed to fabricate the ME composites due to their low cost, high piezoelectric response, and flexibility in modifying the composition to achieve desired properties for targeted applications. For the magnetic component in ME composites, Terfenol-D with high magnetostriction and Metglas (amorphous Fe-alloy) with high magnetic permeability have been the most used materials. Besides magnetic properties, factors such as the processing temperatures, electrical resistance of the material, magnitude of the bias field, phase connectivity have been considered in choosing the magnetostrictive materials.

Table 1. List of well-known piezoelectric and magnetostrictive materials used as constituents of ME composites.

Piezoelectric Phase	Magnetostrictive Phase
Lead-based:	**Metals:**
$Pb(Zr,Ti)O_3$ (PZT)	Fe, Co, Ni
$Pb(Mg_{1/3}Nb_{2/3})O_3$-$PbTiO_3$ (PMN-PT)	**Alloys:**
$Pb(Zn_{1/3}Nb_{2/3})O_3$-$PbTiO_3$ (PZN-PT)	FeNi-based
$Pb(Mg_{1/3}Nb_{2/3})_y (Zr_xTi_{1-x})_{1-y}O_3$ (PMN–PZT)	FeCo-based
$Pb(In_{1/2}Nb_{1/2})O_3$-$Pb(Mg_{1/3}Nb_{2/3})O_3$-$PbTiO_3$ (PIN-PMN-PT)	CoNi-based
Lead-free:	Ni_2MnGa
$BaTiO_3$ (BTO)-based	Permendur (FeCoV)
$(K_{0.5}Na_{0.5})NbO_3$ (KNN)-based	Galfenol (FeGa), FeGaB
$Na_{0.5}Bi_{0.5}TiO_3$ (NBT)-based	Samfenol ($SmFe_2$)
Others:	Terfenol-D ($Tb_{1-x}Dy_xFe_2$)
AlN	Fe-based metallic glasses (FeBSi, FeBSiC, FeCoB,
ZnO	FeCoSi, FeCoSiB, FeCuNbSiB)
$(Sr, Ba)Nb_2O_5$	**Ceramics:**
$Ba_{1-x}Sr_xTiO_3$ (BSTO)	Fe_3O_4
$Bi_{1-x}Sr_xTiO_3$ (BST)	$Zn_{0.1}Fe_{2.9}O_4$ (ZFO)
$La_3Ga_{5.5}SiO_{14}$ (LGS)	$La_xSr_yMnO_3$ (LSMO)
$La_3Ga_{5.5}Ta_{0.5}O_{14}$ (LGT)	$La_xCa_yMnO_3$ (LCMO)
Polyurethane (PU)	Ferrites or doped Ferrites (e.g., $NiFe_2O_4$ (NFO),
Polyvinylidene difluoride (PVDF)	$CoFe_2O_4$ (CFO), Li ferrite, Cu ferrite, Mn ferrite)

The material properties that are considered important when developing ME composites include electrical properties such as dielectric constant or permittivity (ε_r), dielectric loss (tan δ), Curie temperature (T_C), remnant polarization (P_r), coercive electric field (E_c), piezoelectric strain constant (d_{ij}), piezoelectric voltage constant (g_{ij}), electromechanical coupling factor (k_{ij}), mechanical quality factor (Q_m), and acoustic impedance (Z). The magnetic properties of relevance include magnetic permeability (μ), remnant magnetization (M_r), coercive magnetic field (H_c), magnetostriction (λ_{ij}), piezomagnetic coefficient (q_{ij}), and Curie and Neel temperatures (T_C and T_N). Generally, to obtain a better ME response from the composite, the combination of a piezoelectric material displaying a high piezoelectric voltage constant ($g_{ij} = d_{ij}/\varepsilon$) and low dielectric and piezoelectric losses, and a magnetostrictive material exhibiting a high magnetostriction coefficient

(q_{ij}) and good interfacial coupling between these two phase components is desired.

2.3. Fabrication of ME Composites

2.3.1. Bulk ME Composites

A widely used method for synthesizing of 0-3 particulate composites consists of sintering a mixture of ceramic oxide powders at high temperature. This ceramic sintering process is simple and the properties of the composite can be conveniently tailored by selecting the constituent phases, their starting particle sizes, and processing parameters. Using this method, mainly magnetic ferrites have been co-processed with piezoelectric ceramics due to their high temperature stability.Figure 3b shows the microstructure of a $Pb(Zr_{0.56}Ti_{0.44})O_3$–$Ni_{0.6}Zn_{0.2}Cu_{0.2}Fe_2O_4$ sample sintered at 1000 °C [13]. At present, the measured ME coupling of 0-3 particulate composites are far lower than the theoretically predicted ones. This can be attributed to several factors: (a) the misfit strain at the interface arising from the thermal expansion mismatch between the piezoelectric and ferrite phases, which reduces the densification; (b) the interdiffusion and/or chemical reactions between the two phases during high-temperature sintering, which deteriorates the piezoelectricity and/or magnetostriction of constituent phases and the strain transfer between two phases; and (c) current leakage due to the low resistivity in the randomly distributed magnetic phase with low percolation threshold, which makes the electric poling of these ME composites quite difficult [2,3]. Some efforts have been made to solve these issues. Hot pressing and spark plasma sintering (SPS) have been utilized to achieve high density while avoiding possible reaction between the constituent phases [14,15]. A modified precipitation method involving a controlled annealing, quenching, and aging cycle [16] has been employed to obtain good dispersion of ferrite ($NiFe_{1.9}Mn_{0.1}O_4$) particles in the matrix (PZT) phase. Core/shell (ferrite/piezoceramic) structured composites have been synthesized to avoid the contact of the conductive ferrite particles during sintering [17]. Unlike the other cases, Islam *et al.* [18] reported a PZT core/$NiFe_2O_4$ shell ME composite (Figure 3c). To circumvent the problem of high conductivity through the ferrite phase, a 100-mm layer of PZT was used to cover both faces of the sintered pellet.

Very few attempts have been made to fabricate 1-3 type bulk ME composites [19,20,21]. Shi *et al.* [19] prepared a pseudo-1-3 type ME composite consisting of a PZT rod array (with base) and Terfenol-D/epoxy matrix (TDE) using the dice-and-fill technique, where diced PZT sample was filled with an epoxy resin solution containing Terfenol-D particles

(Figure 3e). A peak ME response of 6 V/cm·Oe was obtained between 80 and 110 kHz. Lam *et al.* [20] studied the frequency response of PZT-TDE composite, made using the same dice-and-fill method, under magnetic bias field. It was found that the resonance shifted to a lower frequency with increasing magnetic bias field. Ma and colleagues [21] presented a simple 1-3 composite made up of just a single PZT rod embedded in a TDE mixture (Figure 3f), which avoided the need for dicing of the PZT rod arrays. This composite exhibited a maximum ME response of 18.2 V/cm·Oe at 84 kHz. Such a large ME response was attributed to the enhancement of coupling at electromechanical resonance, which assisted the elastic interaction between the TDE medium and PZT rod.

Preparation of bulk 2-2 laminate composites has been mainly conducted by epoxy bonding and co-firing [5,22,23,24,25,26]. The most common geometries in laminate composites are either bilayer structure of magnetostrictive and piezoelectric layers or a sandwich structure, whereby the piezoelectric layer is arranged between two magnetostrictive ones. Laminate composites made by bonding magnetostrictive and piezoelectric layers with silver epoxy are usually cured at a lower temperature (80–100 °C). The epoxy acts as a conductive bond and strengthens the interface between the composite layers, although the epoxy layer must be kept as thin as possible to obtain good ME coupling. Various combinations of piezoelectric and magnetostrictive materials can be used to fabricate different types of ME composites with this bonding process as it does not involve high temperatures. However, the mechanically soft and viscoelastic epoxy layer will dampen the generated strain and induce a noise floor, suppressing the interface coupling and thus the ME response. Co-firing is an established process used for sintering multilayer capacitors. Besides strengthening the interface in layered ME composites, the co-firing process also provides the possibility of cost-effective mass production. Figure 3h shows the cross-sectional optical image of a NCZF-stack [0.9 PZT-0.1 PZN]-NCZF trilayer co-fired at 900 °C for 3 h [22]. Despite its advantages, the high temperature co-firing process, similar to the ceramic sintering of 0-3 particulate composites, is challenging owing to the different shrinkage rates, thermal expansion mismatch, and interface inhomogeneities.

Piezoelectric macro-fiber composites (MFCs) offer better flexibility and higher performances than the traditional monolithic piezoelectric ceramics. Compared to a PZT plate, MFCs are easier to elongate and shrink when force is applied. Several ME composites have been developed based on these MFC structures, which employ an interdigitated electrode (IDE) [23,24]. Figure 3i displays picture of an IDE/piezofiber core composite and its implementation in a complete Metglas/piezofiber ME sensor. This composite structure yielded a giant ME response with an extremely low equivalent magnetic noise. Some polymer-based laminate ME composites prepared using polyvinylidene difluoride (PVDF), a piezopolymer possessing a low ε and a high g, in combination with Metglas or Vitrovac (Fe-based amorphous alloy) and Terfenol-D, have also shown large ME coupling behavior [25,26].

2.3.2. Film-Based ME Composites

As device design trends towards miniaturization and multifunctionality, thin film conformations are highly desired for application in integrated magnetic/electric devices. In film-based ME composites, optimized interface coupling can be achieved through the direct bonding of components. Moreover, large electric fields can be comfortably applied to thin films since they require relatively smaller bias voltages than those needed for bulk samples. This extends the scope of thin films for devices which have limitations on their operating voltages [27]. The greater freedom and flexibility in the fabrication of film-based ME composites render their property tuning, through interfacial coupling and strain engineering, control of crystal structure and orientation, grain size, and layer thickness, as well as chemical modification with a wide range of substituents, *etc.* Composite films can also facilitate the understanding of the physical phenomena involved in ME effects at lower dimensions, and thus enable the design of new types of magnetoelectrics with novel phase structures. A number of ME composite films have been recently prepared using various film deposition methods. Some of these methods can yield excellent thin film epitaxial growth with atomic scale thickness control and coherent interfaces [12]. However, most ME composite films reported in the literature show very low ME properties due to reduced electromechanical parameters as a consequence of substrate clamping.

Since ME coupling occurs through interfacial strain transfer, the ME response of 0-3 type ME nanocomposites can be enhanced by dispersing magnetic nanoparticles (NPs) with large interfacial area in a piezoelectric matrix. Based on this idea, McDannald *et al.* [28] synthesized PZT-CFO nanocomposite films (Figure 4a) via spin coating, and the ME coupling was modulated by controlling the concentration of CFO NPs dispersed in the PZT matrix. A high ME coefficient of 0.549 V/cm·Oe was obtained with a low concentration of CFO NPs, and the subsequent decrease in ME response with higher CFO concentrations was ascribed to the enhanced agglomeration of NPs.

The 1-3 type ME film composites have mainly been explored at the micro- and nanoscale where they can be easily synthesized through the self-assembly process [29]. Aimon *et al.* [30] developed self-assembled $BiFeO_3$–$CoFe_2O_4$nanocomposites (Figure 4b) that were templated into ordered structures in which the ferrimagnetic $CoFe_2O_4$ pillars formed square arrays having periods of 60–100 nm in a ferroelectric $BiFeO_3$ matrix.

The 1-3 ME composites prepared via self-assembly were reported to exhibit a reduced substrate clamping effect and more efficient strain coupling, as a result of their large interfacial surface area [12]. However, the design and control of such a structure remains a challenge, and their ME coefficients could not be directly measured because of the leakage problem, which resulted from the low resistance of the magnetic pillars penetrating through the films.

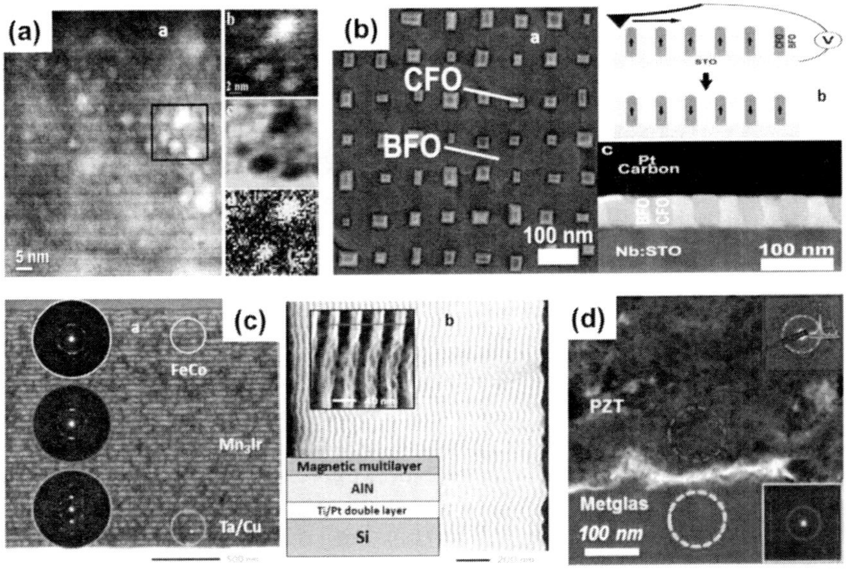

Figure 4. Development of film-based ME composites with (**a**) 0-3 connectivity [28]; (**b**) 1-3 connectivity [30]; (**c**) and (**d**) 2-2 connectivity [33,39].

Deposition methods that have been used to develop thin film-based 2-2 type ME composites include pulsed laser deposition, chemical vapor deposition, sputtering, molecular beam epitaxy, spin coating, spray pyrolysis, and spin-spray techniques [31,32]. Self-biased ME film composites that show high sensitivity to AC magnetic fields were reported by Lage *et al.* [33] These thin film 2-2 ME composites fabricated by magnetron sputtering on silicon-cantilever substrates consisted of piezoelectric AlN and multilayers having the sequence $Ta/Cu/Mn_{70}Ir_{30}/Fe_{50}Co_{50}$ or $Ta/Cu/Mn_{70}Ir_{30}/Fe_{70.2}Co_{7.8}Si_{12}B_{10}$serving as the magnetostrictive component (Figure 4c). Giant ME couplings were reported for ME thin film composites based on $BaTiO_3$-CoFe-$BaTiO_3$, and AlN-

$(Fe_{90}Co_{10})_{78}Si_{12}B_{10}$ systems [34,35]. Thick film-based 2-2 layered ME composites have been prepared by employing several deposition techniques such as tape casting [36], aerosol deposition [37], electrophoretic deposition (EPD) [38], granule spray in vacuum deposition (GSV) [39], electrodeposition [40], and electroless deposition [41]. Recently, a unique fabrication approach involving a combination of room temperature deposition (GSV) and localized annealing of PZT thick film by laser radiation has been demonstrated [39] to synthesize PZT/Metglas ME film composites (Figure 4d). Localized heating through selective absorption of laser irradiation in the PZT film not only induced significant improvement in its dielectric and ferroelectric properties but also contributed to a colossal ME output by avoiding thermal damage to the Metglas substrate and preserving its inherent magnetic properties. Alternatively, photonic sintering or flash light irradiation can also be used to induce localized annealing effects in films deposited on thermally sensitive substrates. This method, commonly employed in printed electronics, uses broadband (ultraviolet to infrared) intense light pulses generated from a Xenon flash lamp to control the thermal diffusion in the nanostructured films.

3. RECENT ADVANCES IN THE DEVELOPMENT OF ME COMPOSITES

3.1. Realization of Broadband ME Response with Piezoelectric Anisotropy

In general, depending upon the material properties, the ME coefficient of laminate composites exhibit maxima at a specific DC bias under off-resonance condition. In the resonant condition, ME laminates exhibit a sharp peak around the resonance frequency. Such behavior limits their ability to sense wide ranges of both DC and AC magnetic fields. The broadband/wideband behavior is characterized by flat ME response over a given AC frequency range and DC magnetic bias. Limited efforts have been made to achieve a wide bandwidth around resonance frequency either by combining several PZT/Terfenol-D bilayers through parallel and series electrical connections [42] or by developing composite structures of PZNT/Metglas with a dimensionally gradient architecture [43,44]. These peak broadening approaches have indeed shown some promise, but they only cover a limited range of frequencies with a non-uniform power distribution. Patil *et al.* [45] demonstrated a resonant ME response with a wide bandwidth by adopting tri-layer ME laminates (Ni/PMN-PZT/Ni) having inherent multiple resonance frequencies, attributable to the anisotropic piezoelectric response of the [011]-oriented PMN-PZT single crystal. The required broadening of the resonance frequencies was achieved by serially connecting three ME laminates with different thickness ratios of magnetostrictive and piezoelectric layers. Such a wide bandwidth, along with the giant ME coupling, could be used in multifunctional devices such as broadband energy harvesters or field sensors. Further, Patil *et al.* [46] demonstrated the presence of giant ME voltage coefficients and magnetic-field-direction dependent ME signals at both low frequencies and resonance frequency in a symmetric ME laminate (Metglas/PMN-PT/Metglas). High performance resulted from the anisotropic transverse piezoelectric coefficients of the [011]-oriented PMN-PT single crystal. In another study, Kambale *et al.* [47] designed the ME rectangular unimorph cantilever beam structure consisting of Ni and PMN-PZT single crystal with <001> and <011>-cut crystallographic directions and investigated their mechanical vibration-based energy harvesting behavior. Both the ME voltage coefficient (α_{ME}) and mechanically harvested power output was found to be strongly dependent on the crystallographic cut directions of the PMN-PZT single crystals.

3.2. ME Composites with Textured Piezoelectric Ceramics

To achieve high ME coefficients, it is necessary to use a piezoelectric material exhibiting a high magnitude of piezoelectric voltage constant ($g = d/\varepsilon$). However, piezoelectric ceramic compositions having a high d_{33} usually show a low g_{33}, while high g_{33} compositions possess a low d_{33}. Thus, the achievement of both high d and g from a single piezoelectric composition has been challenging. Though domain-engineered, relaxor piezoelectric single crystals such as PMN-PT and PZN-PT exhibit excellent piezoelectric properties, their application has been limited due to the high production cost and small size of useful samples. PZT-based compositions cannot be readily grown in single crystal form because of their incongruent melting behavior [48]. Alternately, textured ceramics based on PMN-PT and PMN-PZT have been prepared using the templated grain growth (TGG) method. Here, the texturing produces an engineered domain state resulting in high d values and the use of oriented and platelet shaped template crystals having low permittivity suppresses the increase in ε of the textured ceramic, thereby achieving low cost high performance piezoelectric ceramics suitable for ME composites. Yan *et al.* [49] reported an enhanced α_{ME} for Metglas/textured PMN-PZT/Metglas laminate compared to Metglas/random PMN-PZT/Metglas, where the maximum α_{ME} was found to be 0.348 V/cm·Oe and 1.49 V/cm·Oe, respectively. Because of the environmental concerns of using lead containing materials, recent investigations have focused on developing ME composites employing lead-free piezoelectric ceramics with compositions based on KNN, NBT, and BT. A large value of maximum α_{ME} (1.32 V/cm·Oe) has been reported from a laminate ME composite of Terfenol-D/Mn-doped NBT–BT single crystal/ Terfenol-D [50].

3.3. Self-Biased ME Composites

Most conventional ME composites need a DC bias, to obtain the maximum ME response, and that necessitates the use of permanent magnets or electromagnets resulting in bulky devices and problems of electromagnetic interference. To circumvent these issues, self-biased magnetoelectric (SME) composites that provide sizeable ME coupling under an external AC magnetic field in the absence of a DC magnetic field have been proposed [51]. This would enable device miniaturization and the development of ME composite-based circuit components for integrated electronics and medical applications. The SME composites are categorized into five groups according to their working mechanism: (a) functionally graded FM-based SME; (b) exchange bias-mediated SME; (c) magnetostriction hysteresis-based SME; (d) built-in stress-mediated SME; and (e) non-linear SME composites (Figure 5). Large and tunable SME responses have been reported for both bulk and film-based ME composite systems [33,52]. A recent review by Zhou *et al.* [51] provides a comprehensive coverage of the development and prospects of the SME composites.

3.4. ME Composites with Novel Structures

ME composites with a range of composite connectivity, including new connectivity designs that cannot be readily synthesized by traditional routes are being developed through the fabrication of nanostructured ME materials. In recent years, there have been some studies on the synthesis of ME nanocomposites having core/shell arrangement in the form of nanoparticles, nanowire arrays, and nanotubes. Other 1D composite structures with random and Janus-type arrangements have also been prepared (Figure 6). These nanostructured composites were prepared using different wet chemical synthesis methods. Kalyan *et al.* [53] synthesized $CoFe_2O_4/BaTiO_3$ core/shell nanoparticles (Figure 6a) using a combination of solution processing and high temperature calcination. Rongzheng *et al.* [17] demonstrated the formation of ferrite/perovskite oxide core/shell nanostructures in several multiferroic systems such as $Fe_3O_4/PbTiO_3$, γ-$Fe_2O_3/PbTiO_3$, γ-Fe_2O_3/Pb(Zr,Ti)O_3, $CoFe_2O_4/BaTiO_3$, $CoFe_2O_4/PbTiO_3$ and $CoFe_2O_4$/Pb(Zr,Ti)O_3 using a combined hydrothermal and annealing process. For synthesizing ordered arrays of $NiFe_2O_4$/PZT core/shell nanowires (Figure 6b), Ming *et al.* [54] developed a method involving the combination of a modified sol-gel process, electrochemical deposition, and subsequent oxidization in anodized nanoporous alumina membranes. Recently, Andrew *et al.* [55,56] reviewed the efforts in the development of multiferroic nanofibers with core/shell, random, and Janus-type arrangements (Figure 6c–e), fabricated by electrospinning. Nevertheless, some of the challenges in the practical implementation of these ME nanocomposites include forming isolated multiferroic particles that are free of agglomerates, substrate free assembling of nanofibers into ordered structures, and difficulties in accessing their ME properties. Further, in some cases, these nanostructured ME composites will have to be consolidated into a dense form for device applications which may affect the stability of nanostructures.

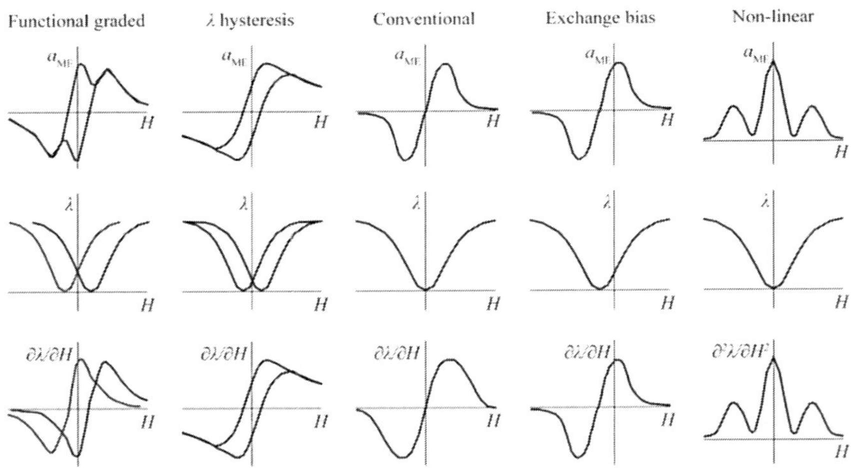

Figure 5. Schematic representation of different types of self-biased ME composites with different α_{ME}-H (top row), λ-H (middle row), and q-H curves (bottom row) [51].

4. CHARACTERIZATION OF ME COUPLING

To understand the effect of multiple parameters and variables (materials, connectivity, operational mode, synthesis method, *etc.*) on the ME coupling, and to predict the performance of ME composites in various applications, investigations on their working behavior are essential. Several characterization tools have been reported in literature on DME and CME coupling, based on the measurement as a function of an applied magnetic field or an electric field over a range of frequencies from DC to 110 GHz. Some of these analysis techniques have been discussed in detail by Srinivasan [59]. The primary tools for studying DME coupling are measurements of the low-frequency ME voltage coefficient, voltage output in response to applied magnetic fields at the bending and electromechanical resonance modes, nonlinear ME effects for large AC magnetic fields, static magnetic field-induced polarization, and magneto-dielectric effects. Studies on the nature of CME coupling include low-frequency ME effects under an applied AC electric field and measurement of the induced magnetic flux in a coil wound around the composite, static electric field E-induced magnetization, E-tuning of inductance, or ferromagnetic resonance. Recently, ME measurements on thin films and nanostructured composites have been made by employing scanning probe microscopy (SPM). Piezoresponse force microscopy (PFM) under an applied magnetic field is used for studies on DME, and magnetic force microscopy (MFM) under an applied voltage is used for analyzing the CME.

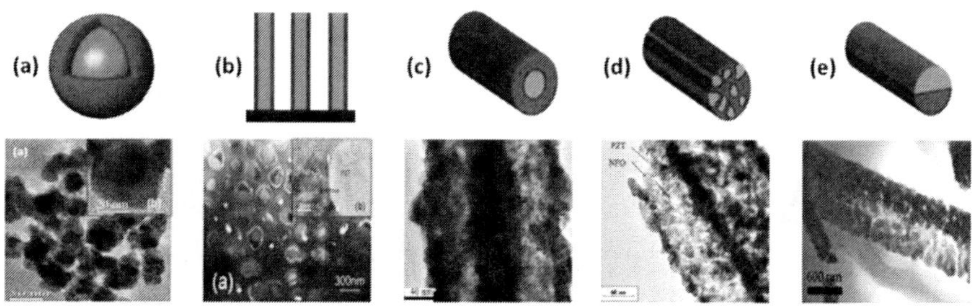

Figure 6. Bi-phasic ME composites with different types of arrangements between the two phases (**a–c**) core/shell arrangement: (**a**) nanoparticles [53]; (**b**) nanowires [54]; and (**c**) nanotubes [57]; (**d**) random arrangement [58]; and (**e**) Janus-type arrangement [56].

From the point of view of functional devices, it is important to control the magnitude of the ME voltage coefficient and its behavior [60]. In a ME composite, the measured DME coupling coefficient is the field conversion ratio between applied $H_{ac}$and induced E_{ac} under a bias field H_{dc}, $\alpha_{ME} = \delta E_{ac}/\delta H_{ac}$. For such measurements, it is necessary to first pole (DC poling or Corona poling) the ferroelectric phase in the composite. In general, under a constant applied AC magnetic field, the ME coupling coefficient first increases with increasing H_{dc}, reaching a maximum at an optimized DC bias (H_{bias}), and then decreases with further increasing H_{dc}. When the bias field is reversed, the ME voltage shows a 180° phase shift. The H-dependence tracks the slope of λ *versus* H. Saturation of λ at high H leads to $\alpha_{ME} = 0$ [27]. For strain-mediated ME composites, it has been shown that $\alpha_{ME} \propto q_{ij} = (d\lambda_{ij})/dH$. This indicates that the ME coefficient is directly related to the nature of the ferromagnetic phase and the effectiveness of elastic coupling between the two phases. Figure 7a demonstrates the typical behavior of the α_{ME}, λ, q, and their relationship as a function of H_{dc}. The variation of magnetostriction as a function of H and the dependence of α_{ME} on q suggests the need for additional H_{dc}. Usually, the α_{ME} value is higher when H and δH are parallel to each other and to the sample plane, compared to the out-of-plane magnetic fields due to demagnetization. So far, giant ME responses have been reported in a variety of composite systems and the highest α_{ME} was reported for composites with ferromagnetic alloys. Figure 8 summarizes the best values of α_{ME} obtained for different combinations of materials in bulk and film-based ME composites having 0-3, 1-3, and 2-2 connectivity.

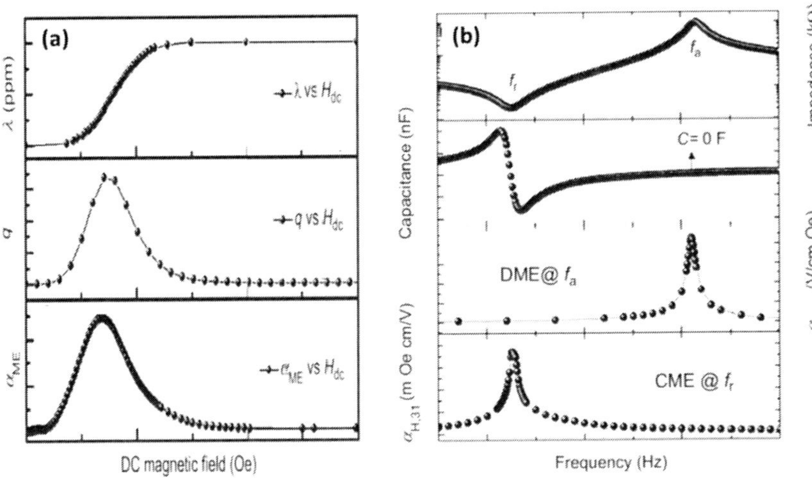

Figure 7. (**a**) Typical behaviors of magnetostriction, piezomagnetic coefficient, and ME coefficient as a function of DC magnetic field [60]; (**b**) Typical behaviors of impedance, capacitance, and DME and CME coefficients as a function of AC frequency [7].

Another parameter of importance is the change in ME coupling as a function of applied AC magnetic field frequency (ME*vs.* f_{ac}), which depends on the sample dimensions. When AC magnetic field is applied at a frequency corresponding to the electromechanical resonance (EMR) of the piezoelectric phase, or the ferromagnetic resonance (FMR) of the magnetic phase in the ME composite, the ME voltage coefficient shows a peak behavior, with as much as a 100-fold increase in its magnitude compared to the off-resonance condition. Cho *et al.* [7] found that the DME coupling was maximized at the anti-resonance frequency (f_a), while the CME coupling exhibited peak at the EMR frequency (f_r) of the piezoelectric layer, as shown inFigure 7b. This phenomenon was further explained by using piezoelectric constitutive equations and combining them with the frequency-dependent capacitance of the piezoelectric layer. Comparison of both DME and CME coefficients have rarely been performed because it has been challenging to achieve strong DME and CME coupling simultaneously [72].

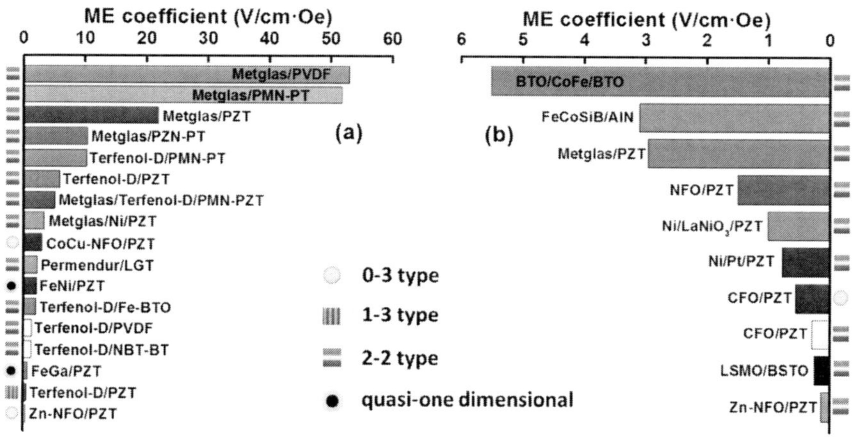

Figure 8. Reported values of off-resonance ME voltage coefficients for various material systems: (**a**) bulk and (**b**) film-based ME composites. Data taken are from Ref. [5,11,21,23,24,25,26,28,34,35,36,37,39,50, 52,61, 62,63,64,65,66,67,68,69,70,71].

5. MAGNETOELECTRIC DEVICES AND APPLICATIONS

Several of the bulk as well as film-based ME composites have been reported to exhibit strong ME responses under both off/on resonance conditions. Based on the type of ME coupling and the mechanisms used to control the order parameters, a variety of applications have been proposed, including magnetic sensors, high-frequency inductors, memory devices, and high-frequency signal processing devices (Table 2). Recent progress on the design and development of some of the ME devices will be discussed in this section.

Table 2. Classification of different ME devices. Adapted from [72].

ME Coupling	Physical Mechanism	ME Devices
Direct ME coupling	H control of electric polarization	Magnetic sensors, current sensors, transformers, gyrators, energy harvesters
Converse ME coupling	E control of magnetization switching	Spintronics, including random access memories, tunnel junctions
	E control of permeability μ	Voltage tunable inductors, tunable band-pass filters, phase shifters
	E control of spin wave	Voltage tunable filters, tunable resonators, phase shifters

5.1. Devices Based on the DME Effect

5.1.1. Magnetic Field Sensors

ME material-based sensor devices are considered to be promising alternatives for conventional Hall sensors and giant magnetoresistive (GMR) devices. Because of their passive nature and self-powered operation at room temperature, ME sensors may be able to replace bulky and expensive superconducting quantum interference devices (SQUIDs). ME sensors have the potential to perform biomagnetic measurements analogous to all of those performed by MEG and fMRI. The key requirements for magnetic field sensors in this application are (i) sensitivity of ~pT to fT per √Hz at low frequencies (10^{-2} to 10^{3} Hz); and (ii) ambient temperature and wide bandwidth (0.1 to 100 Hz) operation [73]. The direct ME coupling effect, where the ME voltage coefficient is dependent on AC and DC magnetic fields, would allow the ME composites to sense either an AC or DC magnetic field by monitoring the output electrical signals. Since a current passing through a wire generates a magnetic field in the surrounding space, the ME composite can also be used as a current probe for detecting current by monitoring the corresponding magnetic flux.

Various ME laminate-based sensors have been reported in the literature over the last decade to detect minuscule magnetic fields which could have applications ranging from medical imaging to oil prospecting. Efforts have focused on: (a) different categories of sensor construction [24,74,75,76], geometry and other features [77,78,79,80]; (b) packaging of the sensor units; (c) improvements in fabrication techniques [81]; (d) signal processing conditions [82,83]; and (e) gradiometric configurations to reduce environmental noise sources [84]. The multi-push-pull configurations, multilayer configuration, and bimorphs, all of which exhibit an improved ME voltage coefficient, have demonstrated considerable potential for sensing low-frequency magnetic field variations. A push-pull mode ME laminate consists of a symmetric longitudinally poled piezoelectric PMN-PT single crystal and two longitudinally magnetized magnetostrictive Terfenol-D layers. The symmetric nature allows for optimized elastic coupling between consecutive layers. As a result, a large ME coefficient of 30 V/cm·Oe at a resonance of 77.8 kHz was obtained with a magnetic field sensitivity of 136 pT/√Hz [74]. The very low sensitivity of 5.1 pT/√Hz, which is very close to the predicted value of 4.2 pT/√Hz, has been achieved for a Metglas/PMNT laminate structure with a multi-push-pull (M-P-P) mode that is longitudinally magnetized and longitudinally poled, coupled with a low noise charge amplifier [24]. Fang *et al.* [85] very recently proposed a Metglas/Mn-PMNT laminate composite consisting of longitudinal magnetized Metglas layers and different numbers (N) of transversely polarized Mn-PMNT fibers connected in series, as illustrated in Figure 9. An ultralow magnetic field sensitivity of 0.87 pT/√Hz at room temperature was reported for these Metglas/Mn-PMNT fiber laminate composites.

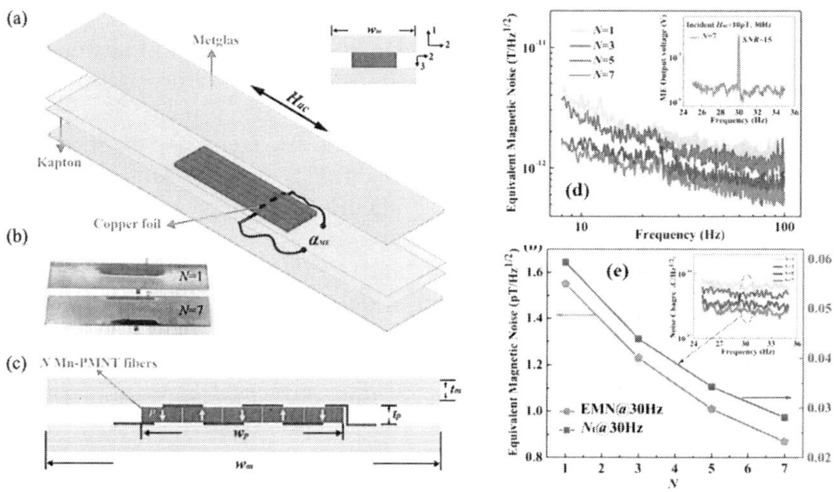

Figure 9. **(a)**–**(c)** 3D structure, photograph, cross-sectional schematic diagram, respectively, of the Metglas/Mn-PMNT composite. **(d)** and **(e)** Equivalent magnetic noise level and total noise charge density of the ME magnetic sensors [85].

Thin film-based ME sensors enable the fabrication of miniaturized low-cost sensor devices with high sensitivity and high spectral resolution [86]. Thin film-based architecture also provides the capability for designing sensor arrays that can be integrated with other circuit components. Zhao *et al.* [87] reported the fabrication of a thin film ME sensor by sputter depositing a $Fe_{0.7}Ga_{0.3}$ film (1.5 μm thick) over a sol-gel derived PZT film (1.5 μm thick) on a micromachined Si (35 μm thick) cantilever. Substantial improvement in ME coupling was observed due to reduction in substrate clamping by decreasing the Si cantilever thickness (from 180 to 35 μm). The sensor device showed a maximum ME coefficient of 1.81 V/cm·Oe at the electromechanical resonance frequency of 333 Hz and could detect a 2.3 × 10^{-8} T (2.3 × 10^{-4} Oe) AC magnetic field with the 50 nV noise floor. Marauska *et al.* [86] developed a ME sensor system with a stack of SiO_2/Ti/Pt/AlN/Cr/FeCoSiB fabricated on a 150 mm thick Si wafer cantilever. The AlN piezoelectric layer (1 μm thick) was grown by reactive pulsed-dc magnetron sputtering, and a $(Fe_{90}Co_{10})_{78}Si_{12}B_{10}$ magnetostricitve layer (2.2 μm thick) was sputter deposited in a magnetic bias field of 10 mT. The ME sensor exhibited a maximum ME coefficient of 1000 (V/m)/(A/m) (= 795.8 V/cm·Oe) under resonance at 2.4 kHz with sensitivity of 780 V/T and noise level above 100 pT/√ Hz. Recently, Lee *et al.* [88] demonstrated all-thin-film ME ac/dc magnetic field sensor arrays made by using PZT and Terfenol-D thin films in the form of 2.98-μm-thick and 300-μm-long microcantilevers (Figure 10a,b). The PZT films, displaying a columnar structure, were deposited by the sol-gel method. Terfenol-D films deposited using the RF magnetron sputtering method formed well crystallized large grains (Figure 10c). The ME voltages were measured at 60 Hz from the cantilevers with the PZT and PZT/Terfenol-D layers with increasing magnetic field between 2 × 10^{-9} T to 20 × 10^{-9} T. It was observed that the PZT/Terfenol-D cantilever produced a significant voltage output, while the PZT-only cantilever showed a negligible output (Figure 10d). There was also a gradual increment in the ME voltage of PZT/Terfenol-D in proportion to the DC magnetic field. The minimum detectable DC magnetic field was estimated by applying a DC current of 0.005 μA corresponding to 1 × 10^{-12} T, and the ME voltage from the sensor measured at 60 Hz AC field in off-resonance mode was found to be 86 μV with a 150-nV noise floor (Figure 10e).

5.1.2. Electric Current Sensors

Conventional current sensors, which operate by detecting the electric current induced by magnetic fields are best represented by Hall effect and reluctance devices [89]. Hall devices need to be powered by highly stable constant current supplies, and their inherently weak Hall voltages (5 to 40 μV/Oe) impose great demands on signal conditioners. Reluctance devices require being interfaced with highly precise integrators, and real-time measurements are generally inhibited at low frequencies (100 Hz). In contrast, current sensors based on ME composites do not suffer from these problems due to the extrinsic ME effect exhibited by the composites. In principle, a straight wire carrying an AC or a DC current will excite an AC or DC vortex magnetic field around this wire according to Ampère's Law. The strength of the magnetic field depends on the current I in the wire and r, the distance from the wire ($H = I/2\pi r$). Ring-type ME laminates, therefore, have the essential configurations of electric current sensors. For example, Leung *et al.* [90] demonstrated a ring-type electric current sensor (Figure 11) operating in vortex magnetic field detection mode. The sensor design was based on a ring-shaped magnetoelectric laminate consisting of an axially polarized PZT piezoelectric ceramic ring bonded between two circumferentially magnetized epoxy-bonded Terfenol-D short-fiber/NdFeB magnet magnetostrictive composite rings. The sensitivity of the electric current sensor was evaluated, both theoretically and experimentally. It was reported that the output voltage had good linear responses to the electric current. The sensor showed a high off-resonance sensitivity of 12.6 mV/A over a flat frequency range of 1 Hz–30 kHz and a large resonance sensitivity of 92.2 mV/A at the resonance of 67 kHz. Excellent linearity and large current sensitivity of 114.2 mV/A was observed in a Metglas/PZT laminate when measuring low-frequency alternating magnetic fields of 50 Hz [91]. This sensor is ideally suited for power-line current measurement. By combining an ME ring with a piezoelectric transformer structure, Zhang *et al.* [92] achieved a high resonance sensitivity of ~157 mV/A at electromechanical resonance frequency of 62 kHz. Yu *et al.* [93] have recently reported the design and implementation of a cantilever-type device for a current sensor based on two layers of longitudinally magnetized Terfenol-D and one layer of transversely polarized piezoelectric PZT material. The layers were epoxy-bonded, and the wire conducting the current was coiled along the device and the output voltage was obtained across the two surfaces of the PZT. Compared with a ring-type current sensor, this sensor could be installed and maintained without any interruption in the power supply, which was very convenient and practical. The high-sensitivity, power-free, bias-free, and wide-bandwidth nature of the ME current sensor also provided great potential for real-time monitoring of the conditions of engineering systems which have electric current-carrying cables or conductors.

5.1.3. Energy Harvesters

Harvesting energy from ambient energy sources such as vibrations, sound, radiofrequency waves, light, temperature gradients, wind, and others is an area of focus for current and next-generation remote monitoring electronic devices and self-powered wireless sensor networks with the goal of improving device lifetime and addressing the limitations of conventional batteries. In addition, the ambient environment is filled with magnetic noise of 50–60 Hz almost everywhere these days. Harvesting this weak and low-frequency magnetic noise (< 1 mT = 10 G) to develop a consistent electricity source remains a difficult challenge. Over the last decade, Ryu group [47,94,95] and other researchers [96,97,98,99,100,101,102,103,104,105] have investigated methods to obtain optimum electricity from the tiny magnetic fields in the surroundings, using magneto-mechano-electric (MME) mechanism. The operation mechanism can be described as follows: When the ME composite is placed in an AC magnetic field, the magnetostrictive layer in the composite responds to the mechanical vibration (magneto-mechano coupling), thereby straining the piezoelectric layer, which results in an output voltage across the electrical load through the direct piezoelectric effect (mechano-electric coupling). Due to the existence of the piezoelectric phase in the ME composite, any mechanical oscillation applied to the composite directly creates electrical voltage. Consequently, the MME generator could be used to harvest energy from both the magnetic field and external vibrations at the same time [95]. This sequential operating process is schematically depicted in Figure 12a.

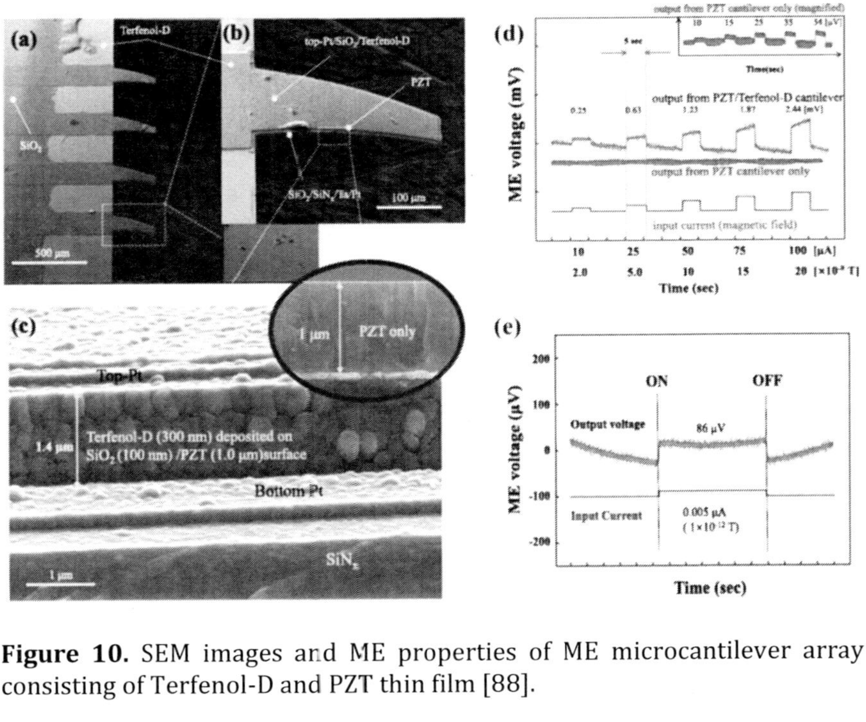

Figure 10. SEM images and ME properties of ME microcantilever array consisting of Terfenol-D and PZT thin film [88].

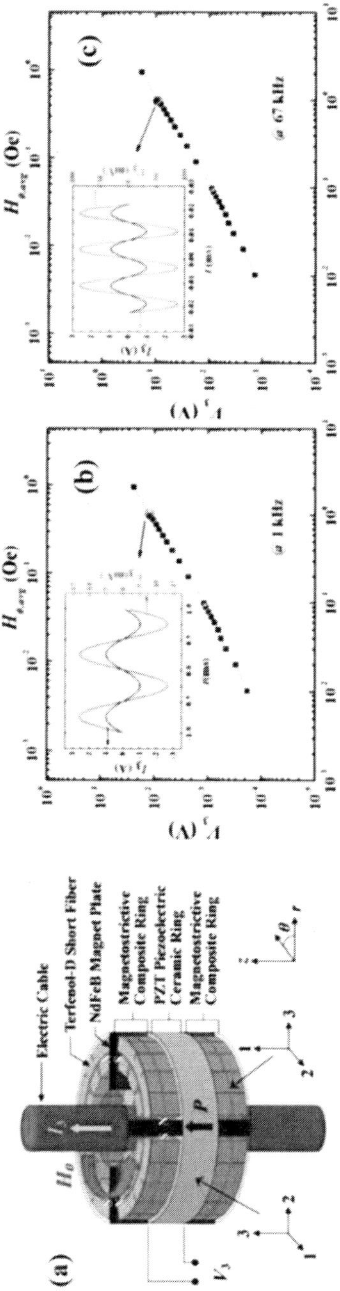

Figure 11. (**a**) Schematic diagram of the proposed ring-type electric current sensor. (**b**) and (**c**) The AC electric voltage (V_3) output from the sensor as a function of both the AC electric current (I_3) and its associated average AC vortex magnetic field ($H_{\theta,avg}$) [90].

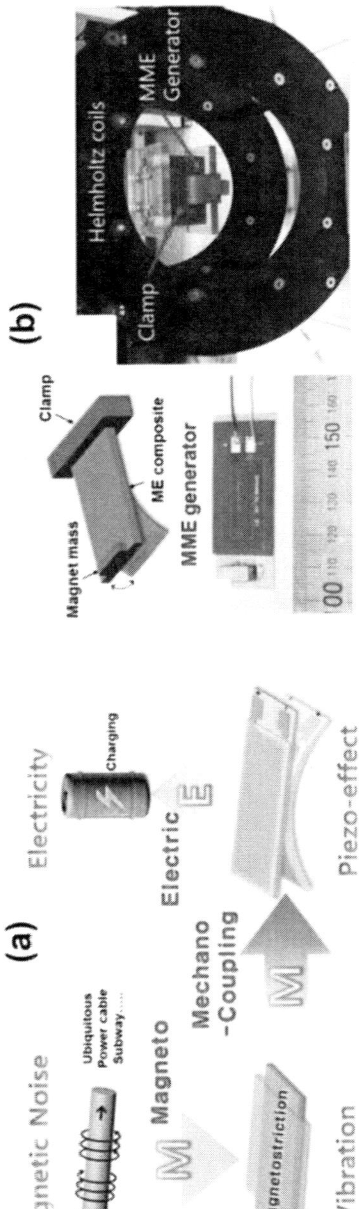

Figure 12. (**a**) Schematic depicting the working principle of Magneto-Mechano-Electric (MME) energy harvester; (**b**) Schematic and photo of cantilever structured MME energy harvester [95].

By selecting high performance piezoelectric and magnetostrictive materials and by optimizing the composite structure, it is expected that high electric power density can be obtained from a low-frequency magnetic field using the MME generator. Gao *et al.* [103] designed an asymmetrical bi-layered push-pull mode Metglas/Pb(Zr,Ti)O_3 laminate, and obtained a giant αME of more than 400 V/cm·Oe with a tunable resonance frequency in the range of 60 Hz to 220 Hz by tip mass loading. The maximum harvested power output was about 16 μW/Oe with a 6 MΩ resistance load, with a corresponding power density of 200 μW/cm³ at 60 Hz. Dong *et al.* [104] achieved a power output of 420 μW/Oe across a 50 kΩ load under an AC magnetic field of 1 Oe at about 21 kHz using an ME cantilever based on the push-pull type Metglas/Pb(Zr,Ti)O_3 laminate. The ME laminate configuration, with an optimized beam aspect ratio of 0.86, exhibited an output power gain of 52.5 mW, and a corresponding power density of about 28.5 mW/cm³ at 30 Hz under 6.9 m/s excitation (0.7 g acceleration) at the resonance frequency, even in the absence of a DC bias magnetic field [105].

Ryu *et al.* [95] demonstrated an MME generator constructed using an anisotropic and flexible piezoelectric $Pb(Mg_{1/3}Nb_{2/3})O_3$–$PbTiO_3$ (PMN–PT) single crystal fiber composite (SFC), a cost-effective magnetostrictive Ni plate and Nd permanent magnetic proof mass (Figure 12b). The flexibility of the SFC ensures the high compliance of the sample, which is ideal for achieving low resonance frequency in a cantilever structure. The flexibility also increases device durability and enables the application of increased strain magnitudes. The Ni plate can be easily self-biased and generates a linear strain response in a low-level magnetic field environment. The performance of the MME generator containing an anisotropic <011> SFC with d_{32} mode under a small noise level magnetic field is shown in Figure 12b. At 60 Hz, and H_{ac}~500 μT, the maximum generated voltage was ~34 Vpp (~12.4 Vrms). The power from the MME generator was high enough to fully charge a 220-μF electrolytic capacitor after rectifying for 3 min. Using the stored power in the charged capacitor, the device was able to turn on 35 commercial high intensity LEDs with a turn on/off frequency of ~1 Hz. Seeking to improve the output power density, many research groups have developed various MME harvesters using different combinations of magnetostrictive and piezoelectric materials; some of these are summarized in Figure 13. These results represent significant advances towards next-generation remote monitoring electronic devices and self-powered wireless sensor networks.

Efforts have also been made to develop energy harvesters based on ME composite films. Onuta and colleagues [101] developed miniaturized energy harvesters using all-thin-film ME structures consisting of piezoelectric $Pb(Zr_{0.52}Ti_{0.48})O_3$(PZT) and magnetostrictive $Fe_{0.7}Ga_{0.3}$ layers (Figure 14). The device was fabricated on a micromachined Si cantilever coated with a silicon oxide/nitride/oxide (ONO) stack (3.8 μm thick). The $Fe_{0.7}Ga_{0.3}$ layer (500 nm thick) was sputtered on the Pt-buffered PZT layer

(500 nm thick) deposited over the ONO stack. The film-stress in the ONO stack was engineered, and a photo-lithographic process was employed to make an unbent, free-standing cantilever beam structure (950 μm long and 200 μm wide). The chip (6.6 mm × 6.6 mm) containing six cantilever devices was mounted in a vacuum chamber placed between a pair of Helmholtz coils, and it was aligned parallel to the magnetic fields. The results of the energy harvesting measurements from a single ME device are shown in Figure 14b,c. The voltage output from the harvesting device became saturated at higher external loading, while the power output was peaked at a load impedance of 12.5 kΩ (Figure 14b) from which the peak power density was determined to be 0.7 mW/cm^3 (RMS). Further, both the voltage and power outputs have shown saturation plateaus with respect to increasing Hac (Figure 14c), and such behavior was attributed to the saturation of the internal stress of magnetic origin that involves the rotation of the magnetization vector.

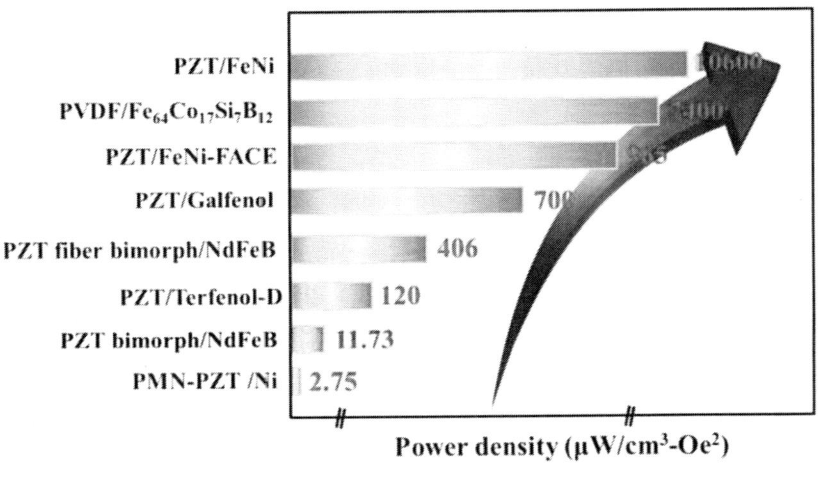

Figure 13. Summary of reported power densities from the MME harvesters made with different composite systems. Data are from references [94,96,97,98,99,100,101,102,103,104,105].

5.1.4. Magnetic Recording Read Head

Magnetic recording read heads in hard disk drives function based on one of the magneto-resistance (MR) effects, *i.e.*, anisotropic magneto-resistance (AMR), giant magneto-resistance (GMR) or tunneling magneto-resistance (TMR). The MR read head contains a sensing stack along with several additional layers and structures that ensures the proper biasing and shielding of the sensor [106,107]. During operation, the internal resistance of the MR sensor changes due to interaction with the stray fields from the recorded bits of a magnetic recording medium. The change in resistance (ΔR) is translated into a read signal as a voltage amplitude change $\Delta V = I \times \Delta R$. However, it is necessary to pass a constant DC test current through the sensor stack to detect the change in the resistance/amplitude. Higher recording densities can be achieved by reducing the shield-to-shield spacing, which is possible by thinning down the sensor stack. But such a trimming of the sensor size inflicts complications in fabrication and in the operation of the sensor and the read head.

Vopson *et al.* [106] proposed a new design for a magnetic recording read head sensor based on the DME, *i.e.* the magnetically induced ME effect, as shown in Figure 15a,b. The proposed sensor stack incorporates a tri-layer ME composite structure and consists of 7 layers, arranged as seed/AFM/FM/FE/FM/AFM/cap, which is much simplified in comparison with the existing TMR sensor stack, which has around 15 layers. The two AFM layers added adjacent to the FM layers produce the required DC bias magnetic field. This design will facilitate reduction in sensor size without sensitivity loss and will eliminate the need for permanent magnets currently used for horizontal biasing. Since the ME-based sensors operate without using any electric power, there will be no need for the DC test current through the sensor, and the data can be read back directly as an induced voltage. This will reduce power consumption as well as minimize the Joule heating issues that occur in the high resistance TMR sensors.

Zhang *et al.* [107] developed a prototype ME read head (Figure 15c) fabricated with BTO-NFO and BTO-CFO bilayered heterostructures that was grown using PLD on 0.7% Nb-doped STO single crystal substrates. The ME sensitivity of this structure was measured to be about 0.5 V/Oe at 1 kHz. To illustrate the reading and writing process, a primal 16-bit input signal (a_k) was written on the disk tracks by a writing current (I_w) in non-return-to-zero format after being encoded into a new signal (b_k). These written bits b_k were able to yield a media field δH in the recording medium. A small in-plane δH (about a few Oersted) that was considered to model the media fields (Figure 15d) from the bits in the recording medium was generated using a set of small Helmholtz coils, and a response signal of 1–2 μV was obtained. In the fully developed device case, greatly enhanced outputs (mV) were expected by applying higher amplitudes of media fields (~100 Oe) and by operating the device at the resonance frequency.

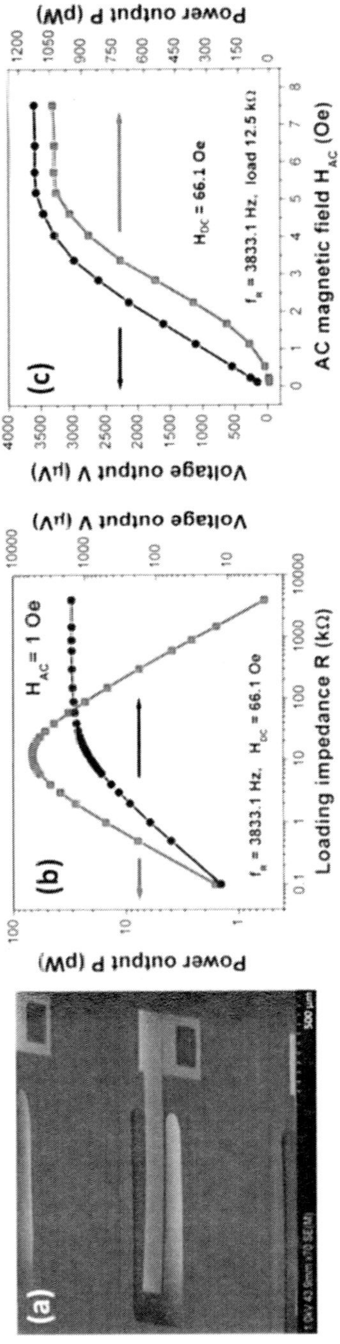

Figure 14. (**a**) SEM micrograph of a thin film ME energy harvester based on PZT/FeGa and (**b,c**) its energy harvesting properties [101].

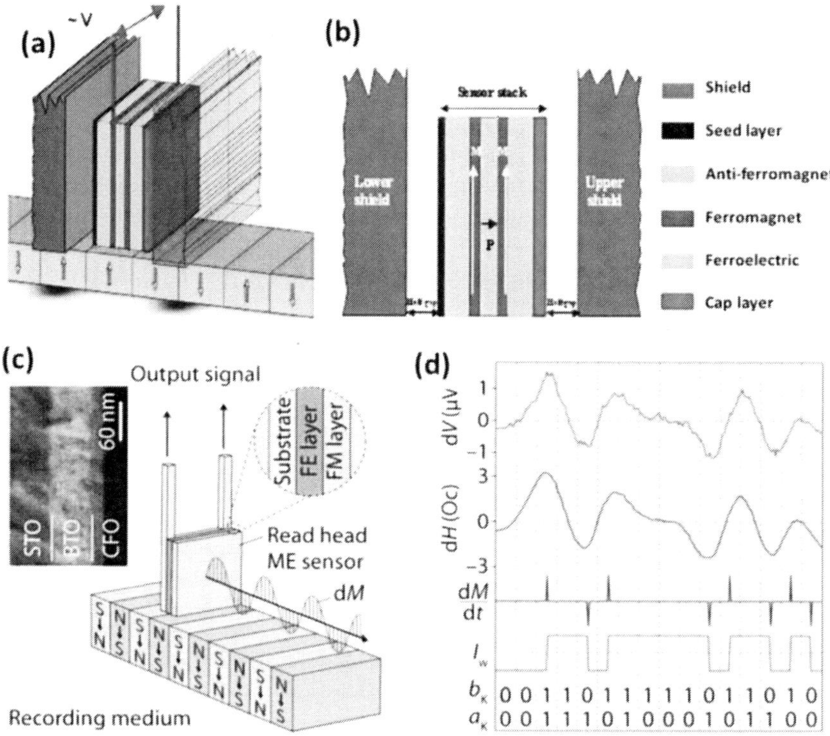

Figure 15. (**a**) and (**b**) 3D diagram and cross-section of the magnetic read head proposed by Vopson *et al.* [106] (**c**) and (**d**) Schematic, TEM image (inset) and voltage output of an ME read-head sensor based on a bilayer CFO/BTO heterostructure grown on a single-crystal STO substrate [107].

5.1.5. Biomedical Applications

Motivated by the advances in multiferroics, ME composites have been suggested for biomedical applications such as wireless endoscopy, minimally invasive surgical tools, and stimulation of functions of living cells [73,108]. The potential use of ME nanoparticles (MENs) as carriers for on-demand drug release and to artificially stimulate the neural activity deep in the brain has also been suggested [109,110].

Wireless capsule endoscopes (WCEs) are often used to examine the gastrointestinal (GI) tract for clinical diagnosis (Figure 16a). Though the WCEs are far less invasive compared to conventional endoscopes, the passive nature of WCEs makes it difficult to control their position and orientation as the capsule moves along the GI tract. Currently, tracking of WCEs is carried out by methods such as radio frequency (RF) triangulation, magnetic tracking, radiation vision, and ultrasound sensing. However, accurate localization and high fidelity tracking of WCEs is still a challenging task. In the magnetic tracking method, a permanent magnet is placed inside the wireless capsule, which is located at an x, y, z location inside the human GI tract with respect to a reference frame. Since the permanent magnet emits a magnetic field, it can be identified using a magnetic sensor, such as a 3-axis magnetic sensor, which can approach nT sensitivity. It has been proposed that multiferroic ME transducers are promising candidates to provide WCE position feedback, by using them in the form of enhanced sensitivity (pT range) magnetic field sensors [73]. By incorporating a MEMS-based ME composite structure into the WCE capsule, power can be generated on demand as required for recharging the battery of the WCE. Further, the use of self-biased ME composites will reduce complexity in the device design. To improve its diagnostic capabilities, focused efforts are needed, particularly to achieve reliable control over the movement and orientation of the capsule within the constraints of the size and shape of the device.

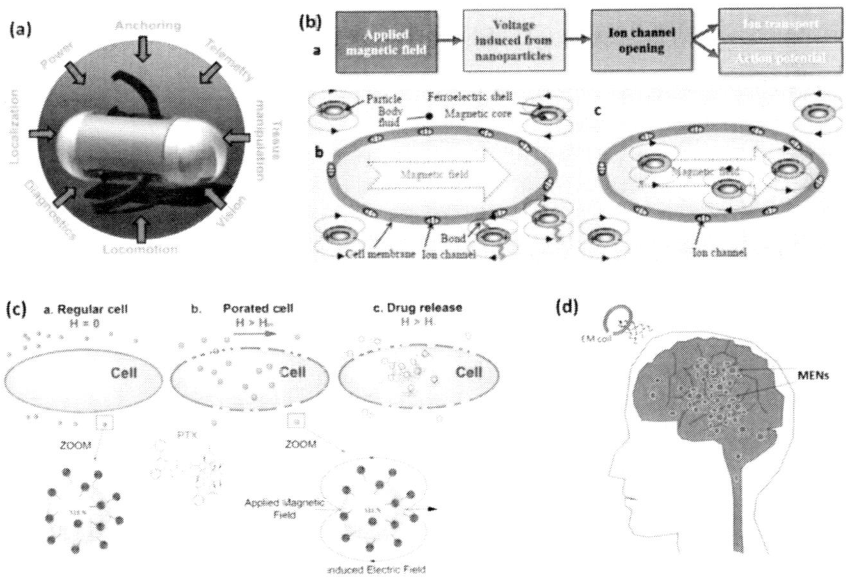

Figure 16. (**a**) Key functionalities that complement the wireless capsule endoscopes [73]; (**b**) Illustration of possible mechanisms of stimulation of ion channels: a chain of actions triggered by the applied magnetic field pulses, external and internal stimulation by uptaken nanoparticles [108]; (**c**) MENs as field-controlled nano-electroporation sites to let the drug through the cancer cell membranes [109]; (**d**) Illustration of the deep brain stimulation approach [110].

Controlling the function of biological macromolecules is of vital importance in health science studies. Approaches used to stimulate cell functions include the use of the heat generated by hysteresis losses in magnetic nanoparticles placed in a high-frequency magnetic field, and the mechanical agitation of magnetic nanoparticles attached to cells using external low-frequency magnetic fields [111,112]. The interaction between electromagnetic fields and biological macromolecules can be understood by studying the ion channels which regulate several cellular processes, such as action potentials in neurons or muscle contraction. Ion channels are membrane proteins that form pores for the controlled exchange of ions across cellular membranes [113]. The common type, voltage-gated ion channels open or close in response to a transmembrane electric field. Ion channel-related disorders have been found to cause several health issues, which are usually treated using pharmacological agents that modify the gating kinetics, or block the channel transiently or permanently. Recently, Kargol *et al.* [108] proposed an innovative approach, based on the use of core/shell structured MENs with a ferromagnetic core and a ferroelectric shell, that would allow the remote control of ion channel gating via externally applied magnetic fields (Figure 16b). In this approach, electric fields in the vicinity of the cells generated by MENs introduced extra- or intracellularly can be locally modified to invoke appropriate conformational changes in the ion channels. The resulting local depolarization or hyperpolarization of the membrane will lead to opening or closing of the ion channels accordingly. Because of the remote way in which the stimulation will be performed, individual cells or selected groups of cells can be targeted, rather than whole tissues.

Targeted drug delivery with adequate high specificity (to tumor cells) remains a formidable task in the treatment of cancer in general, particularly ovarian cancer. Although the survival rates have been improved by intraperitoneal (IP) delivery through a surgically implanted catheter, toxicity and catheter complications have precluded widespread adoption of this invasive means of delivery. Guduru *et al.* [109] addressed this challenge by exploiting the dependence of the membrane's porosity on the electric field, *i.e.* electroporation that can be utilized to trigger drug delivery into the cells. Above a threshold magnetic field (H_{th}), the MENs loaded with the drug and optionally with the biomarker-specific antibodies (for delivery to the tumor cells) can generate localized fields large enough to open up the membrane pores in their proximity and thus allow the delivery of the drug inside the tumor cells. The drug can be released off the MENs by further increasing the field above the second critical value, H_r, necessary for overcoming the drug-MEN binding energy. This hypothesis was testified through*in vitro* studies on human ovarian carcinoma cell (SKOV-3) and healthy cell (HOMEC) lines, where a 30-Oe DC bias was applied to trigger high-specificity uptake of Paclitaxel (PTX) loaded on 30-nm $CoFe_2O_4/BaTiO_3$ core/shell MENs. The drug penetrated through the membrane and completely eradicated the tumor within 24 h without affecting the normal cells.

In the human neural network, chemical and electrical synapses transfer information between adjacent axons and dendrites directly or indirectly through electric field energy. The ability to efficiently control the network at micro- or even nano-scale can enable significant control over important brain functions. Existing non-invasive brain stimulation methods including repetitive transcranial magnetic stimulation (rTMS) and transcranial direct current stimulation (tDCS) are limited in their depth and locality. A potential solution would be to use MENs for non-invasive control of the neural network. In this approach, very low intensity external magnetic field is required to stimulate brain activity at any depth in the brain and the field can be focused to act upon MENs in any particular region of the brain. The external magnetic field generates AC signals in ME nanoparticles that are correlated with the frequency spectrum of the neural activity, which in turn causes neurons in that region to fire at similar frequencies (Figure 16d). Yue *et al.* [110] modeled the effect of MENs to non-invasively stimulate the brain of a patient with Parkinson's disease. Using the optimized values for the concentration of the 20-nm nanoparticles (with α_{ME} of 100 mV/cm·Oe in the aqueous solution) of 3×10^6 particles/cc and excitation frequency of the externally applied 300-Oe magnetic field of 80 Hz, the pulsed sequences of the electric field were brought to the levels comparable to those of healthy people.

5.2. Devices Based on the CME Effect

5.2.1. Magnetoelectric Random Access Memory

Ferroelectric random access memory (FeRAM) stores binary information using ferroeletric polarization states, while magnetic random access memory (MRAM) stores data bits using magnetization states and the property of magneto-resistance in multilayers. Applications of these memory devices are limited by the need for destructive read and reset operations in FeRAM, and by inconveniences in the write process in MRAM caused by the requirement for large currents for magnetization reversal. The mutual control of magnetic order and electric polarization in ME materials can be exploited to combine the best features of FeRAM and MRAM, and thus create a novel type of non-volatile RAM, *i.e.* MeRAM with multiple memory states [114]. This approach would offer greater memory density, reduced power consumption, and improved thermal stability. Bibes *et al.* [115] proposed a hybrid MeRAM device that combines the ME coupling in a multiferroic (consisting of FE and AFM phases) layer with the interfacial exchange bias between the multiferroic AFM phase and a FM layer to induce magnetization switching through voltage control. One challenging task in the realization of practical MeRAM devices is the achievement of reversible 180° deterministic switching of magnetization, on which existing magnetic memories, including hard disk drives and MRAMs, rely upon. Heron *et al.* [116] approached this issue through E-field control of exchange bias in multiferroic/magnetic ($CoFe/BiFeO_3$) heterostructures. Xue *et al.* [117] demonstrated that the E-field induced near 180° magnetization switching in the AFM/FM/FE heterostructures through voltage-controlled tuning of exchange bias and through manipulation of coercive fields. The irreversibility of the E-field induced near 180° magnetization switching in AFM/FM/FE heterostructures was resolved by employing a magnetic impulse, which could switch the magnetization back and lead to a continuous magnetization switching [118]. The heterostructure was formed by epoxy gluing of the magnetron sputtering deposited exchange coupled multilayer structure of NiFe/NiCoO/glass to a (011) cut PZN-PT single crystal, with the magnetic easy axis along the [100] or [01−1] PZN-PT crystal directions (Figure 17a). Switching magnetization was very close to the saturation magnetization (from ~M_s to −0.95 M_s along PZN-PT [01−1]; from 0.76 M_s to −0.9 M_s along PZN-PT [100]), which implies that near 180° magnetization reversal was observed in the NiFe/NiCoO/glass/PZN-PT heterostructure. The reported findings are a significant advancement in realizing the voltage writing of magnetic bits in MeRAM.

5.2.2. Phase Shifters

Microwave phase shifters are important elements for radar applications, telecommunications, oscillators and phased array antenna systems. A wide variety of phase shifters based on semiconductors, ferrites, and ferroelectrics have been developed [119,120,121]. The ferrite-based phase shifters are based on the Faraday rotation of electromagnetic radiation in magnetized ferrite rods in waveguides. Large magnetic bias fields are required for phase tuning, which involves huge power dissipation; consequently they cannot be miniaturized in size or made compatible with integrated circuit technologies. A second class of microwave phase shifters is based on ferroelectric materials. Their distinguishing features are fast electric tunability and low power consumption. However, such phase shifters are very lossy at frequencies above 1–5 GHz. A ferrite-ferroelectric layered structure, such as a ME composite, opens up the possibility of dual-tunable microwave devices, which offers higher efficiency, lower noise, compact size and lightweight compared to conventional microwave deceives. Following theoretical investigations, Ustinov *et al.* [122] reported the development of such ME devices. Their ME dual phase shifter based on

FMR is a bilayer ME composite consisting of a ferrite YIG layer with a thickness of 5.7 μm and a ferroelectric BST layer with a thickness of 500 μm. The schematic structure of the cross-section of a ME dual phase shifter is displayed in Figure 18a. The electric field control of the phase shift arises through a strong ME coupling in between the ferrite and ferroelectric layers. For an applied electric field E = 20 kV/cm, a maximum differential phase shift, Δφ = 650° was achieved. The estimated figure of merit of the phase shifter was about 25 °/dB. An electric field tunable YIG/PZT phase shifter was also designed and characterized by Tatarenko *et al.* [123]. For E = 5–8 kV/cm and Δφ = 90–1800, an insertion loss of 1.5–4 dB was obtained. The observed insertion loss was somewhat closer to the desirable 0.5 dB needed for practical applications. Recently, bilayers of hexaferrites and piezoelectric PZT were used for novel design of phase shifters. For a strip-line Zn_2Y/PMN-PT phase shifter, data on the differential phase shift *versus* E indicated a differential phase shift of 500 for E = 12 kV/cm, a linear variation in the phase shift with E, and an insertion loss of 4–8 dB [124].

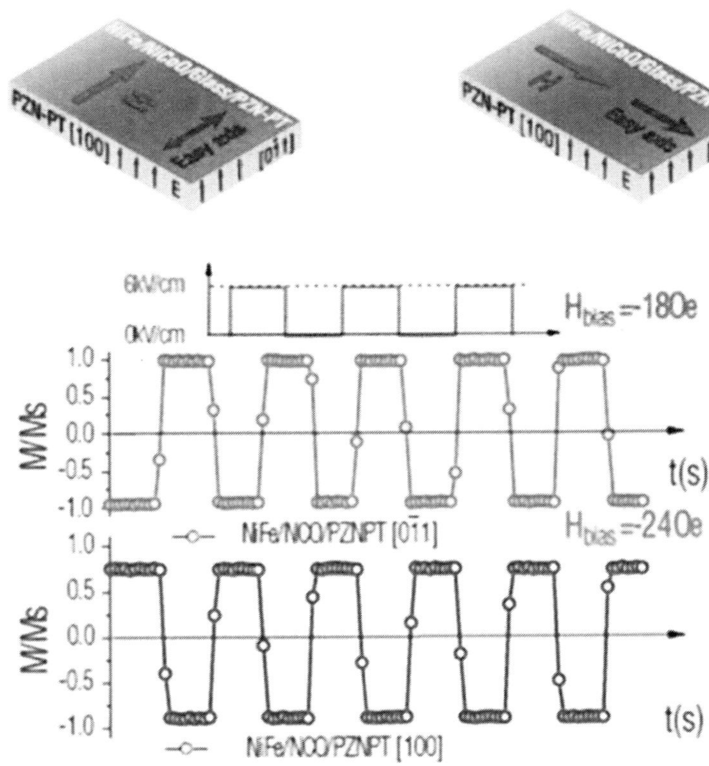

Figure 17. Magnetization switching by E-field in NiFe/NiCoO/glass/PZN-PT heterostructure-based MeRAM [117].

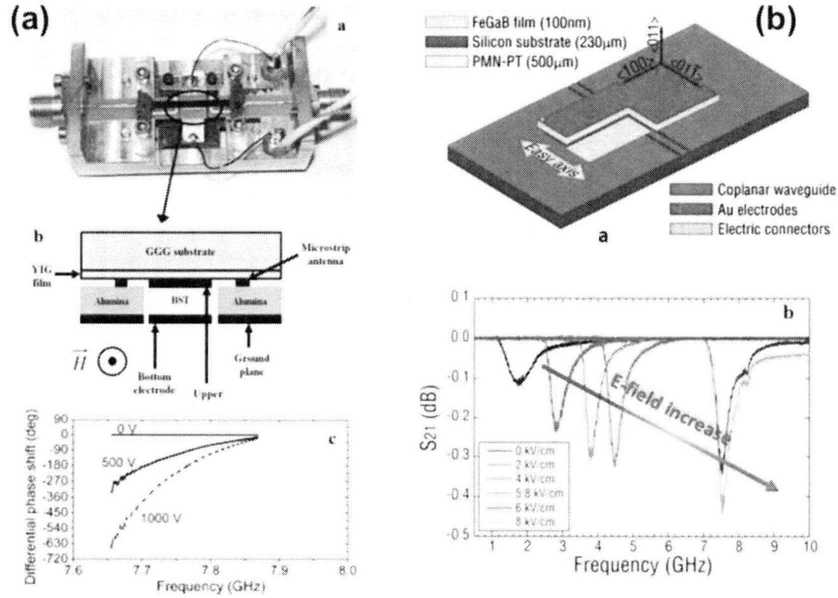

Figure 18. (**a**) Experimental prototype of the YIG/BST phase shifter, schematic of cross-section of the device and the YIG/BST layered structure, and electrically induced differential phase shift [122]; (**b**) Schematic of sample configuration and microwave measurement setup, and electric field dependence of the transmission coefficient (S_{21}) spectra of the FeGaB/PZN-PT resonator [132,133].

5.2.3. Resonators

Resonators are front-end elements in phased array radars, and are also used in filters and phase shifters. At the ferromagnetic resonance (FMR), a ferromagnetic material is subjected to high-frequency excitations [125]. FMR devices based on soft magnetic materials are utilized in microwave signal processing devices, resonators, band-pass/band-stop filters and phase shifters. Such devices require a magnetic field to operate and to achieve frequency tuning, which creates disadvantages like slow operation, high noise levels, large power consumption, and limited miniaturization. By replacing ferrite materials with ME composites, electrical tuning of FMR is possible. Fetisov *et al.* [126] experimentally demonstrated an electric field tunable 1–10 GHz microwave planar resonator using a YIG/PZT composite. The device consisted of a single micro strip (50 mm thick and 3 mm long) deposited onto an Al_2O_3 substrate. For a signal input of 2–10 GHz, 0.1 mW of power was applied to the microstrip transducer. Low input power was chosen to prevent heating of the sample due to power absorption at the

FMR. For zero applied electric field (E = 0), the spectra contained a well-defined FMR absorption peak at about 3.5 GHz, with a maximum insertion loss of 45 dB and a 3-dB line-width of Δf = 3.4 MHz. The off-resonance loss could be partially due to ferroelectric losses in the PZT. When the electric field E = 10 kV/cm was applied, the FMR peak shifted by 40 MHz to a lower frequency. When the electric field E = −10 kV/cm was applied by reversing the voltage applied to the PZT, the peak up-shifted by 38 MHz. The converse ME coupling constant $A = \Delta f/E \approx 4$ MHz cm/kV, and the voltage tuning for E = ±10 kV/cm was only about 2% of the central frequency. Li *et al.* [127] reported X-band resonator studies using epitaxial nickel ferrite ($NiFe_2O_4$) films (thickness of about 2.0 μm) deposited on PZN-PT or PMN-PT substrates using direct liquid injection chemical vapor deposition. A strong ME coupling resulting in large shifts in the FMR profile was observed, and were due to electrostatic field-induced anisotropic magnetic field changes. The data indicated an ME coefficient $A = \delta H/E$ = 20 Oe·cm/kV (or 60 MHz cm/kV) at 9 GHz and 11 GHz, which is an order of magnitude higher than that for YIG/PZT. The main drawback with the use of ferrite material is the large FMR line-width.

Recently, hexagonal ferrites have been used for resonators over a wide frequency range from 15 to 110 GHz [128,129,130]. The Y-($Ba_2Zn_2Fe_{12}O_{22}$:ZnY) and Z-type ($Ba_3Co_2Fe_{24}O_{41}$) hexagonal ferrites with easy-plane magnetic anisotropy are preferred for use at 10–40 GHz [128]. M-type hexaferrites, $MFe_{12}O_{19}$ (M = Ba, Sr) with uniaxial magnetic anisotropy are ideal for the frequency range 40–75 GHz [129]. The Al doped M-type hexaferrites further increase the resonance frequency, and are suitable for 50–110 GHz devices [130]. Tatarenko *et al.* [131] reported the converse ME effect over the 8–25 GHz range in bilayers of single crystal ZnY and polycrystalline PZT or single crystal PMN-PT. The resonator was tuned by 120 MHz with E = 12 kV/cm, and corresponding ME coupling strength A was about 10 MHz cm/kOe. Lou *et al.* [132,133] reported a layered ferroic FeGaB and single crystal piezoelectric PZN-PT-based resonator. They observed that a large electric field induced an effective magnetic anisotropy field of 750 Oe with a narrow FMR line-width of 50 Oe at X-band. This resonator achieved a large electric field tunable FMR frequency range between 1.75 and 7.57 GHz even at zero magnetic bias field, as shown in Figure 18b, which corresponds to a mean tunable frequency per unit electric field of about 970 MHz cm/kOe. The giant tunable magnetic field and FMR frequency of the FeGaB/PZN-PT resonator makes it a promising candidate for wide-band electric field tunable microwave devices. Compared to conventional tunable microwave magnetic devices, which are tuned by magnetic fields, these electrostatically tunable microwave multiferroic devices are much less noisy, and more energy efficient, compact, and lightweight.

5.2.4. Inductors

One of the three fundamental components for electronic circuits, tunable inductors, find widespread use in various applications such as communication systems and power electronics. Most tunable inductors are magnetically tuned by using electromagnets. Since electromagnets are typically bulky, noisy and energy consuming, they are inconvenient for use in such applications. Efforts have been made to develop electronically tunable inductors that have large tunability, high quality factors, and low energy consumption. For example, microelectromechanical systems (MEMS)-based tunable inductors can have very high quality factors, although they have a limited tunable range of 20% and are difficult to fabricate. The use of ME composites as a core material could be advantageous for tunable inductor applications since tuning magnetic properties by electric field is possible even at low fields. A variety of ME composite-based inductors have been made using different structures. Tunability is calculated from inductance (L) using the formula $[\Delta L/L_0 = (L - L_0)/L_0] \times 100$. Fang *et al.* [134] have reported a PZT/MnZn/PZT composite inductor that showed a small tunable inductance range $\Delta L/L_{min}$ of up to 20%. Lou *et al.* [135] reported a composite consisting of a combination of PZT slab and two layers of amorphous Metglas ribbons with a large tunable inductance range $\Delta L/L_{min}$ of up to 450%, together with improved quality factors (Figure 19a). This tunability of inductance and increased quality factor were due to strong ME coupling in the composite core, which led to electric field-induced permeability change. However, due to the relatively large thickness of the Metgals ribbons (~25 μm), excessive eddy current loss limits the operational frequency range of the conventional Metglas-based inductors to be <100–200 kHz, and leads to low quality factors of these inductors at microwave frequencies. Recently, Lin *et al.* [136] investigated the effect of ferromagnetic layer thickness on the performance of ME inductors. The inductance tunable range of $\Delta L/L_{min}$ of 370% was achieved, together with a significant 3-fold enhancement in quality factor, with an increase in the operational frequency range. To miniaturize the inductor, a ring inductor was designed by Mandal *et al.* [137] using insulating single phase $Bi_{0.7}Dy_{0.3}FeO_3$ multiferroic material, but the tunability of the inductor was only about 18%.

5.2.5. ME Antenna

In recent years, reduction in antenna size, which is suitable for lower frequency bands where wavelength is larger, has received significant attention because of the demands of numerous mobile communication systems. Several approaches have been devised to achieve antenna size reductions [138,139]. The first is to use capacitive or inductive loadings and/or meandered lines to obtain slow-wave resonance. The main drawbacks to these approaches are increased ohmic losses and narrow bandwidths. The second is to reduce the wavelength of the structure using dielectric materials, but this is prone to surface-wave excitations and corresponding losses of efficiency. A new approach was recently found that uses a novel engineered ME composite material. A half-wavelength resonant microstrip antenna on an ME composite of nickel zinc ferrites ($Ni_{1-x}Zn_xFe_2O_4$:NZFO) and ferroelectric BST was made by Petrov *et al.* [140]. They concentrated on the miniaturization of the 100-MHz antenna, and its effects on impedance matching, bandwidth and efficiency (Figure 19b). The ME composite disc with a diameter of 22 cm and thickness of 0.85 cm was metallized on one side, and the microstrip had a length of 22 cm and width of 0.65 cm. A vector network analyzer was used for measurements of return and transmission losses. The measured SWR at resonance was close to 1.3 and indicated good impedance matching with free space. The absence of reflection at the boundary between the substrate of the antenna and the surrounding medium reduced the energy absorbed in the substrate. The helical antennas were fabricated using Co_2Z type hexaferrite and ferrite substrates [141,142]. The resonant frequency and bandwidth at 10 dB for the antennas were measured to be 195 MHz and 27 MHz, and 209 MHz and 41 MHz, respectively. A helical antenna was also fabricated on (Co/Ti) doped BaM substrate (45 × 11 × 3.8 mm^3), and the antenna was then mounted on an FR4 board (10 × 5 mm^2) with an 8 × 5 mm^2 copper ground [143]. A 50-Ω coaxial cable was used to feed the antenna. The antenna performance was characterized with a network analyzer and a Wheeler cap. The antenna resonant frequency shifted from 231 to 201 MHz with increasing permeability from 4.51 to 8.58. This data indicates that the factor of miniaturization of the antenna is about 7% with a permeability of 8.58. It is worth noting here that the low-loss (Co/Ti) doped BaM is an excellent soft magnetic material for miniature antenna applications in a very high-frequency (30–300 MHz) range.

6. SUMMARY

In recent years, much attention has been paid towards exploiting the promising technological potential of magnetoelectric coupling in materials. Although there has been tremendous progress in the development of both bulk and film-based ME composites, efforts are likely to continue in the pursuit of desired levels of performance. To realize strong ME coupling and high performance ME composites it is essential to select the appropriate combination of piezoelectric and magnetostrictive materials with better properties; adopt a suitable fabrication approach and configuration of the composite constituents; optimize interfacial coupling; reduce the noise contributions to the output signal; tune and control other dynamic parameters; and achieve a comprehensive understanding of the contributing factors and physical phenomena by theoretical modeling. To effectively employ the coupling across interfaces in ME composites and to achieve giant ME effects, the nature of the interfaces should be unraveled. ME-based devices are compact, light-weight, fast in response, less noisy, and energy-efficient and thus are viable alternatives for some existing conventional electrical and magnetic devices. Although some prototype devices based on bulk ME composites have been demonstrated, challenges still exist in terms of obtaining control over materials properties and fabrication, and optimization of the device performance. Further, the reliability of the ME composite systems needs to be evaluated in practical device designs. While much work remains to be done to accomplish on-chip integration of the ME devices, there is significant scope, and many avenues can be explored.

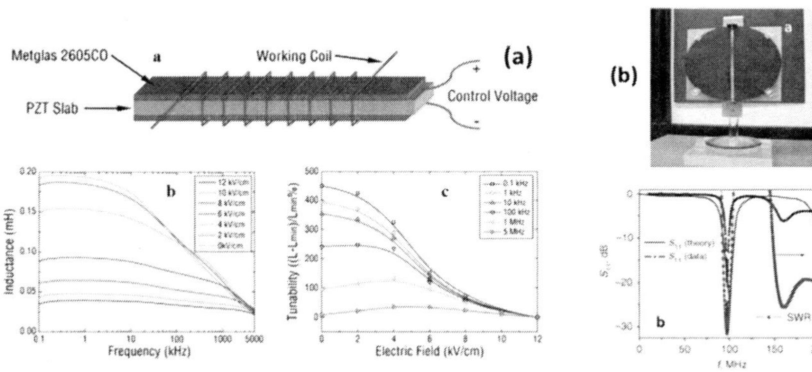

Figure 19. (**a**) Schematic of the magnetoelectric inductor, inductance spectra, and inductance tunability at different frequencies and electric field [135]; (**b**) Miniature microstrip antenna on NZFO/BST composite substrate, measured and estimated return loss and measured SWR for the antenna [140].

ACKNOWLEDGMENTS

This work was supported by the Global Frontier R&D Program (Grant Nos. 2013M3A6B1078872) on Center for Hybrid Interface Materials (HIM) of the National Research Foundation (NRF) of Korea funded by the Ministry of Science, ICT & Future Planning, Korea and internal R&D program of Korea Institute of Materials Science (KIMS). S.P. would like to acknowledge the financial support from Office of Basic Energy Science, Department of Energy, USA through Grant No. DE-FG02-06ER46290.

CONFLICTS OF INTEREST

The authors declare no conflict of interest.

REFERENCES

1. Wang, Y.; Hu, J.; Lin, Y.; Nan, C.-W. Multiferroic magnetoelectric composite nanostructures. NPG Asia Mater. 2010, 2, 61–68.

2. Priya, S.; Islam, R.; Dong, S.; Viehland, D. Recent advancements in magnetoelectric particulate and laminate composites. J. Electroceram. 2007, 19, 149–166.

3. Nan, C.-W.; Bichurin, M.I.; Dong, S.; Viehland, D.; Srinivasan, G. Multiferroic magnetoelectric composites: Historical perspective, status, and future directions. J. Appl. Phys. 2008, 103, 031101.

4. Vaz, C.A.F.; Hoffman, J.; Ahn, C.H.; Ramesh, R. Magnetoelectric coupling effects in multiferroic complex oxide composite structures. Adv. Mater. 2010, 22, 2900–2918.

5. Ryu, J.; Priya, S.; Uchino, K.; Kim, H.-E. Magnetoelectric Effect in Composites of Magnetostrictive and Piezoelectric Materials. J. Electroceram. 2002, 8, 107–119.

6. Fang, Z.; Zhang, Q.; Datta, S. Ultra sensitive magnetic sensors integrating the giant magnetoelectric effect with advanced microelectronics. Ph.D. Thesis, Pennsylvania State University, University Park, PA, USA, December 2011.

7. Cho, K.-H.; Priya, S. Direct and converse effect in magnetoelectric laminate composites. Appl. Phys. Lett. 2011, 98, 232904.

8. Newnham, R.E.; Skinner, D.P.; Cross, L.E. Connectivity and piezoelectric-pyroelectric composites. Mater. Res. Bull.1978, 13, 525–536.

9. Nan, C.-W.; Jia, Q. Obtaining ultimate functionalities in nanocomposites: Design, control, and fabrication. MRS Bull.2015, 40, 719–724.

10. Park, C.S.; Ryu, J.; Choi, J.J.; Park, D.S.; Ahn, C.W.; Priya, S. Giant Magnetoelectric Coefficient in 3–2 Nanocomposite Thick Films. Jpn. J. Appl. Phys. 2009, 48, 080204.

11. Gillette, S.M.; Fitchorov, T.; Obi, O.; Jiang, L.; Hao, H.; Wu, S.; Chen, Y.; Harris, V.G. Effects of intrinsic magnetostriction on tube-topology magnetoelectric sensors with high magnetic field sensitivity. J. Appl. Phys. 2014,115, 17C734.

12. Yan, Y.; Priya, S. Multiferroic Magnetoelectric Composites/Hybrids. In Hybrid and Hierarchical Composite Materials; Kim, C.-S., Randow, C., Sano, T., Eds.; Springer International Publishing: Cham, Switzerland, 2015; pp. 95–160.

13. Islam, R.; Rong, C.-B.; Liu, J.P.; Priya, S. Effect of gradient composite structure in cofired bilayer composites of Pb(Zr0.56Ti0.44)O3–Ni0.6Zn0.2Cu0.2Fe2O4 system on magnetoelectric coefficient. J. Mater. Sci. 2008, 43, 6337–6343.

14. Srinivasan, G.; DeVreugd, C.P.; Flattery, C.S.; Laletsin, V.M.; Paddubnaya, N. Magnetoelectric interactions in hot-pressed nickel zinc ferrite and lead zirconante titanate composites. Appl. Phys. Lett. 2004, 85, 2550–2552.

15. Jiang, Q.H.; Shen, Z.J.; Zhou, J.P.; Shi, Z.; Nan, C.-W. Magnetoelectric composites of nickel ferrite and lead zirconnate titanate prepared by spark plasma sintering. J. Eur. Ceram. Soc. 2007, 27, 279–284.

16. Islam, R.A.; Priya, S. Synthesis of high magnetoelectric coefficient composites using annealing and aging route. Int. J. Appl. Ceram. Tech. 2006, 3, 353–363.

17. Liu, R.; Zhao, Y.; Huang, R.; Zhao, Y.; Zhou, H. Multiferroic ferrite/perovskite oxide core/shell nanostructures. J. Mater. Chem. 2010, 20, 10665–10670.

18. Islam, R.A.; Bedekar, V.; Poudyal, N.; Liu, J.P.; Priya, S. Magnetoelectric properties of core-shell particulate nanocomposites. J. Appl. Phys. 2008, 104, 104111.

19. Shi, Z.; Nan, C.W.; Zhang, J.; Cai, N.; Li, J.-F. Magnetoelectric effect of Pb(Zr,Ti)O3 rod arrays in a (Tb,Dy)Fe2/epoxy medium. Appl. Phys. Lett. 2005, 87, 012503.

20. Lam, K.H.; Lo, C.Y.; Chan, H.L.W. Frequency response of magnetoelectric 1–3-type composites. J. Appl. Phys. 2010,107, 093901.

21. Ma, J.; Shi, Z.; Nan, C.W. Magnetoelectric properties of composites of single Pb(Zr,Ti)O3 rods and Terfenol-D/Epoxy with a single-period of 1–3-type structure. Adv. Mater. 2007, 19, 2571–2573.

22. Islam, R.A.; Ni, Y.; Khachaturyan, A.G.; Priya, S. Giant magnetoelectric effect in sintered multilayered composite structures. J. Appl. Phys. 2008, 104, 044103.

23. Dong, S.; Zhai, J.; Li, J.; Viehland, D. Near-ideal magnetoelectricity in high-permeability magnetostrictive/piezofiber laminates with a (2–1) connectivity. Appl. Phys. Lett. 2006, 89, 252904.

24. Wang, Y.; Gray, D.; Berry, D.; Gao, J.; Li, M.; Li, J.; Viehland, D. An extremely low equivalent magnetic noise magnetoelectric sensor. Adv. Mater. 2011, 23, 4111–4114.

25. Mori, K.; Wuttig, M. Magnetoelectric coupling in Terfenol-D/polyvinylidenedifluoride composites. Appl. Phys. Lett.2002, 81, 100–101.

26. Silva, M.; Reis, S.; Lehmann, C.S.; Martins, P.; Lanceros-Mendez, S.; Lasheras, A.; Gutiérrez, J.; Barandiarán, J.M. Optimization of the Magnetoelectric Response of Poly(vinylidene fluoride)/Epoxy/Vitrovac Laminates. ACS Appl. Mater. Inter. 2013, 5, 10912–10919.

27. Lawes, G.; Srinivasan, G. Introduction to magnetoelectric coupling and multiferroic films. J. Phys. D Appl. Phy. 2011,44, 243001.

28. McDannald, A.; Staruch, M.; Sreenivasulu, G.; Cantoni, C.; Srinivasan, G.; Jain, M. Magnetoelectric coupling in solution derived 3–0 type PbZr0.52Ti0.48O3:xCoFe2O4 nanocomposite films. Appl. Phys. Lett. 2013, 102, 122905.

29. Zheng, H.; Wang, J.; Lofland, S.E.; Ma, Z.; Mohaddes-Ardabili, L.; Zhao, T.; Salamanca-Riba, L.; Shinde, S.R.; Ogale, S.B.; Bai, F.; et al. Multiferroic BaTiO3-CoFe2O4 nanostructures. Science 2004, 303, 661–663.

30. Aimon, N.M.; Kim, D.H.; Sun, X.; Ross, C.A. Multiferroic behavior of templated BiFeO3-CoFe2O4 self-assembled nanocomposites. ACS Appl. Mater. Interf. 2015, 7, 2263–2268.

31. Srinivasan, G. Layered multiferroic composites. In Composite Magnetoelectrics; Srinivasan, G., Priya, S., Sun, N.X., Eds.; Woodhead Publishing: Cambridge, UK, 2015; pp. 55–70.

32. Zhou, Z.; Sun, N.X. Multiferroic nanostructures. In Composite Magnetoelectrics; Srinivasan, G., Priya, S., Sun, N.X., Eds.; Woodhead Publishing: Cambridge, UK, 2015; pp. 71–86.

33. Lage, E.; Kirchhof, C.; Hrkac, V.; Kienle, L.; Jahns, R.; Knöchel, R.; Quandt, E.; Meyners, D. Exchange biasing of magnetoelectric composites. Nat. Mater. 2012, 11, 523–529.

34. Greve, H.; Woltermann, E.; Quenzer, H.-J.; Wagner, B.; Quandt, E. Giant magnetoelectric coefficients in (Fe90Co10)78Si12B10-AlN thin film composites. Appl. Phys. Lett. 2010, 96, 182501.

35. Nair, S.S.; Pookat, G.; Saravanan, V.; Anantharaman, M.R. Lead free heterogeneous multilayers with giant magneto electric coupling for microelectronics/microelectromechanical systems applications. J. Appl. Phys. 2013, 114, 064309.

36. Srinivasan, G.; Rasmussen, E.T.; Gallegos, J.; Srinivasan, R.; Bokhan, Y.I.; Laletin, V.M. Magnetoelectric bilayer and multilayer structures of magnetostrictive and piezoelectric oxides. Phys. Rev. B. 2001, 64, 214408.

37. Kambale, R.C.; Ryu, J.; Patil, D.R.; Chai, Y.S.; Kim, K.H.; Yoon, W.-H.; Jeong, D.-Y.; Park, D.-S.; Kim, J.-W.; Choi, J.-J.; et al. Colossal magnetoelectric response of PZT thick films on Ni substrates with a conductive LaNiO3 electrode. J. Phys. D Appl. Phy. 2013, 46, 092002.

38. Zhou, D.; Jian, G.; Zheng, Y.; Gong, S.; Shi, F. Electrophoretic deposition of BaTiO3/CoFe2O4 multiferroic composite films. Appl. Surf. Sci. 2011, 257, 7621–7626.

39. Palneedi, H.; Maurya, D.; Kim, G.-Y.; Priya, S.; Kang, S.-J.L.; Kim, K.-H.; Choi, S.-Y.; Ryu, J. Enhanced off-resonance magnetoelectric response in laser annealed PZT thick film grown on magnetostrictive amorphous metal substrate.Appl. Phys. Lett. 2015, 107, 012904.

40. Pan, D.A.; Bai, Y.; Chu, W.Y.; Qiao, L.J. Ni–PZT–Ni trilayered magnetoelectric composites synthesized by electro-deposition. J. Phys. Condens. Matter. 2008, 20, 025203.

41. Bi, K.; Wang, Y.G.; Wu, W.; Pan, D.A. The large magnetoelectric effect in Ni–lead zirconium titanate–Ni trilayers derived by electroless deposition. J. Phys. D Appl. Phy. 2010, 43, 132002.

42. Yu, H.; Zeng, M.; Wang, Y.; Wan, J.G.; Liu, J.-M. Magnetoelectric resonance-bandwidth broadening of Terfenol-D/epoxy-Pb(Zr,Ti)O3 bilayers in parallel and series connections. Appl. Phys. Lett. 2005, 86, 032508.

43. Park, C.-S.; Ahn, C.-W.; Ryu, J.; Yoon, W.-H.; Park, D.-S.; Kim, H.-E.; Priya, S. Design and characterization of broadband magnetoelectric sensor. J. Appl. Phys. 2009, 105, 094111.

44. Park, C.-S.; Priya, S. Broadband/wideband magnetoelectric response. Adv. Condens. Matter Phys. 2012, 2012, 323165.

45. Patil, D.R.; Kambale, R.C.; Chai, Y.; Yoon, W.-H.; Jeong, D.-Y.; Park, D.-S.; Kim, J.-W.; Choi, J.-J.; Ahn, C.-W.; Hahn, B.-D.; et al. Multiple broadband magnetoelectric response in thickness-controlled Ni/[011] Pb(Mg1/3Nb2/3)O3-Pb(Zr,Ti)O3 single crystal/Ni laminates. Appl. Phys. Lett. 2013, 103, 052907.

46. Patil, D.R.; Chai, Y.; Kambale, R.C.; Jeon, B.-G.; Yoo, K.; Ryu, J.; Yoon, W.-H.; Park, D.-S.; Jeong, D.-Y.; Lee, S.-G.; et al. Enhancement of resonant and non-resonant magnetoelectric coupling in multiferroic laminates with anisotropic piezoelectric properties. Appl. Phys. Lett. 2013, 102, 062909.

47. Kambale, R.C.; Yoon, W.-H.; Park, D.-S.; Choi, J.-J.; Ahn, C.-W.; Kim, J.-W.; Hahn, B.-D.; Jeong, D.-Y.; Lee, B.-C.; Chung, G.-S.; et al. Magnetoelectric properties and magnetomechanical energy harvesting from stray vibration and electromagnetic wave by Pb(Mg1/3Nb2/3)O3-Pb(Zr,Ti)O3 single crystal/Ni cantilever. J. Appl. Phys. 2013, 113, 204108.

48. Priya, S.; Yang, S.C.; Maurya, D.; Yan, Y. Recent advances in piezoelectric and magnetoelectric materials phenomena. In Composite Magnetoelectrics; Srinivasan, G., Priya, S., Sun, N.X., Eds.; Woodhead Publishing: Cambridge, UK, 2015; pp. 103–157.

49. Yan, Y.; Cho, K.-H.; Maurya, D.; Kumar, A.; Kalinin, S.; Khachaturyan, A.; Priya, S. Giant energy density in [001]-textured Pb(Mg1/3Nb2/3)O3-PbZrO3-PbTiO3 piezoelectric ceramics. Appl. Phys. Lett. 2013, 102, 042903.

50. Wang, Y.; Zhao, X.; Jiao, J.; Zhang, Q.; Di, W.; Luo, H.; Leung, C.M.; Or, S.W. Lead-free magnetoelectric laminated composite of Mn-doped Na0.5Bi0.5TiO3–BaTiO3 single crystal and Tb0.3Dy0.7Fe1.92 alloy. J. Alloy Compd. 2010, 496, L4–L6.

51. Zhou, Y.; Maurya, D.; Yan, Y.; Srinivasan, G.; Quandt, E.; Priya, S. Self-biased magnetoelectric composites: An overview and future perspectives. Energy Harvest. Syst. 2016, 3, 1–42.

52. Mandal, S.K.; Sreenivasulu, G.; Petrov, V.M.; Srinivasan, G. Magnetization-graded multiferroic composite and magnetoelectric effects at zero bias. Phys. Rev. B. 2011, 84, 014432.

53. Raidongia, K.; Nag, A.; Sundaresan, A.; Rao, C.N.R. Multiferroic and magnetoelectric properties of core-shell CoFe2O4@BaTiO3 nanocomposites. Appl. Phys. Lett. 2010, 97, 062904.

54. Liu, M.; Li, X.; Imrane, H.; Chen, Y.; Goodrich, T.; Cai, Z.; Ziemer, K.S.; Huang, J.Y.; Sun, N.X. Synthesis of ordered arrays of multiferroic NiFe2O4-Pb(Zr0.52Ti0.48)O3 core-shell nanowires. Appl. Phys. Lett. 2007, 90, 152501.

55. Starr, J.D.; Andrew, J.S. Janus-type bi-phasic functional nanofibers. Chem. Comm. 2013, 49, 4151–4153.

56. Andrew, J.S.; Starr, J.D.; Budi, M.A.K. Prospects for nanostructured multiferroic composite materials. Scripta Mater.2014, 74, 38–43.

57. Xie, S.; Ma, F.; Liu, Y.; Li, J. Multiferroic CoFe2O4-Pb(Zr0.52Ti0.48)O3 core-shell nanofibers and their magnetoelectric coupling. Nanoscale 2011, 3, 3152–3158.

58. Xie, S.H.; Li, J.Y.; Liu, Y.Y.; Lan, L.N.; Jin, G.; Zhou, Y.C. Electrospinning and multiferroic properties of NiFe2O4–Pb(Zr0.52Ti0.48)O3 composite nanofibers. J. Appl. Phys. 2008, 104, 024115.

59. Srinivasan, G. Magnetoelectric characterization techniques. In Composite Magnetoelectrics; Srinivasan, G., Priya, S., Sun, N.X., Eds.; Woodhead Publishing: Cambridge, UK, 2015; pp. 41–54.

60. Zhou, Y.; Apo, D.J.; Sanghadasa, M.; Bichurin, M.; Petrov, V.M.; Priya, S. Magnetoelectric energy harvester. InComposite Magnetoelectrics; Srinivasan, G., Priya, S., Sun, N.X., Eds.; Woodhead Publishing: Cambridge, UK, 2015; pp. 161–207.

61. Dong, S.; Zhai, J.; Xing, Z.; Li, J.; Viehland, D. Giant magnetoelectric effect (under a dc magnetic bias of 2 Oe) in laminate composites of FeBSiC alloy ribbons and Pb(Zn1⁄3Nb2⁄3)O3–7%PbTiO3 fibers. Appl. Phys. Lett. 2007, 91, 022915.

62. Park, C.-S.; Cho, K.-H.; Arat, M.A.; Evey, J.; Priya, S. High magnetic field sensitivity in Pb(Zr,Ti)O3–Pb(Mg1/3Nb2/3)O3single crystal/Terfenol-D/Metglas magnetoelectric laminate composites. J. Appl. Phys. 2010, 107, 094109.

63. Ramana, M.V.; Reddy, N.R.; Sreenivasulu, G.; Kumar, K.V.S.; Murty, B.S.; Murthy, V.R.K. Enhanced mangnetoelectric voltage in multiferroic

particulate Ni0.83Co0.15Cu0.02Fe1.9O4−δ/PbZr0.52Ti0.48O3 composites—dielectric, piezoelectric and magnetic properties. Curr. Appl. Phys. 2009, 9, 1134–1139.

64. Sreenivasulu, G.; Fetisov, L.Y.; Fetisov, Y.K.; Srinivasan, G. Piezoelectric single crystal langatate and ferromagnetic composites: Studies on low-frequency and resonance magnetoelectric effects. Appl. Phys. Lett. 2012, 100, 052901.

65. Chen, Y.; Gillette, S.M.; Fitchorov, T.; Jiang, L.; Hao, H.; Li, J.; Gao, X.; Geiler, A.; Vittoria, C.; Harris, V.G. Quasi-one-dimensional miniature multiferroic magnetic field sensor with high sensitivity at zero bias field. Appl. Phys. Lett. 2011,99, 042505.

66. Zhang, N.; Fan, J.; Rong, X.; Cao, H.; Wei, J. Magnetoelectric effect in laminate composites of Tb1−xDyxFe2−y and Fe-doped BaTiO3. J. Appl. Phys. 2007, 101, 063907.

67. Islam, R.A.; Kim, H.; Priya, S.; Stephanou, H. Piezoelectric transformer based ultrahigh sensitivity magnetic field sensor. Appl. Phys. Lett. 2006, 89, 152908.

68. Feng, M.; Wang, J.-J.; Hu, J.-M.; Wang, J.; Ma, J.; Li, H.-B.; Shen, Y.; Lin, Y.-H.; Chen, L.-Q.; Nan, C.-W. Optimizing direct magnetoelectric coupling in Pb(Zr,Ti)O3/Ni multiferroic film heterostructures. Appl. Phys. Lett. 2015, 106, 072901.

69. Tahmasebi, K.; Barzegar, A.; Ding, J.; Herng, T.S.; Huang, A.; Shannigrahi, S. Magnetoelectric effect in Pb(Zr0.95Ti0.05)O3 and CoFe2O4 heteroepitaxial thin film composite. Mater. Des. 2011, 32, 2370–2373.

70. Martínez, R.; Kumar, A.; Palai, R.; Srinivasan, G.; Katiyar, R.S. Observation of strong magnetoelectric effects in Ba0.7Sr0.3TiO3/La0.7Sr0.3MnO3 thin film heterostructures. J. Appl. Phys. 2012, 111, 104104.

71. Park, C.-S.; Khachaturyan, A.; Priya, S. Giant magnetoelectric coupling in laminate thin film structure grown on magnetostrictive substrate. Appl. Phys. Lett. 2012, 100, 192904.

72. Sun, N.X.; Srinivasan, G. Voltage control of magnetism in multiferroic heterostructures and devices. SPIN 2012, 2, 1240004.

73. Paluszek, M.; Avirovik, D.; Zhou, Y.; Kundu, S.; Chopra, A.; Montague, R.; Priya, S. Magnetoelectric composites for medical application. In Composite Magnetoelectrics; Srinivasan, G., Priya, S., Sun, N.X., Eds.; Woodhead Publishing: Cambridge, UK, 2015; pp. 297–327.

74. Dong, S.; Zhai, J.; Bai, F.; Li, J.-F.; Viehland, D. Push-pull mode magnetostrictive /piezoelectric laminate composite with an enhanced magnetoelectric voltage coefficient. Appl. Phys. Lett. 2005, 87, 062502.

75. Gao, J.; Das, J.; Xing, Z.; Li, J.; Viehland, D. Comparison of noise floor and sensitivity for different magnetoelectric laminates. J. Appl. Phys. 2010, 108, 084509.

76. Li, M.; Wang, Z.; Wang, Y.; Li, J.; Viehland, D. Giant magnetoelectric effect in self-biased laminates under zero magnetic field. Appl. Phys. Lett. 2013, 102, 082404.

77. Wang, Y.; Hasanyan, D.; Li, M.; Gao, J.; Li, J.; Viehland, D. Equivalent magnetic noise in multi-push-pull configuration magnetoelectric composites: Model and experiment. IEEE Trans. Ultrason. Ferroelectr. Freq. Control. 2013, 60, 1227–1233.

78. Das, J.; Gao, J.; Xing, Z.; Li, J.F.; Viehland, D. Enhancement in the field sensitivity of magnetoelectric laminate heterostructures. Appl. Phys. Lett. 2009, 95, 092501.

79. Gao, J.; Gray, D.; Shen, Y.; Li, J.; Viehland, D. Enhanced dc magnetic field sensitivity by improved flux concentration in magnetoelectric laminates. Appl. Phys. Lett. 2011, 99, 153502.

80. Wang, Y.; Li, M.; Hasanyan, D.; Gao, J.; Li, J.; Viehland, D. Geometry-induced magnetoelectric effect enhancement and noise floor reduction in Metglas/piezofiber sensors. Appl. Phys. Lett. 2012, 101, 092905.

81. Li, M.; Berry, D.; Das, J.; Gray, D.; Li, J.; Viehland, D. Enhanced sensitivity and reduced noise floor in magnetoelectric laminate sensors by an improved lamination process. J. Amer. Ceram. Soc. 2011, 94, 3738–3741.

82. Wang, Y.J.; Gao, J.Q.; Li, M.H.; Shen, Y.; Hasanyan, D.; Li, J.F.; Viehland, D. A review on equivalent magnetic noise of magnetoelectric laminate sensors. Philos. Trans. R. Soc. London Ser. A 2014, 372.

83. Gao, J.; Wang, Y.; Li, M.; Shen, Y.; Li, J.; Viehland, D. Quasi-static (f<10−2 Hz) frequency response of magnetoelectric composites based magnetic sensor. Mater. Lett. 2012, 85, 84–87.

84. Shen, Y.; Gao, J.; Shen, L.; Gray, D.; Li, J.; Finkel, P.; Viehland, D.; Zhuang, X.; Saez, S.; Dolabdjian, C. Analysis of the environmental magnetic noise rejection by using two simple magnetoelectric sensors. Sens. Actuators A Phys. 2011,171, 63–68.

85. Fang, C.; Jiao, J.; Ma, J.; Lin, D.; Xu, H.; Zhao, X.; Luo, H. Significant reduction of equivalent magnetic noise by in-plane series connection in magnetoelectric Metglas/Mn-doped Pb(Mg1/3Nb2/3)O3-PbTiO3 laminate composites. J. Phys. D Appl. Phy. 2015, 48, 465002.

86. Stephan, M.; Robert, J.; Henry, G.; Eckhard, Q.; Reinhard, K.; Bernhard, W. MEMS magnetic field sensor based on magnetoelectric composites. J. Micromech. Microeng. 2012, 22, 065024.

87. Zhao, P.; Zhao, Z.; Hunter, D.; Suchoski, R.; Gao, C.; Mathews, S.; Wuttig, M.; Takeuchi, I. Fabrication and characterization of all-thin-film magnetoelectric sensors. Appl. Phys. Lett. 2009, 94, 243507.

88. Lee, D.G.; Kim, S.M.; Yoo, Y.K.G.; Han, J.H.; Chun, D.W.; Kim, Y.-C.; Kim, J.; Hwang, K.S.; Kim, T.S.; Jo, W.W.; et al. Ultra-sensitive magnetoelectric microcantilever at a low frequency. Appl. Phys. Lett. 2012, 101, 182902.

89. Ramsden, E. Hall-Effect Sensors: Theory and Applications; Newnes Publications: Burlington, NJ, USA, 2006; pp. 1–9.

90. Leung, C.M.; Or, S.W.; Zhang, S.; Ho, S.L. Ring-type electric current sensor based on ring-shaped magnetoelectric laminate of epoxy-bonded Tb0.3Dy0.7Fe1.92 short-fiber/NdFeB magnet magnetostrictive composite and Pb(Zr, Ti)O3piezoelectric ceramic. J. Appl. Phys. 2010, 107, 09D918.

91. Lu, C.; Li, P.; Wen, Y.; Yang, A.; Yang, C.; Wang, D.; He, W.; Zhang, J. Magnetoelectric composite Metglas/PZT-based current sensor. IEEE Trans. Magn. 2014, 50, 1–4.

92. Zhang, S.; Leung, C.M.; Kuang, W.; Or, S.W.; Ho, S.L. Concurrent operational modes and enhanced current sensitivity in heterostructure of magnetoelectric ring and piezoelectric transformer. J. Appl. Phys. 2013, 113, 17C733.

93. Yu, X.; Lou, G.; Chen, H.; Wen, C.; Lu, S. A slice-type magnetoelectric laminated current sensor. IEEE Sens. J. 2015, 15, 5839–5850.

94. Patil, D.R.; Zhou, Y.; Kang, J.-E.; Sharpes, N.; Jeong, D.-Y.; Kim, Y.-D.; Kim, K.H.; Priya, S.; Ryu, J. Anisotropic self-biased dual-phase low frequency magneto-mechano-electric energy harvesters with giant power densities. APL Mater.2014, 2, 046102.

95. Ryu, J.; Kang, J.-E.; Zhou, Y.; Choi, S.-Y.; Yoon, W.-H.; Park, D.-S.; Choi, J.-J.; Hahn, B.-D.; Ahn, C.-W.; Kim, J.-W.; et al. Ubiquitous magneto-mechano-electric generator. Energy Environ. Sci. 2015, 8, 2402–2408.

96. Zhou, Y.; Apo, D.J.; Priya, S. Dual-phase self-biased magnetoelectric energy harvester. Appl. Phys. Lett. 2013, 103, 192909.

97. Liu, G.; Ci, P.; Dong, S. Energy harvesting from ambient low-frequency magnetic field using magneto-mechano-electric composite cantilever. Appl. Phys. Lett. 2014, 104, 032908.

98. Lasheras, A.; Gutiérrez, J.; Reis, S.; Sousa, D.; Silva, M.; Martins, P.; Lanceros-Mendez, S.; Barandiarán, J.M.; Shishkin, D.A.; Potapov, A.P. Energy harvesting device based on a metallic glass/PVDF magnetoelectric laminated composite.Smart Mater. Struct. 2015, 24, 065024.

99. Li, P.; Wen, Y.; Liu, P.; Li, X.; Jia, C. A magnetoelectric energy harvester and management circuit for wireless sensor network. Sens. Actuators A. Phys. 2010, 157, 100–106.

100. Han, J.; Hu, J.; Wang, S.X.; He, J. Magnetic energy harvesting properties of piezofiber bimorph/NdFeB composites.Appl. Phys. Lett. 2014, 104, 093901.

101. Onuta, T.-D.; Wang, Y.; Long, C.J.; Takeuchi, I. Energy harvesting properties of all-thin-film multiferroic cantilevers.Appl. Phys. Lett. 2011, 99, 203506.

102. Bian, L.; Wen, Y.; Li, P.; Gao, Q.; Zheng, M. Magnetoelectric transducer with high quality factor for wireless power receiving. Sens. Actuators A Phys. 2009, 150, 207–211.

103. Gao, J.; Hasanyan, D.; Shen, Y.; Wang, Y.; Li, J.; Viehland, D. Giant resonant magnetoelectric effect in bi-layered Metglas/Pb(Zr,Ti)O3 composites. J. Appl. Phys. 2012, 112, 104101.

104. Dong, S.; Zhai, J.; Li, J.F.; Viehland, D.; Priya, S. Multimodal system for harvesting magnetic and mechanical energy.Appl. Phys. Lett. 2008, 93, 103511.

105. Cho, K.-H.; Park, H.-Y.; Heo, J.S.; Priya, S. Structure-performance relationships for cantilever-type piezoelectric energy harvesters. J. Appl. Phys. 2014, 115, 204108.

106. Vopsaroiu, M.; Blackburn, J.; Muniz-Piniella, A.; Cain, M.G. Multiferroic magnetic recording read head technology for 1Tbit/in.2 and beyond. J. Appl. Phys. 2008, 103, 07F506.

107. Zhang, Y.; Li, Z.; Deng, C.; Ma, J.; Lin, Y.; Nan, C.-W. Demonstration of magnetoelectric read head of multiferroic heterostructures. Appl. Phys. Lett. 2008, 92, 152510.

108. Kargol, A.; Malkinski, L.; Caruntu, G. Biomedical applications of multiferroic nanoparticles. In Advanced Magnetic Materials; Malkinski, L., Ed.; InTech: Rijeka, Croatia, 2012; pp. 89–118.

109. Guduru, R.; Liang, P.; Runowicz, C.; Nair, M.; Atluri, V.; Khizroev, S. Magneto-electric nanoparticles to enable field-controlled high-specificity drug delivery to eradicate ovarian cancer cells. Sci. Rep. 2013, 3, 2953.

110. Yue, K.; Guduru, R.; Hong, J.; Liang, P.; Nair, M.; Khizroev, S. Magneto-electric nano-particles for non-invasive brain stimulation. PLoS One 2012, 7, e44040.

111. Dobson, J. Remote control of cellular behaviour with magnetic nanoparticles. Nat. Nano 2008, 3, 139–143.

112. Huang, H.; Delikanli, S.; Zeng, H.; Ferkey, D.M.; Pralle, A. Remote control of ion channels and neurons through magnetic-field heating of nanoparticles. Nat. Nano. 2010, 5, 602–606.

113. Ashcroft, F.M. Ion Channels and Disease; Academic Press: San Diego, CA, USA, 2000; pp. 1–2.

114. Vopson, M.M. Fundamentals of multiferroic materials and their possible applications. Crit. Rev. Solid State Mater. Sci.2015, 40, 223–250.

115. Bibes, M.; Barthelemy, A. Multiferroics: Towards a magnetoelectric memory. Nat. Mater. 2008, 7, 425–426.

116. Heron, J.T.; Trassin, M.; Ashraf, K.; Gajek, M.; He, Q.; Yang, S.Y.; Nikonov, D.E.; Chu, Y.H.; Salahuddin, S.; Ramesh, R. Electric-field-induced magnetization reversal in a ferromagnet-multiferroic heterostructure. Phys. Rev. Lett. 2011, 107, 217202.

117. Xue, X.; Zhou, Z.; Peng, B.; Zhu, M.; Zhang, Y.; Ren, W.; Ren, T.; Yang, X.; Nan, T.; Sun, N.X.; et al. Electric field induced reversible 180° magnetization switching through tuning of interfacial exchange bias along magnetic easy-axis in multiferroic laminates. Sci. Rep. 2015, 5, 16480.

118. Liu, M.; Lou, J.; Li, S.; Sun, N.X. E-field control of exchange bias and deterministic magnetization switching in AFM/FM/FE multiferroic heterostructures. Adv. Funct. Mater. 2011, 21, 2593–2598.

119. White, J.F. Review of semiconductor microwave phase shifters. Proc. IEEE 1968, 56, 1924–1931.

120. Boyd, C.R. A 60 GHz dual-mode ferrite phase shifter. In Proceedings of the 1982 IEEE MTT-S International Microwave Symposium Digest, Dallas, TX, USA, 15–17 June 1982; pp. 257–259.

121. Zhao, Z.; Wang, X.; Choi, K.; Lugo, C.; Hunt, A.T. Ferroelectric Phase Shifters at 20 and 30 GHz. IEEE Trans. Microw. Theory Tech. 2007, 55, 430–437.

122. Ustinov, A.B.; Srinivasan, G.; Kalinikos, B.A. Ferrite-ferroelectric hybrid wave phase shifters. Appl. Phys. Lett. 2007, 90, 031913.

123. Tatarenko, A.S.; Srinivasan, G.; Bichurin, M.I. Magnetoelectric microwave phase shifter. Appl. Phys. Lett. 2006, 88, 183507.

124. Tatarenko, A.S.; Srinivasan, G. A strain engineered voltage tunable millimeter-wave ferrite phase shifter. Microwave Opt. Technol. Lett. 2011, 53, 261–264.

125. Kittel, C. On the theory of ferromagnetic resonance absorption. Phys. Rev. 1948, 73, 155–161.

126. Fetisov, Y.K.; Srinivasan, G. Electric field tuning characteristics of a ferrite-piezoelectric microwave resonator. Appl. Phys. Lett. 2006, 88, 143503.

127. Li, N.; Liu, M.; Zhou, Z.; Sun, N.X.; Murthy, D.V.B.; Srinivasan, G.; Klein, T.M.; Petrov, V.M.; Gupta, A. Electrostatic tuning of ferromagnetic resonance and magnetoelectric interactions in ferrite-piezoelectric heterostructures grown by chemical vapor deposition. Appl. Phys. Lett. 2011, 99, 192502.

128. Srinivasan, G. Microwave and millimeter-wave multiferroic devices. In Composite Magnetoelectrics; Srinivasan, G., Priya, S., Sun, N.X., Eds.; Woodhead Publishing: Cambridge, UK, 2015; pp. 241–264.

129. Wang, S.G.; Yoon, S.D.; Vittoria, C. Microwave and magnetic properties of double-sided hexaferrite films on (111) magnesium oxide substrates. J. Appl. Phys. 2002, 92, 6728–6732.

130. Ustinov, A.B.; Tatarenko, A.S.; Srinivasan, G.; Balbashov, A.M. Al substituted Ba-hexaferrite single-crystal films for millimeter-wave devices. J. Appl. Phys. 2009, 105, 023908.

131. Tatarenko, A.S.; Murthy, D.V.B.; Srinivasan, G. Hexagonal ferrite-piezoelectric composites for dual magnetic and electric field tunable 8–25 GHz microstripline resonators and phase shifters. Microw. Opt. Technol. Lett. 2012, 54, 1215–1218.

132. Lou, J.; Reed, D.; Pettiford, C.; Liu, M.; Han, P.; Dong, S.; Sun, N.X. Giant microwave tunability in FeGaB/lead magnesium niobate-lead titanate multiferroic composites. Appl. Phys. Lett. 2008, 92, 262502.

133. Lou, J.; Liu, M.; Reed, D.; Ren, Y.; Sun, N.X. Giant electric field tuning of magnetism in novel multiferroic FeGaB/Lead Zinc Niobate–Lead Titanate (PZN-PT) Heterostructures. Adv. Mater. 2009, 21, 4711–4715.

134. Fang, X.; Zhang, N.; Wang, Z.L. Converse magnetoelectric effects on heterotype electrostrain-piezopermeability composites. Appl. Phys. Lett. 2008, 93, 102503.

135. Lou, J.; Reed, D.; Liu, M.; Sun, N.X. Electrostatically tunable magnetoelectric inductors with large inductance tenability. Appl. Phys. Lett. 2009, 94, 112508.

136. Hwaider, L.; Jing, L.; Yuan, G.; Hasegawa, R.; Ming, L.; Howe, B.; Jones, J.; Brown, G.; Sun, N.X. Voltage tunable magnetoelectric inductors with improved operational frequency and quality factor for power electronics. IEEE Trans. Magn. 2015, 51, 1–5.

137. Mandal, M.; Duttagupta, S.P.; Palkar, V.R. Study of multiferroic Bi0.7Dy0.3FeO3 based tunable ring inductor. J. Phys. D Appl. Phy. 2013, 46, 325001.

138. Garg, R. Microstrip Antenna Design Handbook; Artech House: Boston, MA, USA, 2001.

139. Kim, D. Planar magneto-dielectric metasubstrate for miniaturization of a microstrip patch antenna. Microw. Opt. Technol. Lett. 2012, 54, 2871–2874.

140. Petrov, R.V.; Tatarenko, A.S.; Pandey, S.; Srinivasan, G.; Mantese, J.V.; Azadegan, R. Miniature antenna based on magnetoelectric composites. Electron. Lett. 2008, 44, 506–508.

141. Bae, S.; Hong, Y.K.; Lee, J.J.; Jalli, J.; Abo, G.S.; Lyle, A.; Seong, W.M.; Kum, J.S. Low loss Z-type barium ferrite (Co2Z) for terrestrial digital multimedia broadcasting antenna application. J. Appl. Phys. 2009, 105, 07A515.

142. Bae, S.; Hong, Y.K.; Lee, J.J.; Abo, G.S.; Jalli, J.; Park, J.H.; Seong, W.M.; Kum, J.S.; Ahn, W.K.; Park, S.H. Miniaturized broadband ferrite T-DMB antenna for mobile-phone applications. IEEE Trans. Magn. 2010, 46, 2361–2364.

143. Lee, J.; Hong, Y.-K.; Lee, W.; Abo, G.S.; Park, J.; Neveu, N.; Seong, W.-M.; Park, S.-H.; Ahn, W.-K. Soft M-type hexaferrite for very high frequency miniature antenna applications. J. Appl. Phys. 2012, 111, 07A520.

6

POSITION ESTIMATION OF RFID-BASED SENSORS USING SAW COMPRESSIVE RECEIVERS

M. Brandl, T. Posnicek, K. Kellner

Center for Integrated Sensor Systems, Danube University Krems, Krems A-3500, Austria

ABSTRACT

Wireless data acquisition and data fusion in medical and environmental applications is mainly based on additional information like the position of the sensor device. An accurate and easy to implement method for wireless position estimation of RFID devices is described in this paper. For RFID communication and data transmission, the tags operate within the 13.56 MHz ISM-band with a standardized protocol defined in ISO 14443. The RFID-tag position estimation is done by transmitting linear chirp signals from the RFID tag to base stations on RF frequencies. The chirp signals are generated by SAW devices which are triggered by a low power consumption pulse generator based on an avalanche transistor. The performance of position estimation under AWGN conditions was simulated and compared with common methods. An improvement in the mean square position error in the range of the chirp compression gain was found.

KEYWORDS

RFID tags; Position measurement; Chirp modulation; Surface acoustic wave devices; Correlation receiver

1. INTRODUCTION

Radio frequency identification (RFID) tags are used in numerous applications like wirelessly readable product labels, automation and manufacturing processes, product tracking, animal identification, and so on. The miniaturization process of RFID chips by enhanced semiconductor processes has led to RFID chips having very tiny dimensions. Hitachi has launched a highly integrated RFID device with a footprint of 0.4 mm in a square operating in the 2.45 GHz microwave band [1]. This kind of miniaturized device helps to equip most everyday products with cheap RFID tags for identification and tracking.

An RFID set consists of two parts: the RFID device (tag, transponder) and a base station or reader. Such RFID tags gain their powers for operation from the RF-field of base stations in their vicinity. In passive operation, the transponder is only activated if the received signal strength from the base station exceeds the power-up threshold of the tag. Therefore, for passive devices, the communication distance between the RFID tag and the base station is limited by the antenna size, the transmitted RF power, and the receiver sensitivity [2]. The data transmission format as well as the modulation formats for up- and down-link between the RFID tag and reader station for chip- and smart-cards are defined in ISO/IEC 14443 [3]. Four different frequency bands are currently in use depending on the application: Low frequency RFID systems operate at 125 kHz, high frequency (HF) systems operate at 13.56 MHz, ultra-high frequency (UHF) RFID devices use 433 or 868 MHz (Europe), and microwave RFID tags use the 2.4 or 5.8 GHz band. Apart from the application, all RFID devices utilize a bandwidth of only a few kilohertz or megahertz, and therefore they are classical narrowband transmission systems.

Positioning and tracking of objects are among the leading applications of RFID devices and can be categorized by operation principle into two groups:

First, for a RFID position estimation, one or more base stations are placed at fixed and known locations. If an RFID tag approaches a region where the RF field of the base station powers up the device, the tag starts a repeated transmission of its ID. Assuming ID reception in line-of-sight (LOS) scenarios, a rough position estimation can be carried out because the object must be located within a circle which is centered at the base station and has a maximum radius given by the communication range. More precision in position estimation can be gained using triangulation and trilateration methods with three or more base stations [4], [5] and [6].

A second positioning scenario that can be used especially for tracking moving objects and robots should be discussed. A base station with a directional antenna that receives the transmitted signals (IDs) from a subset of RFID devices located within the response range is fixed on the moving device. The positions of the replaying RFIDs are known and therefore the actual location of the moving object can be estimated [7] and [8]. This method can be combined with RSS (Received Signal Strength) information management and optimization [9]. In general, when using the RSS signal for position estimation of passive RFIDs by one or more base stations, the results can be imprecise because of time varying transmission channels and multipath propagation. Several positioning concepts based on RSS sensing are summarized in [10] and [11]. An example of multiple target localization sparse networks based on compressive sampling (CS)-based algorithms for reducing the computational complexity for localization formulations is given in [12].

In this paper, a system for position estimation of passive RFID devices based on ToA (time of arrival) measurements of a transmitted chirp signal are presented (Fig. 1). The advantage of this principle is that spread spectrum signals like a chirp signals enhances the signal to noise ratio by the signal correlation gain if they are received by a signal matched filter. For chirp signal generation and RFID position estimation, the passive signal processing and filtering capability of surface acoustic wave (SAW) devices are used. In the RFID-tag to base station direction (downlink), the linear chirp filter will be excited by a short and broadband pulse which generates a chirp signal as its impulse response on the output of the SAW filter. In the base station, a linear chirp filter is used as a matched filter for the transmitted signal, allowing a position estimation of the RFID tag by trilateration. In contrast to [13], where chirp signals for position estimation of active RFID chirp transceivers based on the frequency modulated continuous wave (FMCW) radar principle are introduced, the method proposed in this paper is based on low power RFID tags without the need for digital signal processing capabilities and FMCW detectors. References [14] and [15] discuss SAW-based reflectors with different codes for position estimation by signal correlation but without the advantage of broadband signals and high compression gains like in chirp based systems.

2. SYSTEM DESCRIPTION

In general, the developed RFID tag can be split into two functional groups, where one part is a standardized RFID transceiver at 13.56 MHz for wireless ID and data transmission and the second part is a SAW-filter-based chirp generator operating at a center frequency of 250 MHz for position measurement of the RFID tag. The core of the RFID unit and the pulse generator was built with a low power microcontroller MSP430F149 (Texas Instruments, USA).

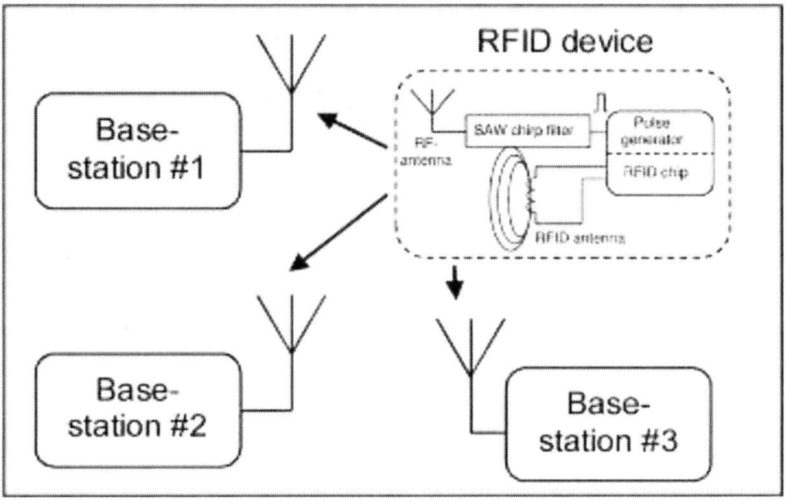

Figure. 1. Transmission of linear chirp signals for ToA measurement.

The concept of RFID position estimation is based on ToA measurement of a linear chirp signal matched filter impulse response called the autocorrelation function (ACF) (Fig. 1,Fig. 2 and Fig. 3). On the RFID tag, a linear chirp is generated by a SAW filter as its impulse response by exciting the SAW device with a short pulse. To excite the full bandwidth of the chirp filter, the corresponding spectrum of the trigger pulse must cover at least the bandwidth of the SAW chirp filter. The chirp signal (impulse response of the SAW filter) is transmitted via an RF antenna to the base stations for ToA measurement. The received chirp signal is correlated in the base station with a signal-matched filter which delivers the chirp autocorrelation function (Fig. 2 and Fig. 3). After subsequent incoherent demodulation of the chirp ACF, the ToA information can be gained by threshold or peak detection. The differences in ToA measurement of the transmitted chirp signal at each base station are denoted as Δt_{mn}, where m and n represent the individual numbers of the base stations.

The wireless powering, ID, and data transmission as well as the modulation formats of the RFID unit of the device are performed in separate channels at different frequencies which are defined in international standards such as ISO/IEC 14443 for proximity and smart cards [3]. By separating RFID-related functions (wireless powering, data transmission, etc.) and the chirp signal transmission for tag position estimation into different frequency bands, possible jamming and interference can be avoided. In our concept, the RFID unit of the device was operating in the 13.56 MHz band, where the power for the RFID device is transmitted via an RF-carrier and the data transmission is done by load modulation on the carrier. The position measurement of the unit was done by chirp signals with a center frequency of 250 MHz and a bandwidth of 80 MHz.

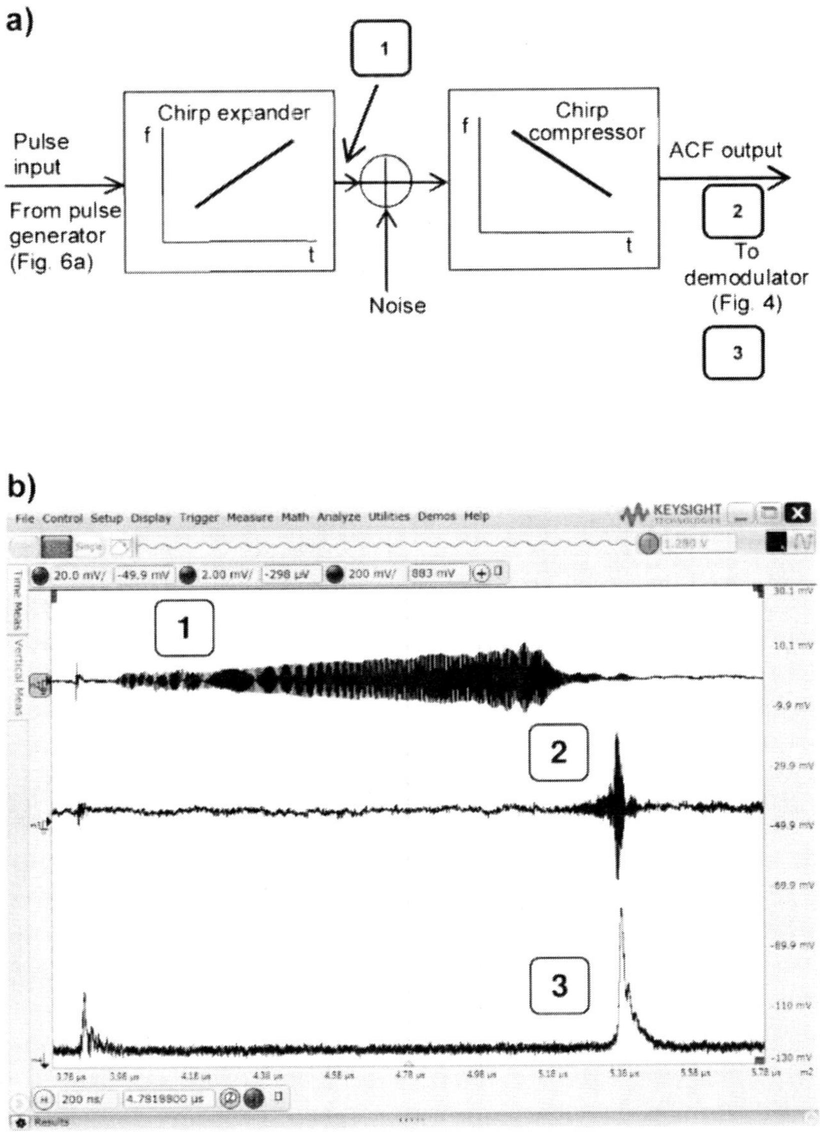

Figure. 2. (a) Principle of chirp transmission and correlation by a matched filter. (b) Time signals at different positions of the transmission system (1) generated chirp signal, (2) ACF at filter output and (3) after incoherent demodulation.

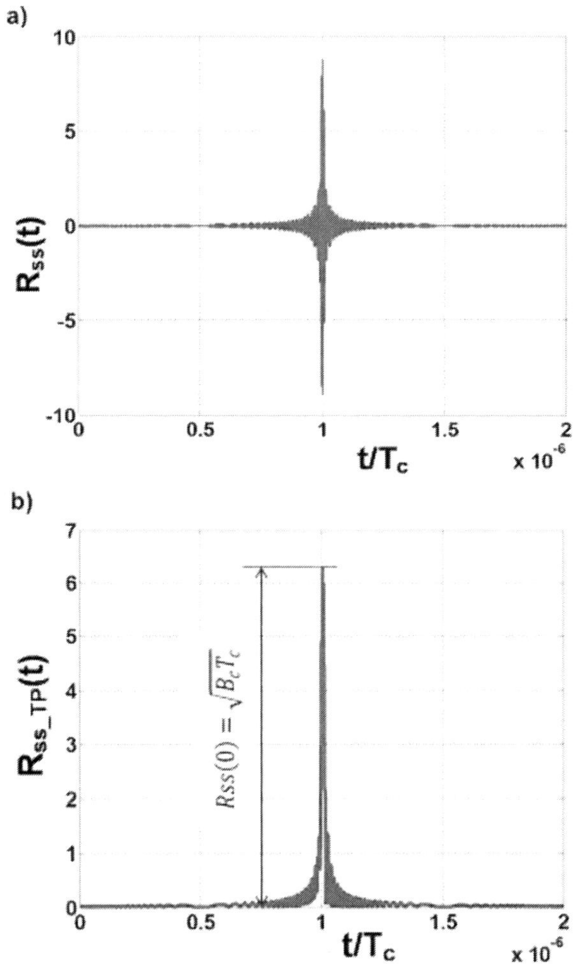

Figure. 3. (a) Bandpass signal of a compressed chirp pulse $R_{ss}(t)$ and (b) its envelope after incoherent demodulation with the peak amplitude at center.

2.1. Linear chirp signals

For precise position estimation of objects, the shortest possible pulses should be used for high range resolution. In radar technology, linear frequency modulated signals (chirp signals) are in widespread use because of their high time-bandwidth product, the resulting high correlation gain, and therefore their excellent signal-to-noise ratio (SNR) at the receiver.

A linear chirp signal with amplitude A (and without amplitude weighting) can be written in the time domain as

$$s(t) = A \times \cos\left(\omega_0 \times t + \frac{\mu \times t^2}{2}\right) \text{ for } -\frac{T_c}{2} < t < \frac{T_c}{2}. \quad (1)$$

The signal s(t) in Eq. (1) roams linearly in a bandwidth B_c centered around the frequency ω_0 in a time interval T_c. The dispersion coefficient μ represents the change of frequency over time. If the frequency increases with time ($\mu > 0$), the signal is called an up-chirp; otherwise it is called a down-chirp.

For chirp-based RFID position estimation, a short pulse generated by the RFID unit excites the SAW chirp filter for signal transmission. An optimal receiver regarding the SNR utilizes matched filter correlative signal processing. In Fig. 2a,b, the principle of chirp transmission and impulse compression including the all relevant signals is sketched.

The auto-correlation function (ACF) $R_{ss}(t)$ of a chirp signal, that is, the response of the signal-matched filter with the normalized impulse response h(t).

$$h(t) = 2\sqrt{\frac{B_c}{T_c}} s^*(-t) = 2\sqrt{\frac{B_c}{T_c}} \cos\left(\omega_0 t - \frac{\mu t^2}{2}\right)$$

$$\text{for } -\frac{T_c}{2} < t < \frac{T_c}{2}, \quad (2)$$

The ACF $R_{ss}(t)$ is given by the convolution of s(t) with h(t), with the dispersion coefficient $\mu = \frac{2\pi B_c}{T_c}$ [17] (Fig. 3a).

$$R_{ss}(t) = 2A\sqrt{\frac{B_c}{T_c}} \frac{\sin\left[\frac{\mu t}{2}(T_c - |t|)\right]}{\mu t} \cos\omega_0 t \text{ for } -T_c < t < T_c \quad (3)$$

The envelope of the signal at the center frequency ω_0 is similar to a sin(x)/x function (Fig. 3a,b). The transmitted chirp signal's energy E_c can be calculated from the chirp signal s(t) by

$$E_c = \int_{-\frac{T_c}{2}}^{\frac{T_c}{2}} s^2(t)\,dt = \frac{A^2 T_c}{2}. \quad (4)$$

The transmitted power P_c is the transmitted signal energy divided by the chirp duration T_c:

$$P_c = \frac{A^2}{2}. \quad (5)$$

In the receiver, the signal is compressed in a main lobe of duration $1/B_c$. If the energy is assumed to be constant, the peak value of the ACF after pulse compression $R_{ss}(0)$ is given by

$$R_{ss}(0) = A \times \sqrt{T_c \times B_c} \tag{6}$$

The factor $T_c \times B_c$ is the gain in peak signal power due to impulse compression. It is equal to the enhancement of the peak SNR in the case of additive white Gaussian noise. If the amplitude of the chirp is almost constant for its duration T_c, sidelobes with a first peak amplitude of approximately −13 dB relative to the main peak occur after impulse compression and may cause errors in signal detection; for example, they may produce ghost targets in radar technology and non-unique and potentially wrong position information in the proposed application. To avoid this and to achieve a better main-to-side-lobe ratio, a signal weighting function $W(\omega)$ is applied in the frequency domain. The generalized cosine square weighting function for a signal with center frequency ω_0 and bandwidth B is given by

$$W(\omega) = k + (1 - k) \times \cos^n\left(\frac{\pi(\omega - \omega_0)}{B}\right). \tag{7}$$

By applying the commonly used Hamming weighting function ($k = 0.08$, $n = 2$ for Eq. (7)), a theoretical maximum of the main-to-side-lobe ratio of approximately 42 dB can be gained, but the compressed impulse is widened by a factor of approximately 1.5[17] and [18].

2.2. Surface acoustic wave devices for chirp signal generation

In principle, chirp signal synthesis and impulse compression can be performed in a fast digital signal processor. By implementing such signal synthesis in a digital direct synthesizer, chirps with a bandwidth of up to a few hundreds of megahertz can be generated for transmission.

An analog method of generating chirps is several decades old and is still applied to impulse compression of chirps with a bandwidth of a few hundreds of megahertz is the utilization of dispersive SAW delay lines [19] and [20]. On a piezoelectric SAW substrate, metallic interdigital transducers (IDTs) convert an electrical RF signal into a SAW and reconvert it into an electric signal. The propagation velocity of the SAW is smaller by a factor of 10e5 than that of an electromagnetic wave in free space. The maximum acoustoelectric interaction occurs at that frequency, where the acoustic wavelength is equal to the geometric period of the IDT's fingers [20]. The interaction will occur at lower frequencies for a wider geometric period than for a narrower one. In Fig. 2, chirp generation and impulse compression in a SAW device are drawn schematically. If the input IDT is excited by a signal covering the full frequency band of the filter, the spectral components with lower frequencies will occur first at the output and the highest ones will occur last; the group delay is dispersive. In Fig. 2, an up-chirp SAW filter is drawn, responding to an applied impulse with an up-chirp. By definition, the impulse response of a signal matched filter $h_m(t)$ is the conjugated complex time reverse of the received signal s(t) and delivers the signal ACF at the filter output.

$$h_m = s^*(-t) \tag{8}$$

As indicated in Fig. 2 and Eq. (2), a chirp matched filter can be gained by reversing the IDT of the dispersive SAW delay line. This means that the matched filter impulse response $g_m(t)$ is a down-chirp for a received up-chirp signal s(t). The spectral components of the chirp signal excite an electric signal at the corresponding IDT finger pairs. For a short time interval, this electric charge distribution is in phase and can be summed up constructively by the bus bars to the compressed output impulse. If a part of the chirp signal does not arrive at the receiver, the corresponding electric charge is absent at the bus bars and the output signal contains less energy but is still the sum of the remaining in-phase signal components compressed to a short time interval. This is an explicit advantage of broadband signals regarding reliability and robustness. Since the SAW device is strictly linear, the impulse compression also operates for negative SNRs at the input. So the signal can still be detected even when the received chirp is buried in noise and interference at the receiver's input.

3. RFID POSITION ESTIMATION

As shown in Fig. 1 and Fig. 2, a SAW chirp filter is excited by the RFID unit of the tag with a short pulse to generate a chirp signal as the filter's impulse response. In the base station, the ACF with its maximum $R_{ss}(0)$ (Eqs. (2) and (6)) of the received chirp signal is built by correlation in a signal matched filter having the time inverse impulse response to the received signal. The proposed principle for RFID position detection can be used for RFID devices or RFID-based sensors which are powered by battery or for passive devices [21]. In passive devices, the energy for the SAW excitation pulse is stored in a capacitor which is charged via an RF-carrier at timepoints when the RFID device is in the vicinity of a base station.

The position estimation is done by trilateration based on the time differences of the received chirp signal at the base stations (Fig. 7). A detailed block diagram of a base station with a standardized RFID interface and supporting chirp-based position estimation of RFID tags is depicted in Fig. 4.

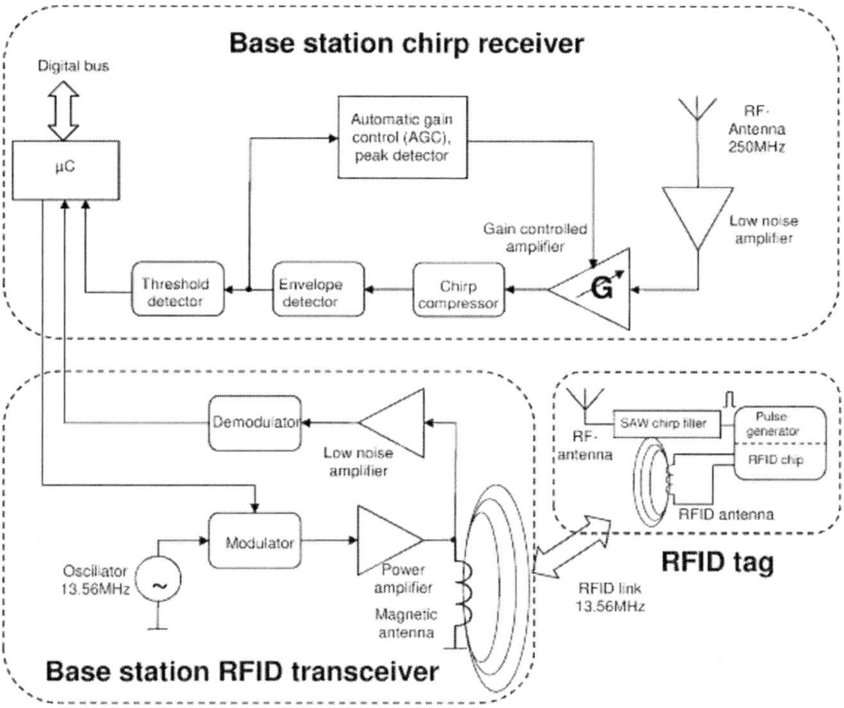

Figure. 4. Detailed block diagram of an RFID base station and an RFID tag with the support of chirp-based position estimation.

3.1. Design of the RFID device

In a periodic interval, the RFID unit or a low power microcontroller generates a short impulse that triggers the SAW chirp filter to generate and transmit a chirp signal to the base stations for position estimation. The energy for operation of the microcontroller is from an onboard coin battery or is stored in a high grade capacitor (Gold-Cap, Super-Cap) [22], which can be charged from time to time by the RF-field of the base station or by energy-harvesting methods [23]. But all the proposed energy storage units have limited capacity and supply voltage. For a high voltage trigger pulse, which is necessary for exciting the SAW chirp filter, an amplification stage should be used.

A highly efficient method of generating short and broadband pulses with rise and fall times in the nanosecond range is the use of transistors operating in the avalanche-breakdown region. An avalanche breakdown can be generated in a bipolar junction transistor by applying a reverse voltage beyond the collector-emitter breakdown voltage[24]. At very high reverse voltages, hot electrons with high kinetic energy are generated in the semiconductor crystal. These electrons can knock out other electrons, which results in a rapidly growing electron avalanche, and the crystal becomes conducting in sub-nanoseconds. This principle is well suited to the generation of pulses that are of high energy, short, and therefore broadband in the frequency domain.

The RFID unit is typically powered by a lithium coin battery with a nominal voltage of 3.3 V or by a high grade capacitor with a typical maximum operating voltage below 5 V. To reach the avalanche-breakdown region of a commercial bipolar junction transistor, a reverse voltage above about 150 V must be applied. To generate such high voltages in an easy way, we have used a field effect transistor (FET) as a switch (T1) together with a 47 μH inductivity to build an ignition circuit for the avalanche transistor T2 (Fig. 5a). The FET is turned on for 4 μs, where during this time, an exponentially increasing current limited by the on-resistance of the FET switch flows through the inductivity L1 (mean current during 4 μs of T1 on-time: $I_{L\text{-}mean}$ ~ 100 mA at U_b = 3.0 V, Fig. 6). After switching the FET off (Fig. 5b), a high voltage pulse (U_{DS}) which is above the breakdown voltage of the bipolar transistor T2 is generated by the inductivity L1. Therefore an avalanche breakdown is generated in transistor T2 and a short current pulse flows through the RF-balun ETC1-1-13TR (M/A-COM Technology Solutions, USA) on which the SAW chirp filter (DS1804C, Nanotron Technologies GmbH, Germany) is connected. The RF-balun is used for balancing and impedance matching. For an optimal chirp generation, the pulse rise time and duration of the trigger pulse must at least cover the chirp bandwidth in the frequency domain and should be more than 300 MHz which corresponds to pulses with a rise time below 3 ns. The measured chirp output power at the antenna was −20 dBm for a supply voltage U_b = 3.0 V. The current and energy consumption of the chirp generation unit are depicted in Fig. 6. The trigger circuit from Fig. 5a is switched on for T_{on} = 4 μs and the mean current consumption during this time is about 80 mA for U_b = 2.5 V and rises to a mean current of 140 mA at U_b = 4 V.

The mean charge necessary for the generation of one chirp signal is the product of the current consumption and the time duration. At a supply voltage of 3 V, the charge consumption of the trigger circuit is about 1.2e−7 mAh for the generation of one chirp signal (Fig. 6). The current consumption of the microcontroller which delivers the trigger pulse to the FET switch can be assumed to be about 10 mA for a time duration of 10 μs, resulting in a withdrawn charge capacity of 2.8e–8 mAh per period. The typical useable capacity of a miniature lithium coin battery CR1216 with a diameter of 12 mm and a height of 1.6 mm is 25 mAh [25]. Therefore, the total charge withdrawn from the coin battery per period is 1.48e–7 mAh, and this means that a total of about 1.7e8 chirp signals can be generated and transmitted. The simple calculation shows that the battery capacity should not be the limiting factor of the sensor tag life cycle and a very high number of position estimations can be done. Because of the low amount of energy necessary for supplying the RFID device, low-power energy-harvesting methods can be used.

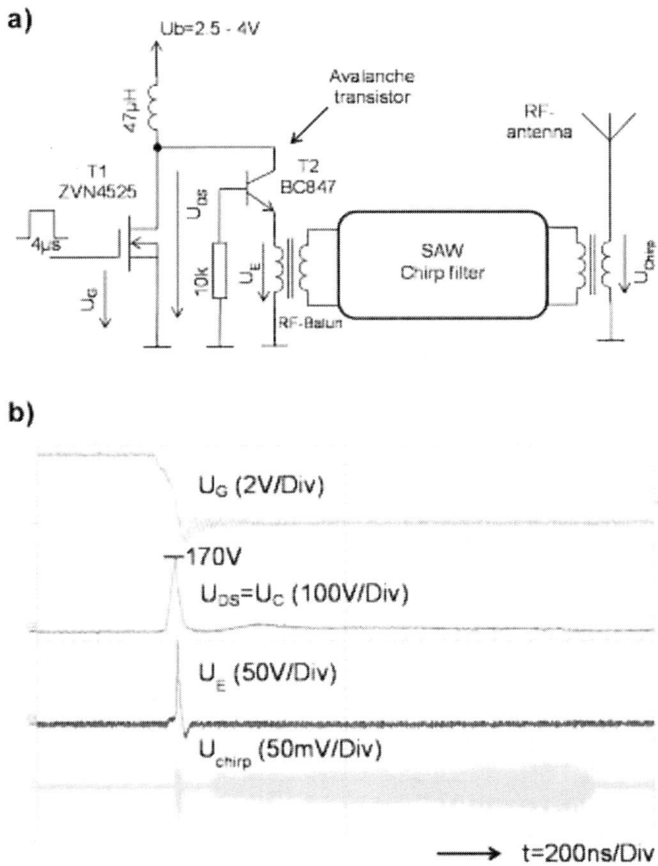

Figure. 5. (a) Pulse generator based on avalanche transistor for the generation of chirp signals. (b) Measured time signals during operation.

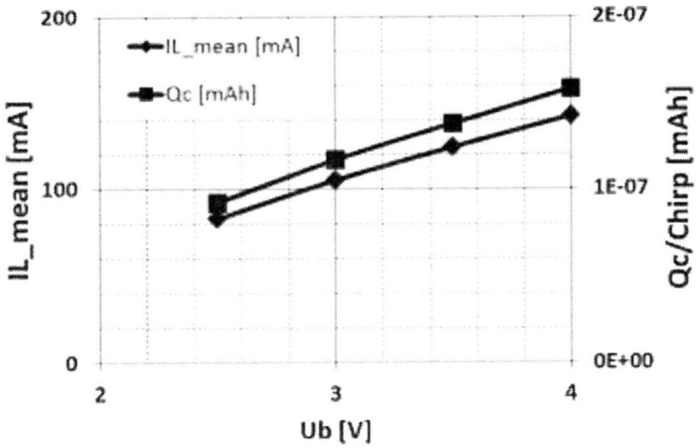

Figure. 6. Current and charge consumption of the chirp trigger circuit for T_{on} = 4 μs.

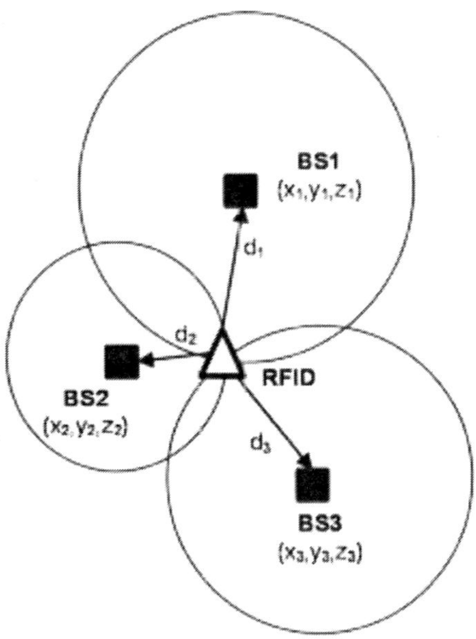

$$d1 = c\Delta t_{12} = \sqrt{(x_1 - x)^2 + (y_1 - y)^2 + (z_1 - z)^2}$$
$$d2 = c\Delta t_{23} = \sqrt{(x_2 - x)^2 + (y_2 - y)^2 + (z_2 - z)^2}$$
$$d3 = c\Delta t_{31} = \sqrt{(x_3 - x)^2 + (y_3 - y)^2 + (z_3 - z)^2}$$

Figure. 7. Trilateration principle for 3D RIFD position estimation.

3.2. Position estimation and simulation results

The RFID is not continuously in the active mode but will normally be set in the low-power sleep state and powered up only at scheduled time points or on request by a received wake-up signal. In the sleep mode, the RFID receiver is powered down and no active amplification of the wake-up signal can be achieved. Therefore, the tag can only be woken up at defined time points by a watchdog clock system implemented on the on-board microcontroller or at any time point by an external trigger, for example, when the RFID tag is in the transmission range and powered by the base station. After wake-up, the RFID tag starts a periodic transmission of chirp signals. The tag position estimation is done by ToA measurements at the receiving base stations and subsequent trilateration using the equations depicted in Fig. 7.

The coordinates of the three base stations are denoted by (x_1, y_1, z_1), (x_2, y_2, z_2), and (x_3, y_3, z_3). The tag position with coordinates (x, y, z) can be estimated from the ToA differences of the received chirp signal. The localization accuracy of the proposed method was determined under the assumption of an AWGN (additive white Gaussian noise) transmission channel where the received signal is corrupted with noise (Eq. (9)and Fig. 9). The accuracy of position estimation based on chirp signals under LOS conditions is mainly given by the peak amplitude of the chirp ACF (Eq. (6)) and the noise on the transmission channel. For a theoretical model of the time-position error caused by white additive noise, it will be assumed that the noise has a power spectral density G(f) = η/2 and zero mean value [26]. A selective low-pass filter in the receiver limits the noise bandwidth to $B_N \leq B_c/2$. The output signal v(t) of the filter sampled at the peak value of the ACF at $t = T_s$ is

$$v(T_s) = R_{ss}(T_s) + n(T_s) = A \times \sqrt{T_c \times B_c} + \varepsilon_A \qquad (9)$$

where $\varepsilon_A = n(T_s)$ is the amplitude error with variance

$$\sigma_A^2 = n^2(T_s) = \eta B_C/2 \qquad (10)$$

Any filter bandwidth B_N that is less than the half chirp bandwidth $B_c/2$ lowers the noise bandwidth but reduces the output ACF amplitude and the pulse rise time $t_r = 1/2B_n = 1/B_c$, assuming that $B_N = B_c/2$ for the maximal pulse rise time. The SNR at the threshold detector is

$$SNR = 10 \times \log\left(\frac{R_{ss}^2(0)}{\sigma_A^2}\right). \qquad (11)$$

In the receiver of the base station, a threshold detector is used for ToA measurement of the transmitted chirp pulse. For incoherent detection of the received chirp signal, the probability density function (pdf) of the noise floor is Rayleigh distributed and the pdf of the ACF-peak is Rican distributed at the threshold detector, respectively. For this case, the optimal threshold detector level V_{opt} is given by [26],

$$V_{opt} = \frac{R_{ss}(0)}{2}\sqrt{1+\frac{1}{SNR}} \text{ and}$$

$$V_{opt} \approx \frac{R_{ss}(0)}{2} \text{ for } (SNR \gg 1). \tag{12}$$

For reliable position estimation accuracies, the SNR must be much greater than 1 and therefore the threshold level is set to half of the ACF maximum $R_{ss}(0)/2$. To obtain a constant amplitude of the ACF at the threshold detector, a control loop with a peak-detection circuit and a gain-controlled amplifier is used (Fig. 4). The noise perturbation $n(t_b)$ causes a time-position error ε_t, and from Fig. 8, using an approximation with similar triangles, it can be found that

$\varepsilon_t/n(t_b)=t_r/R_{ss}(0)$ and $\varepsilon_t[t_r/R_{ss}(0)]n(t_b)$

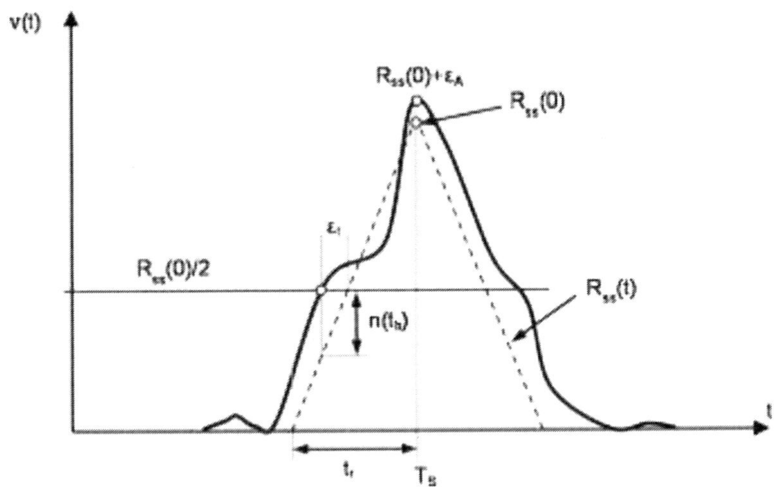

Figure. 8. Time-position error caused by noise.

The variance of the time-position error is therefore

$$\sigma_t^2 = [t_r/R_{ss}(0)]^2\,\overline{n^2(t_b)} = [t_r/R_{ss}(0)]^2\,\eta B_N \tag{14}$$

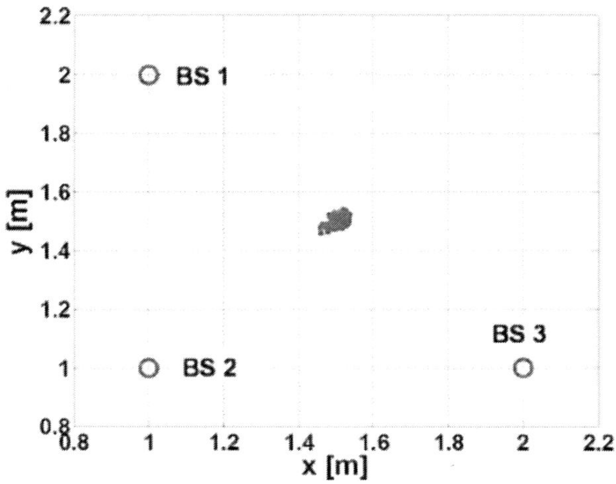

Figure. 9. Estimated position of the RFID tag. N = 100, SNR = 10 dB.

and using Eqs. (4) and (6) it can be found that

$$\sigma_t^2 = \frac{\eta}{2A^2 B_c^2 T_c} = \frac{\eta}{4B_c^2 E_c}. \tag{15}$$

It follows that the variance of the time-position error can be reduced by choosing chirp signals with high bandwidth coverage B_c and long time-duration T_c, resulting in high correlation gain, which is the product T_cB_c.

Fig. 9 shows the simulation results for an RFID tag located at x = y = 1.5 m, z = 0, and the estimated positions calculated by ToA under AWGN conditions, with an SNR of 10 dB on the transmission channel. The chirp signal was defined with a chirp bandwidth of B_c = 80 MHz, a chirp duration of T_c = 1.2 μs, and a center frequency of f_0 = 250 MHz.

The distribution of the localization error is depicted in Fig. 10, from which the mean position estimation accuracy was calculated within a SNR of −10 to +40 dB (Fig. 11).

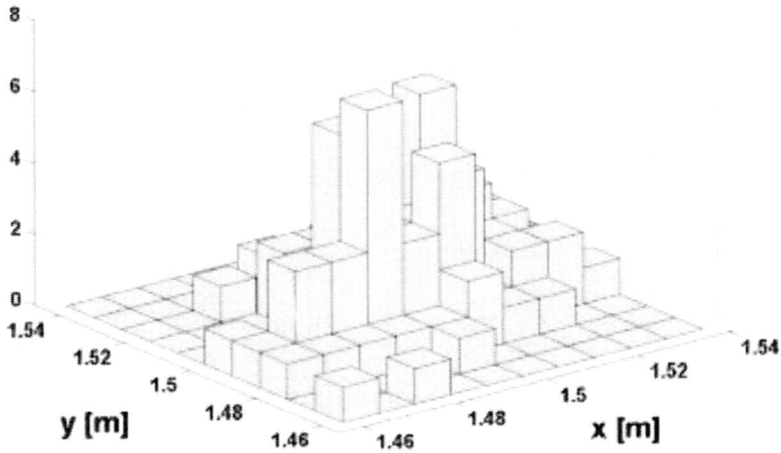

Figure. 10. Distribution of the estimated RFID position. N = 100, SNR = 10 dB.

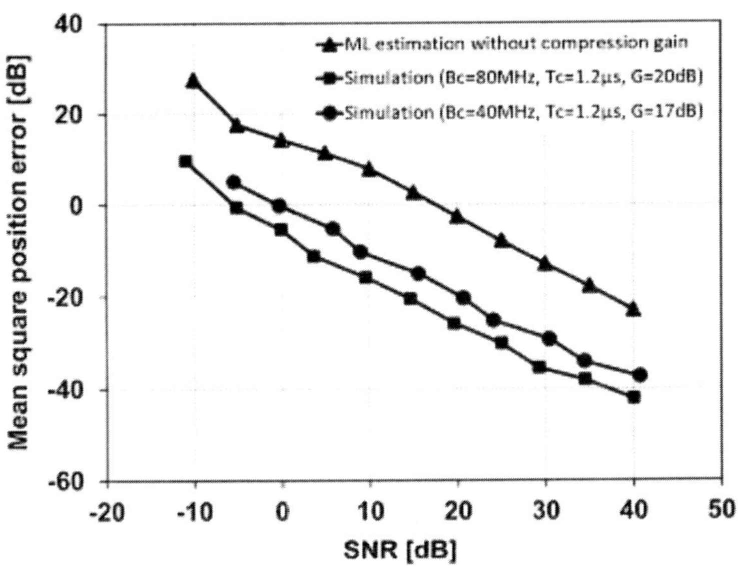

Figure. 11. MSPE comparison of position estimation methods based on signals with and without correlation gain (N = 100).

The proposed ranging method using linear chirp signals with a correlation gain $G = 20 \times \log(\sqrt{T_c B_c})$ is compared with classical ToA principles using signals without compression gain. Both methods are compared under AWGN and LOS conditions. The classical maximum likelihood (ML) estimation method is based on algorithms described in [27]. In Fig. 11 the mean square position error (MSPE) in dependency on the transmission channel SNR is depicted. The MSPE is defined as $E\left\{(\hat{x}-x)^2+(\hat{y}-y)^2\right\}$. The difference in MSPE found between signals with and without correlation gain is about the signals correlation gain G. This is shown for linear chirp signals with a bandwidth B_c = 40 MHz and 80 MHz, respectively. Both chirp signals have a duration of T_c = 1.2 μs. The calculated correlation gain G is ~17 dB for a chirp bandwidth of 40 MHz and ~20 dB for a chirp bandwidth of 80 MHz.

For scenarios of position estimation with more than one RFID tag, the transmitted chirp signals must have a unique code for differentiation. This can be accomplished from the RFID tag by transmitting in quick succession a fixed number of chirp signals in different time slots within a transmission frame. This principle is called pulse position modulation (PPM) [28] and is able to assign the received chirp pulses to a dedicated RFID tag. Probable multi-tag-collision scenarios of RFID devices can be solved by random multiple access approaches implemented in the MAC layer.

4. CONCLUSIONS

An accurate and power-saving principle for wireless position estimation of RFID devices was proposed. For correlative signal processing with a high immunity against interference in the transmission channel, linear chirp signals for locating RFID devices are used. The signal generation is carried out on the RFID tag by triggering a SAW-based chirp filter with short and broadband pulses. For pulse generation, a low power consumption trigger circuit consisting of a transistor, which is operated in the avalanche mode, was used. Due to the low power consumption of the positioning principle, the RFID device can be supplied by a miniature lithium coin battery or by energy-harvesting methods. Several simulations of the chirp-based position estimation method under AWGN conditions underlie the accuracy and robustness of the proposed principle and show a reduction of the mean square position error in range of the chirp compression gain.

ACKNOWLEDGMENTS

The authors thank the government of Lower Austria for funding this work and the European Commission (EFRE) for supporting our activities under project ID no. WST3-T-91/026-2012.

Some of the content of the presented work was partly pre-published as work in progress at IEEE Eurosensors 2014 [16].

REFERENCES

1. Hitachi, Hitachi μ-Chip–The World's smallest RFID IC Hitachi Mu-Solutions. Available: http://www.hitachi.co.jp (2004).

2. P.V. Nikitin, K.V.S. Rao **Performance Limitations of Passive UHF RFID Systems** IEEE, Piscataway, N.J (2006)

3. ISO14443. Available: http://www.openpcd.org/ISO14443

4. M. Bouet, A.L. dos Santos, RFID tags: Positioning principles and localization techniques. Wireless Days, 2008. WD 08. 1st IFIP (2008) 1–5.

5. C. Wang, H. Wu, N.-F. RFID-based 3-D positioning schemes. IEEE INFOCOM -26th IEEE International Conference on Computer Communications (2007) 1235–1243.

6. N. Bulusu, J. Heidemann, D. Estrin **GPS-less low-cost outdoor localization for very small devices** IEEE Pers. Commun., 7 (5) (2000), pp. 28–34

7. P. Youngsu, W.L. Je, K. SangWoo, Improving position estimation on RFID tag floor localization using RFID reader transmission power control. Robotics and Biomimetics, 2008. ROBIO 2008. IEEE International Conference (2009) 1716–1721.

8. M. Bouet, G. Pujolle, L-VIRT: A 3-D range-free localization method for RFID tags based on virtual landmarks and mobile readers. Consumer Communications and Networking Conference, CCNC 2009. 6th IEEE (2009) 1–5.

9. F. Manzoor, H. Yi, K. Menzel, Passive RFID-based indoor positioning system: an algorithmic approach. RFID-Technology and Applications (RFID-TA), IEEE International Conference (2010) 112–117.

10. H. Koyuncu, S.H. Yang **A survey of indoor positioning and object locating systems** IJCSNS Int. J. Comput. Sci. Network Secur., 10 (5) (2010), pp. 121–128

11. H. Liu, H. Darabi, P. Banerjee, J. Liu **Survey of wireless indoor positioning techniques and systems** IEEE Trans. Syst. Man Cybern. Part C (Appl. Rev.), 37 (6) (2007), pp. 1067–1080

12. M. Banitalebi-Dehkordi, J. Abouei, K.N. Plataniotis **Compressive-sampling-based positioning in wireless body area networks** Biomed. Health Inf. IEEE J., 18 (1) (2014), pp. 335–344

13. H. Yanchuan, P.V. Brennan, A. Seeds **Active RFID location system based on time-difference measurement using a linear FM chirp tag signal** Personal, Indoor and Mobile Radio Communications, 2008. PIMRC 2008. IEEE 19th International Symposium (2008), pp. 1–5

14. T.F. Bechteler, H. Yenigun **2-D localization and identification based on SAW ID-tags at 2.5 GHz** IEEE Trans. Microwave Theory Tech., 51 (5) (2003), pp. 1584–1590

15. C. Hausleitner, R. Steindl, A. Pohl, M. Brandl, F. Seifert **State of the art radio interrogation system for passive surface acoustic wave sensors** EUROCOMM 2000. Information Systems for Enhanced Public Safety and Security. IEEE/AFCEA (2000), pp. 158–161

16. M. Brandl, K. Kellner **Position estimation of RFID based sensors using passive SAW compressive receivers** Procedia Eng., 87 (2014), pp. 1263–1265

17. C.E. Cook, M. Bernfeld **JT Radar Signals** Artech House (1993)

18. M. Kowatsch, H.R. Stocker, J. Lafferl **Time sidelobe performance of low time-bandwidth product linear FM pulse compression systems** Sonics Ultrason. IEEE Trans., 28 (4) (1981), pp. 285–286

19. W. Hirt, S. Pasupathy **Continuous phase chirp (CPC) signals for binary data communication—Part I: coherent detection and Part II: noncoherent detection** IEEE Trans. Commun., 29 (6) (1981), pp. 836–858

20. D.P. Morgan **Surface-Wave Devices for Signal Processing** Elsevier, Amsterdam New York (1985)

21. M. Brandl, J. Grabner, K. Kellner, F. Seifert, J. Nicolics, S. Grabner, G. Grabner **A low-cost wireless sensor system and its application in dental retainers** Sens. J. IEEE, 9 (3) (2009), pp. 255–262

22. B.E. Conway **Electrochemical Supercapacitors Scientific Fundamentals and Technological Applications** Plenum Press, New York (1999)

23. S. Sudevalayam, P. Kulkarni **Energy harvesting sensor nodes: survey and implications** Commun. Surv. Tutorials IEEE, 13 (3) (2011), pp. 443–461

24. R.J. Baker **High voltage pulse generation using current mode second breakdown in a bipolar junction transistor** Rev. Sci. Instrum., 62 (4) (1991), pp. 1031–1036

25. Lithium coin battery CR1216 data sheet. Available: http://data.energizer.com/PDFs/cr1216.pdf.

26. A.B. Carlson **Communication Systems An Introduction to Signals and Noise in Electrical Communication** McGraw-Hill, New York (1986)

27. S.A. Zekavat, R.M. Buehrer **Handbook of Position Location: Theory, Practice and Advances** John Wiley & Sons (2012)

28. Springer, W. Gugler, M. Huemer, L. Reindl, C. Ruppel, R. Weigel **Spread spectrum communications using chirp signals** EUROCOMM 2000. Information Systems for Enhanced Public Safety and Security. IEEE/AFCEA (2000), pp. 166–170

7

DIAGNOSIS OF CARBONATION INDUCED CORROSION INITIATION AND PROGRESSION IN REINFORCED CONCRETE STRUCTURES USING PIEZO-IMPEDANCE TRANSDUCERS

V. Talakokula[a, 1], S. Bhalla[b, ,], R.J. Ball[c], C.R. Bowen[d], G.L. Pesce[c], R. Kurchania[e], B. Bhattacharjee[b], A. Gupta[b], K. Paine[c]

a Department of Civil Engineering, ABES Engineering College, Ghaziabad, 201009, India

b Department of Civil Engineering, Indian Institute of Technology Delhi, Hauz Khas, New Delhi 110016, India

c Department of Architecture and Civil Engineering, University of Bath, Bath BA2 7AY, United Kingdom

d Department of Mechanical Engineering, University of Bath, Bath BA2 7AY, United Kingdom

e Department of Physics, Maulana Azad National Institute of Technology (MANIT), Bhopal 462051, India

ABSTRACT

In addition to chloride induced corrosion, the other commonly occurring type of rebar corrosion in reinforced concrete structures is that induced by the ingress of atmospheric carbon dioxide into concrete, commonly referred to as 'carbonation induced corrosion'. This paper presents a new approach for detecting the onset and quantifying the level of carbonation induced rebar corrosion. The approach is based on the changes in the mechanical impedance parameters acquired using the electro-mechanical coupling of a piezoelectric lead zirconate titanate (PZT) ceramic patch bonded to the surface of the rebar. The approach is non-destructive and is demonstrated though accelerated tests on reinforced concrete specimens subjected to controlled carbon dioxide exposure for a period spanning over 230 days. The equivalent stiffness parameter, extracted from the frequency response of the admittance signatures of the PZT patch, is found to increase with penetration of carbon dioxide inside the surface and the consequent carbonation, an observation that is correlated with phenolphthalein staining. After the onset of rebar corrosion, the equivalent stiffness parameter exhibited a reduction in magnitude over time, providing a clear indication of the occurrence of corrosion and the results are correlated with scanning electron microscope images and Raman spectroscopy measurements. The average rate of corrosion is determined using the equivalent mass parameter. The use of PZT ceramic transducers, therefore, provides an alternate and effective technique for diagnosis of carbonation induced rebar corrosion initiation and progression in reinforced concrete structures non-destructively.

KEYWORDS

SEM; Chloride; Corrosion; Carbonation; Reinforced concrete; Electro-mechanical impedance (EMI) technique

1. INTRODUCTION

Rebar corrosion is widely regarded as the most common cause of deterioration and premature failure of reinforced concrete (RC) structures [1]. In general, concrete is expected to provide a high degree of corrosion protection to the embedded steel rebars owing to its highly alkaline nature. A well compacted and cured concrete, with low water-cement ratio, has a permeability that is low enough to inhibit the penetration of corrosion inducing agents from the environment. However, such ideal conditions are seldom met in practice, leading to corrosion of the reinforcement, which is the major tensile load resisting element in RC structures.

Rebar corrosion is of two types, namely (i) chloride induced and (ii) carbonation induced. In the first type, chloride ions, either present in the raw materials or introduced from outside, due to insufficient cover or high permeability, lead to the breakdown of the alkaline passive film around the rebar [2] and [3]. The second type of rebar corrosion, namely carbonation induced corrosion, results from the ingress of atmospheric carbon dioxide

(CO_2) into the concrete. The penetrated CO_2 dissolves in the pore solution to form carbonic acid, which reacts with calcium hydroxide and unhydrated silicates in the cement paste, forming mainly calcium carbonate [4]. Although calcium carbonate fills up the concrete pores leading to a somewhat lower permeability, the associated drop in the pore fluid alkalinity from a pH value of over 12 to as low as 8 disrupts the passivating layer surrounding the steel reinforcement, leading to corrosion [5]. Irrespective of whether the primary mechanism is chloride or carbonation induced, rebar corrosion is generally accompanied by the loss of cross section and the accumulation of corrosion products, which occupy much larger volume than the original steel. Hence, this process generates tensile stresses, which leads to cracking and spalling of the concrete [3]. Rebar corrosion may not necessarily imply structural collapse but it causes significant loss of structural serviceability.

A wide range of electro-chemical/non electro-chemical techniques and sensors based techniques have been reported in the literature for the purpose of diagnosing and quantification of the rebar corrosion. Table 1 summarises some of the techniques along with their advantages and limitations [6]. Recently Miller et al. [7] described an inductive sensor to image steel reinforcement bars in concrete and visualise surface corrosion. The present research uses the electro-mechanical impedance (EMI) technique which has recently emerged as a promising tool for structural health monitoring (SHM) and non-destructive evaluation (NDE) for a wide variety of materials and structures. This technique works by comparing the sensor response in a healthy (undamaged) structure to the response measured at a specific time point of interest. Differences are then associated with chemical change, such as corrosion, or a physical change such as cracking.

Talakokula et al. [6] recently proposed an EMI based approach for detecting and measuring the level of chloride induced rebar corrosion, validated through accelerated corrosion tests on embedded rebars. This paper focuses on extending the approach to carbonation induced corrosion, which operates with a somewhat different mechanism and reports on a study involving accelerated carbonation tests carried out on cylindrical specimens, spanning over 230 days of exposure. The corrosion assessment is based on the mechanical impedance parameters determined from EMI signatures of a piezoelectric lead zirconate titanate (PZT) ceramic patch bonded to the surface of the rebar. The PZT based equivalent stiffness parameter (ESP) and equivalent mass parameter (EMP) has been primarily employed for corrosion assessment. The results are correlated with phenolphthalein stain test and high resolution microscopic images. The following sections of the paper cover the physical principles of the EMI technique followed by experimental results and analysis.

Table 1. Comparison of various corrosion detection techniques [6].

Detection technique	Advantage	Disadvantage
Half-cell potential measurements	• Common procedure for routine inspection of RC structures • Simple and cost-effective	• Provides only information on probability of corrosion, cannot indicate corrosion rates • Values can be misleading • Only a first approach to corrosion detection and must be complimented by other techniques for corrosion rates • Requires the knowledge of potential in non-corroded areas
Polarization resistance technique	• Provides a measurement of the corrosion rate in terms of current density • An instantaneous measuring system	• In most cases, e.g., when the cover concrete is too thick, the confinement of the polarization area may not be achieved.
Gravimetric mass loss measurements	• Most effective method for calculating corrosion rates	• Destructive method • Requires the knowledge of the mass of rebar in healthy state
EMI technique	• The assessment is non-destructive in nature • Detection is based on the structural response • Equivalent structural parameters can be extracted thus circumventing the determination of actual parameters • Reduces number of sensors by allowing self-sensing actuation • Sensitive to damage induced due to carbonation related corrosion	• Maintaining sensor health and bonding is important • Requires knowledge of a healthy structure

2. FUNDAMENTALS OF THE ELETRO-MECHANICAL IMPEDANCE (EMI) TECHNIQUE

Piezoelectric materials are categorized as 'smart' materials owing to their 'stimulus-response' behaviour, exhibited with reference to mechanical and electrical domains. When subjected to mechanical stresses, a PZT patch develops surface charges, a phenomenon called the 'direct piezoelectric effect'. Similarly, when subjected to an electric field, it undergoes mechanical strain, often referred to as the 'converse piezoelectric effect'. The direct and the converse effects are mathematically represented by Eqs. (1) and (2) respectively [8]:

$$D_3 = \overline{\varepsilon_{33}^T} E_3 + d_{31} T_1 \quad (1)$$

$$S_1 = \frac{T_1}{\overline{Y^E}} + d_{31} E_3 \quad (2)$$

where S_1 is the strain along axis '1', D_3 the electric displacement over the surface of the PZT patch, d_{31} the piezoelectric strain coefficient and T_1 the axial stress in the patch along axis '1'. It is assumed that the axis '3' points along the thickness of the PZT patch and the axes '1' and '2' lie in the plane of the patch as shown in Fig. 1(a) and (b) for a typical PZT patch. $\overline{Y^E} = Y^E(1 + \eta j)$ represents the Young's modulus of the patch at constant electric field and $\overline{\varepsilon_{33}^T} = \varepsilon_{33}^T(1 - \delta j)$ is its electric permittivity at constant stress, with η and δ denoting the mechanical loss factor and the dielectric loss factor respectively.

The EMI technique, which is the tool used in this paper for diagnosing the corrosion initiation and progression, exploits the direct and the converse effects simultaneously and has been demonstrated to be capable for SHM of a wide variety of engineering structures, ranging from laboratory sized components to large prototype structures [9],[10], [11] and [12]. Several new cost-effective variants of the technique have also been proposed [13], [14] and [15]. The EMI technique is essentially vibration based in nature, however operating in a somewhat high frequency range of 30–400 kHz. Such high frequencies of vibrations are achieved by electrically exciting the PZT patch (in turn bonded to the surface of the structure to be monitored), by means of an impedance analyzer/inductance capacitance and resistance (LCR) meter. Under such external field excitation, the bonded patch induces deformations in the host structure (converse effect), whose response is transferred back to the patch (direct effect) in the form of an admittance signature, consisting of the conductance '*G*' (real part) and the susceptance '*B*' (imaginary part). Frequencies in excess of 500 kHz are not favourable due to a strong possibility of the sensing region becoming extremely localized relative to the PZT patch[7] and [16]. It is of interest to note that in this context, the conductance/susceptance relate to the piezoceramic sensor and its electro-mechanical response due to the mechanical properties of its surrounding environment; it is therefore not possible to specifically relate the sensor conductance to that of the galvanic corrosion process on the surface of the rebar. Any damage to the host structure affects the dual process (direct and converse effect) and leads to a change in the admittance spectrum, that is, the plot of conductance/susceptance as a function of frequency. This in turn provides the necessary Indication for detection of damage. Liang et al. [17] proposed the impedance based analytical formulations to model PZT-structure electro-mechanical interaction in 1D structures that are relevant to the EMI technique. Bhalla and Soh [18] extended Liang's impedance formulations to 2D structures, (Fig. 1c), by introducing the concept of 'effective impedance'. They had taken into account the signatures of the PZT patch prior to integration into the structure, often termed signatures under 'free-free'condition. Use of these signatures in the governing 2D wave equation, allowed derivation of the following expression for the complex electro-mechanical admittance $\overline{Y}$ as

$$\overline{Y} = G + Bj = 4\omega j\frac{l^2}{h}\left[\overline{\varepsilon_{33}^T} - \frac{2d_{31}^2\overline{Y^E}}{(1-\nu)} + \frac{2d_{31}^2\overline{Y^E}}{(1-\nu)}\left(\frac{Z_{a,eff}}{Z_{s,eff}+Z_{a,eff}}\right)\overline{T}\right] \quad (3)$$

where, l is the half-length of the patch, ν the Poisson's ratio of the patch and ω the angular frequency. The term $\overline{T}$ is the complex tangent ratio, ideally equal to tan $(kl)/kl$, with $k = \omega\sqrt{\rho(1-\nu^2)/\overline{Y^E}}$ being the wave number. Correction factors C_1 and C_2(determined from the signatures of the PZT patch in the 'free-free' condition) were introduced by Bhalla and Soh [18] to realistically represent the behaviour of an actual PZT patch, introducing the modifying expression for $\overline{T}$ as

$$\overline{T} = \frac{1}{2}\left(\frac{\tan(C_1 kl)}{C_1 kl} + \frac{\tan(C_2 kl)}{C_2 kl}\right) \quad (4)$$

Further, the modified mechanical impedance of the PZT patch was derived as

$$Z_{a,eff} = \frac{2h\overline{Y^E}}{j\omega(1-\nu)\overline{T}} \quad (5)$$

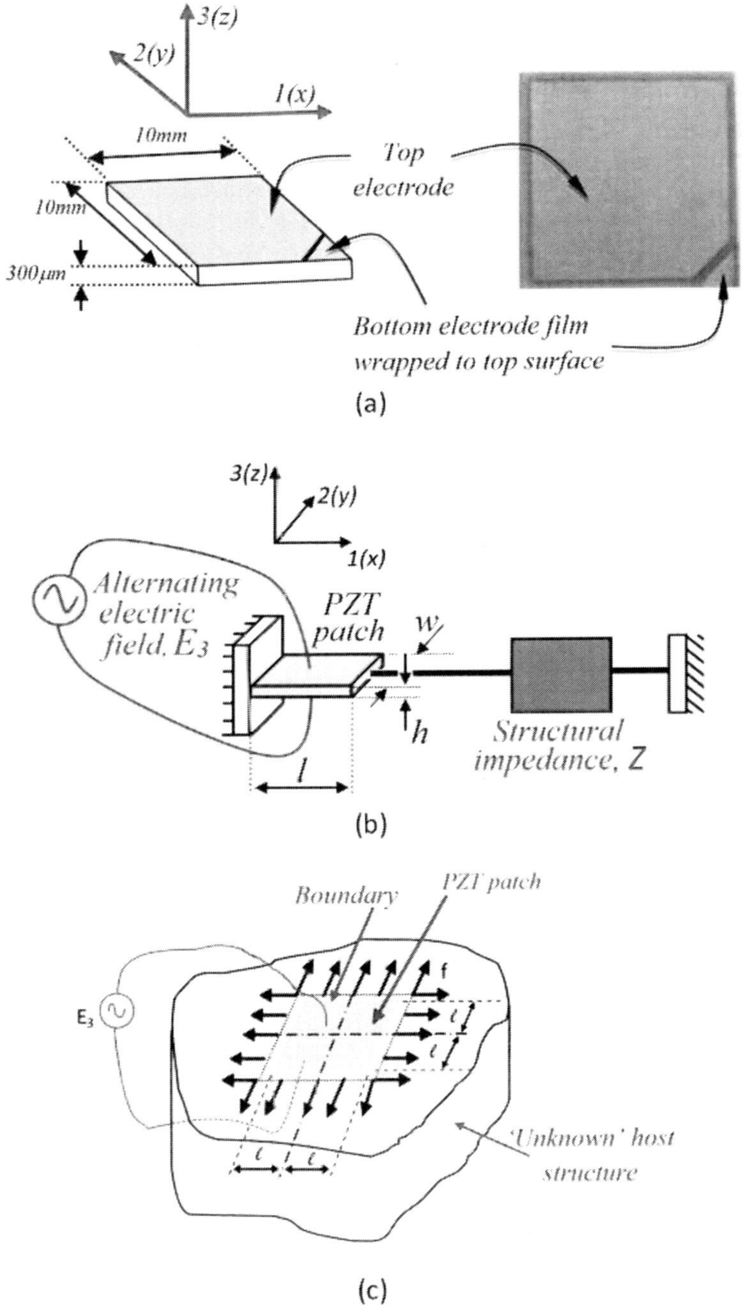

Figure. 1. PZT patch and its interaction with host structure. (a) Geometric details of a typical PZT patch (b) Modelling scheme of PZT-structure interaction 1D (c) A PZT patch bonded to the host structure 2D.

It can be observed from Eq. (3) that any damage to the host structure (i.e. any change in mechanical impedance '$Z_{s,eff}$ ') will induce a deviation in $\overline{Y}$, thereby providing an indication of the damage. Using the computational procedure outlined by Bhalla and Soh[19], Eq. (3) can be utilized to obtain the real and the imaginary components (x and y respectively) of the structural impedance $Z_{s,eff}=x+yj$ at a particular angular frequencyω from the measured G and B values. Thus, bonding a PZT patch to a structure and measuring its admittance signature enables the determination of the drive point mechanical impedance of the structure, $Z_{s,eff} = x + yj$, from the measured data alone, without demanding any *a priori* information governing the phenomenological nature of the structure. Depending upon the variation of 'x' and 'y' with frequency and the associated values, the inherent elements constituting the host structure can be identified[20]. Soh and Bhalla [11] and Bhalla et al. [12] found in their experiments that the PZT patches identified the host structure as a *Kelvin-Voigt* system, consisting of a parallel combination of spring and the damper elements. The stiffness element, termed as the ESP, has been found to be related to the actual stiffness of the host structure. Monitoring the ESP at different times provides an accurate condition of the residual stiffness of the structure and can be employed for SHM.

The next sections describe the identification of structural system based on the values of 'x' and 'y' and utilization of the ESP for monitoring and quantifying the severity of the structural damage induced by carbonation on rebars embedded in concrete.

3. ACCELERATED CARBONATION TESTS ON RC SPECIMENS: EXPERIMENTAL PROCEDURE

Four RC cylinders of grade M30 [21] were cast with diameters of 25, 35, 45 and 60 mm respectively and length 100 mm, as described in Table 2. A single piece of 16 mm diameter high yield deformed (HYD) steel rebar, grade Fe415 [22], was cast at the centre of each cylinder. The rebar's length was 130 mm, allowing 30 mm to project out at one end of the cylinder, as shown in Fig. 2. The four different diameters provided concrete cover depths of 4.5, 9.5, 14.5 and 22 mm respectively. Specimen number 1, with a cover depth 22 mm, was employed as a reference for the purpose of monitoring carbonation penetration depth through phenolphthalein staining. No PZT patch was attached to the reference sample as this was to be used for destructive testing such as for phenolphthalein staining to determine the carbonation depth and imaging analysis. On the other samples, a PZT patch measuring 10 × 10 × 0.3 mm (Fig. 1a), of ferroelectric grade PIC 151 [23] was bonded to the rebar at the mid length, where a small flat surface was achieved through machining to allow for the bonding of the patch, as shown in Fig. 2. A thin layer of epoxy was applied on the machined rebar surface first and the PZT patch was placed on it; the whole set-up was then kept undisturbed at room temperature for 24 h to enable full curing of the adhesive, after which wires were soldered to the PZT patch's electrodes. The PZT patch was then covered with an additional thin layer of epoxy adhesive as a protective measure. The concrete cylindrical specimens were cast using ordinary Portland cement (OPC) 53 Grade, fine aggregate of Zone 1 grading [24]and crushed coarse aggregate of nominal size 10 mm. Table 3 presents the details of the composition of the concrete mix. The moulds were filled with concrete in three layers along with the rebar (with the PZT patch bonded on it), taking particular care in pouring the concrete to avoid damage to the sensor. The concrete was compacted carefully using a table vibrator. For the 25 mm diameter specimen, it was necessary to remove some of the larger pieces of the aggregate from the mortar at the time of casting to accommodate the thin concrete cover of 4.5 mm. The specimens were demoulded after 24 h of casting. The wires of the PZT patch were connected to an LCR meter (Agilent E4980 [25]). In this manner, the electro mechanical admittance signature, consisting of the real part (G) and the imaginary part (B) were acquired in the frequency range of 50–400 kHz. A frequency interval of 100 Hz was used for each impedance measurement.

Table 2. Specimen identification number, dimensions and cover to reinforcement.

Specimen	Specimen size (mm)	Cover to reinforcement (mm)
1	60 × 100(Reference specimen)	22
2	25 × 100	4.5
3	35 × 100	9.5
4	45 × 100	14.5

Compared to chloride-induced corrosion, carbonation is a slower process. In high-quality concrete, it has been estimated that carbonation will proceed at a rate of not more than 1.0 mm per year [26]. The carbonation rate is significantly increased in concrete with a high water-cement ratio (more than specified by IS 10262, 2009 [27]), low cement content, short curing period, lower strength, and highly permeable or porous paste. In addition, the carbonation rate is highly dependent on the relative humidity of the concrete. The highest rates of carbonation occur when the relative humidity ranges between 50% and 75%. Carbonation-induced corrosion typically occurs on areas of building facades that are exposed to rainfall, shaded from sunlight, and have smaller concrete cover over the reinforcing steel [27]. Since the carbonation process is too slow under natural conditions, accelerated carbonation is preferred in the laboratory investigations since the 1960s [28]. In the present study, the accelerated corrosion experiments were initiated after a period of 28 days from the day of casting, assuming that any further concrete strength gain on account of concrete curing is negligible.

After the baseline signatures were acquired, the specimens were placed in a carbonation chamber. The CO_2 concentration in the chamber was maintained at 5% under a relative humidity of 75 ± 5% and temperature of 25 °C. Studies of the chemical changes and phase analysis of Portland cement pastes under conditions of accelerated carbonation in excess of 3% (CO_2 concentration) have been conducted by Castellote and colleagues[29]. They demonstrated that, when CO_2 concentrations were in excess of 3%, microstructural changes were observed compared to ambient conditions. This should be considered when translating results from this study to systems carbonating under ambient conditions. Wax was applied to the circular ends of each cylindrical samples to allow CO_2 ingress radially from the sides only. During the accelerated carbonation exposure, the admittance signatures were acquired at frequent intervals throughout the experimental exposure period of 230 days. In order to measure the carbonation penetration depth, a 20 mm thick slice was cut from the bottom of the reference specimen, as shown in Fig. 3 and the fracture surface was sprayed with a saturated solution of phenolphthalein indicator in ethanol. The slicing was done on the same face in dry conditions. However, the ends were covered immediately with wax after cutting, so that any CO_2 ingress or diffusion of moisture takes place only radially. Maximum and minimum depth of carbonation was measured and numerically averaged. During the exposure period, the carbonation penetration depth was periodically measured using the phenolphthalein staining by cutting fresh slices from the reference specimen.

At high pH (uncarbonated), the indicator has a bright pink colour. However, below a pH of 8.6 (carbonated condition), it becomes colourless. The interface between the pink and the colourless regions indicates the carbonation front. Maximum and minimum penetration depths were measured and numerically averaged. Depths of 2.5, 12 and 15 mm were observed at 15, 30 and 40 days respectively. The low values of carbonation depth observed during the initial days may be attributed to the initial high moisture content in the specimens. Based on the concrete covers described previously (see Table 2), it is estimated that the carbonation front reached the rebar after 18, 26 and 39 days for specimens 2, 3 and 4 respectively. The numbers of days has been calculated assuming that carbonation follows a square root time law. The next section correlates these observations (which are made through destructive tests) with ESP obtained non-destructively via the EMI technique.

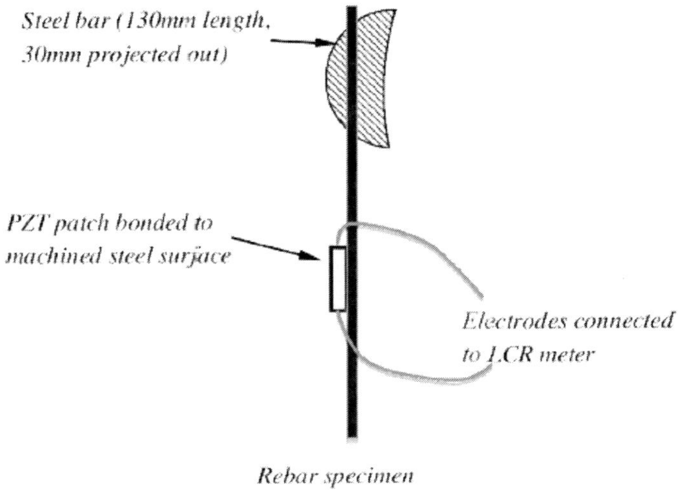

(a)

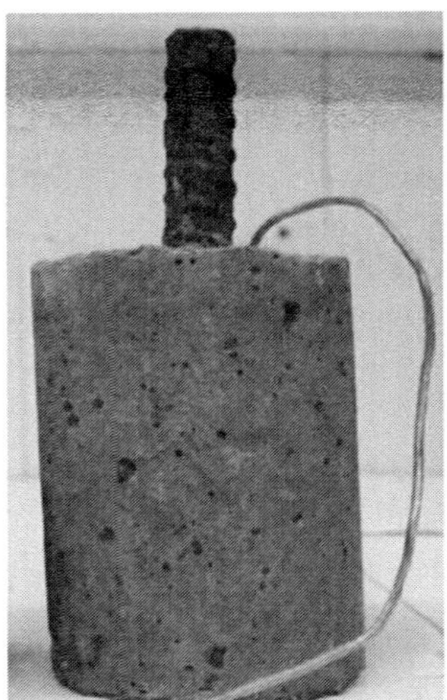

(b)

Figure. 2. Rebar specimen. (a) Typical specimen geometry (b) typical sample.

Table 3. Concrete mix design specification.

Materials	Quantity
Water cement ratio	0.4
OPC (kg/m^3)	500
Fine aggregate (kg/m^3)	570.5
Coarse aggregate (kg/m^3) 10 mm	1062.2
Slump (mm)	100

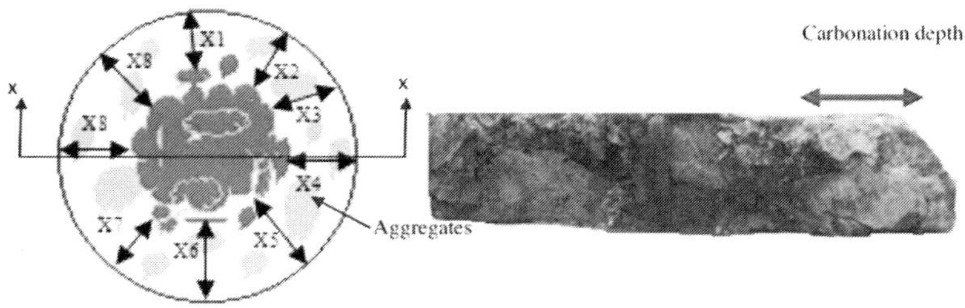

(a) (b)

Figure. 3. Phenolphthalein indicator solution applied to a fresh fracture slice of concrete. (a) Schematic diagram showing reading locations. (b) Section xx.

The baseline signatures of the specimens were first acquired as reference prior to placing in the carbonation chamber. Fig. 4 shows the baseline conductance spectrum of the PZT patches of three specimens, acquired after 28 days of curing, before being subjected to accelerated carbonation. Even though the initial baseline signatures of the specimens are different, when applying the EMI technique it is the variation of the signature of interest with the baseline signature that is important as it is used here to deduce changes in the specimen.

4. CHANGES IN IMPEDANCE BASED EQUIVALENT PARAMETERS DURING CARBONATION

Fig. 5 (a–c) show the conductance signatures of the bonded PZT patches for the three test specimens (Nos. 2–4) acquired during various times. It is worth noting that all the specimens were 28 days old before the initiation of the experiment and hence the majority of the hydration process was already completed. The figures show that the resonance peaks of the conductance signature have changed their magnitude and frequency for all the three specimens with increasing carbonation.

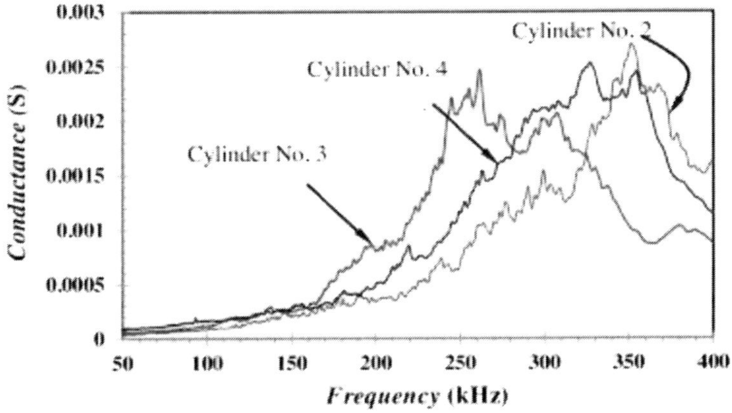

Figure. 4. Baseline conductance signatures of PZT patch embedded in RC cylindrical specimens.

To quantify damage, at first crudely, root mean square deviation (RMSD) metric was utilized, which is defined as [12]

$$RMSD = \sqrt{\frac{\sum_{1}^{N}(G_i - G_i^0)^2}{\sum_{1}^{N}(G_i^0)^2}} \times 100 \tag{6}$$

where G_i is the conductance of the PZT patch at any stage during the test and $G_i°$ is the baseline value (in pristine condition), i representing the frequency index (100–400 kHz). Fig. 6(a–c) shows the variation of the RMSD index of the three specimens against the exposure period. The RMSD values increased during the exposure period, which provides a useful indicator that the PZT sensors are sensitive enough to identify changes in the system. This is not always the case, as previous studies on other systems such as reinforced concrete structures and steel bolted joints [6] and [12]. RMSD sometimes exhibit a random response, hence under these conditions the RMSD approach is unreliable. Additionally RMSD is a non - parameter based statistical index to quantify the changes in the signature acquired by the PZT patch. Previous studies by the authors [6] and [12] have confirmed that it is suitable for preliminary analysis not very reliable for in depth analysis. It just provides the deviation of the raw signature with respect to the baseline, thereby not directly linking to any structure related parameter

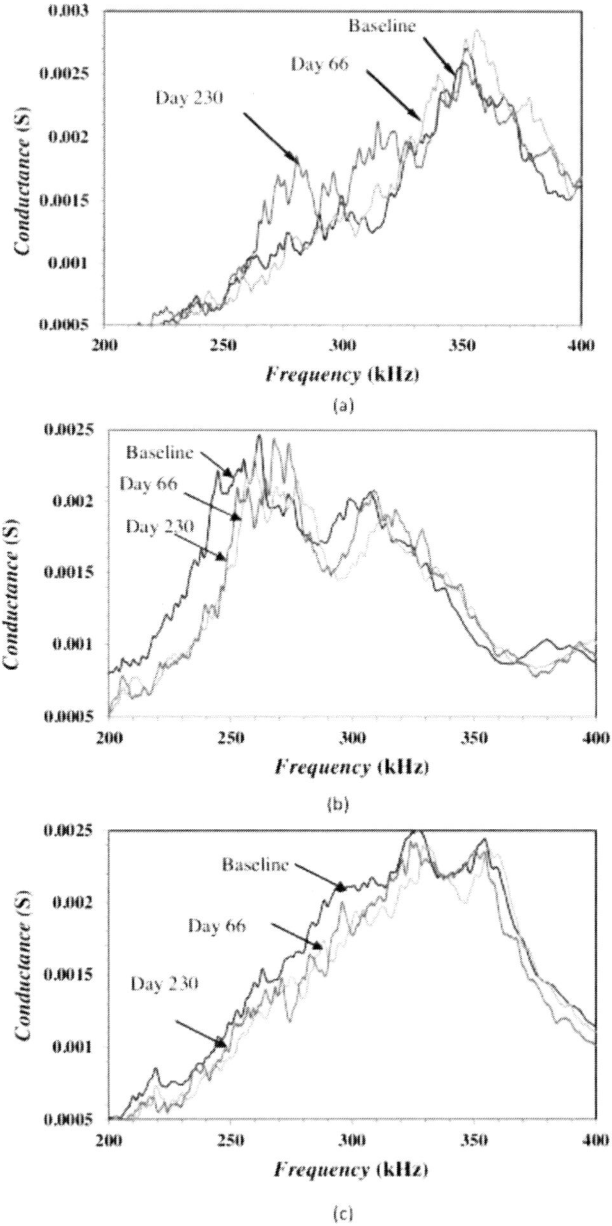

Figure. 5. Variation of conductance signatures during accelerated carbonation corrosion procress(a) Cylinder No. 2 (b) Cylinder No. 3 (c) Cylinder No. 4.

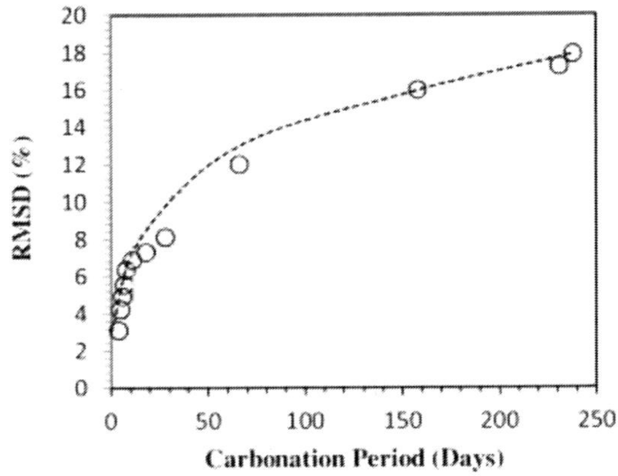

(a)

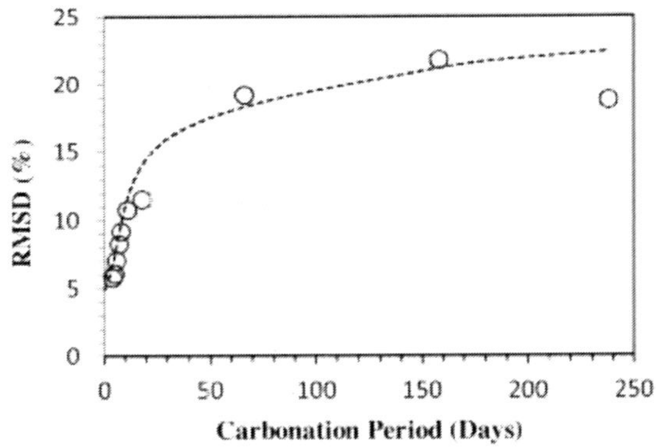

(b)

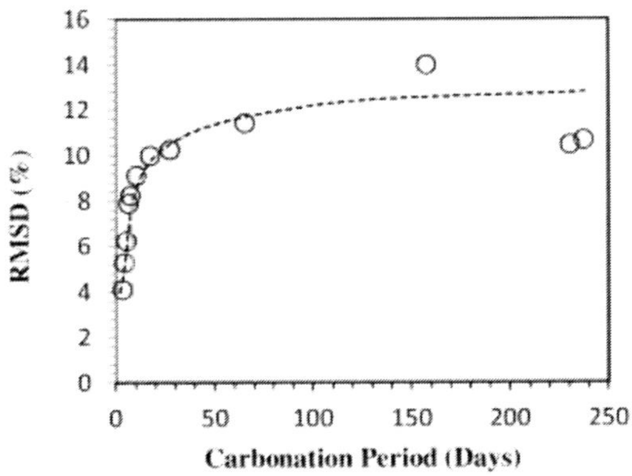

(c)

Figure. 6. Variation of Variation of RMSD index during accelerated carbonation corrosion procress. (a) Cylinder No. 2 (b) Cylinder No. 3 (c) Cylinder No. 4.

4.1. Equivalent stiffness parameters (ESP)

To gain deeper insight into the corrosion mechanism, equivalent structural parameters were extracted from the impedance spectrum using the computational procedure outlined by Bhalla and Soh [19]. The extracted mechanical impedance $Z_{s,eff}$ consists of the real and imaginary components *x* and *y* respectively. Observing the variation of *x* and*y* closely in the frequency range of 150–250 kHz revealed that the variation is similar to the variation of *x* and *y* of a parallel spring-damper-mass (k-c-m) combination [20] as shown in Fig.7(a). For this system,

$$x = c \text{ and } y = m\omega - \frac{k}{\omega} \tag{7}$$

where c is the equivalent damping parameter, k the equivalent stiffness parameter (ESP), m the equivalent mass parameter (EMP) and ω the angular frequency. The variation of '*x*' and'*y*' for both equivalent system (parallel spring-damper-mass) and that of the experimental is shown in Fig. 7(b and c) for a typical specimen (no.2), confirming the observation that the host structure has been identified as a parallel spring-damper-mass system.

For this system, the values of the ESP and the EMP of the host structure were determined using Eq. (7) (refer [6], for details). Fig. 8 displays the effect of the accelerated carbonation exposure on the identified ESP, which shows an initial increase in ESP for all the three specimens. This increase is attributed to calcium carbonate formed during carbonation, filling the pores and leading to an increase in stiffness. However, once the carbonation front reached the steel/concrete interface, a reduction in alkalinity and subsequent break down of the passive film is expected to trigger corrosion. The greater volume of the corrosion products would induce stresses in the adjacent concrete material, leading to the formation of micro-cracks and a decrease in the ESP values. Hence, the phase where the ESP value begins to decrease indicates the initiation of corrosion. From these plots, based on the trend of the ESP, this period can be projected to be lie somewhere between 50 and 90 days of exposure, the exact time differs from one specimen to other. It should also be emphasised that the ESP values are not directly related to the depth of carbonation. Depths of 2.5, 12 and 15 mm were observed at 15, 30 and 40 days respectively by means of destructive tests conducted on the reference specimen as described before. The carbonation process lowers the local pH around the rebar, allowing corrosion to initiate, which in turn influences the ESP as discussed before. When considered together with the cover depths reported in Table 2, a sequential pattern is noted in relation to the carbonation front, which reached the rebar after 18, 26 and 39 days for specimens 2, 3 and 4 respectively. It can be noted from Fig. 8that the peak of the ESP plot does indeed occur after these specific periods. The ESP based diagnosis clearly proves its leverage over the conventional statistical diagnostics such as RMSD as covered before.

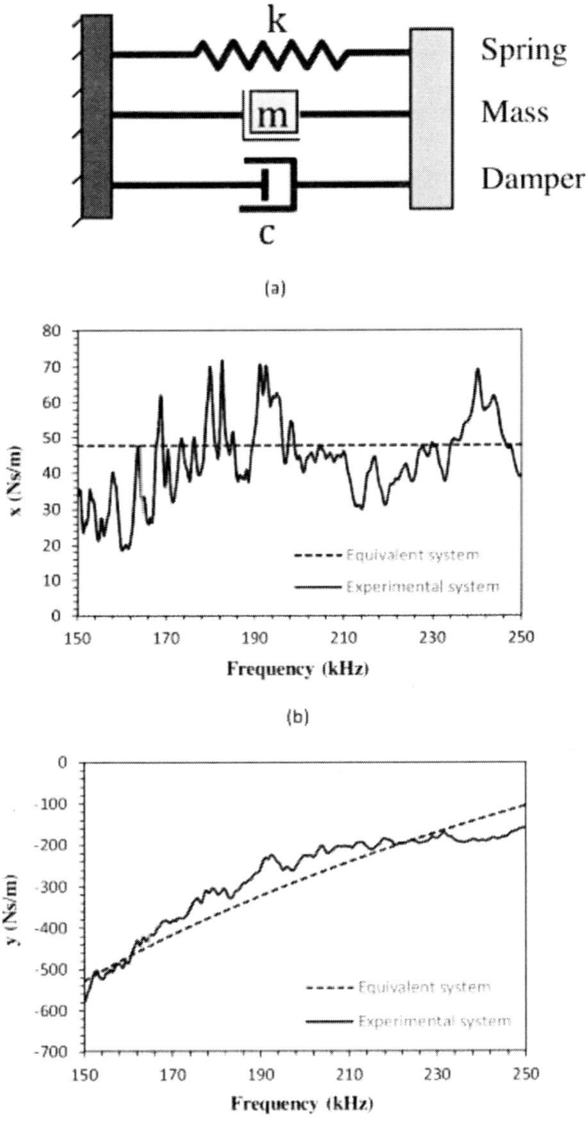

Figure. 7. Equivalent stiffness parameters. (a) Identified Parallel combination of spring-mass-damper system (b) Comparison of experimental plots of x with those of identified system for typical cylinder (No. 2) (c) Comparison of experimental plots of y with those of identified system for typical cylinder (No. 2).

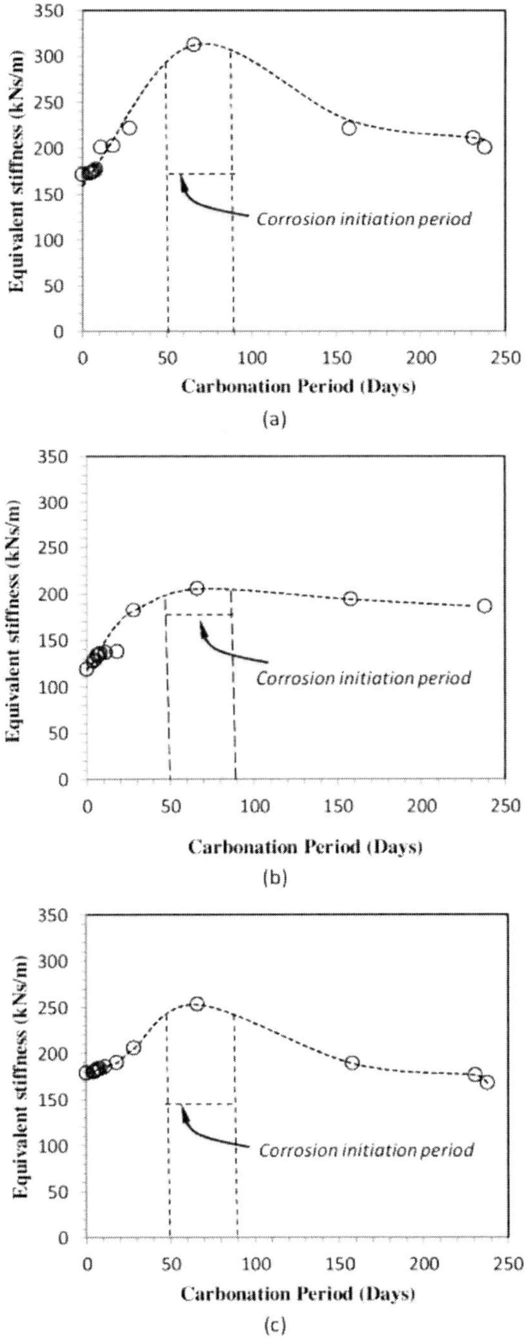

Figure. 8. Variation of equivalent stiffness during accelerated carbonation corrosion process. (a) Cylinder No. 2 (b) Cylinder No. 3 (c) Cylinder No. 4.

In addition, the non-dimensional stiffness loss *($\Delta k/k$)* was determined to quantify the level of corrosion damage on account of carbonation. Fig. 9(a–c) shows the variation of *($\Delta k/k$)*with the exposure period for the three test specimens. From the figures, it can be observed that for smaller diameter specimen (No. 2), the downward trend of the curve started when the Δk/k value reached 0.45. As the diameter of the specimens increased (Nos. 3 and 4), thereby increasing the cover to steel, it is noted that the peak $\Delta k/k$ values are lower at 0.4 and 0.3 respectively. Further, relative to the peak value of the ESP, the decrease in the $\Delta k/k$ values of specimens 2, 3 and 4 are found to be 0.3, 0.2 and 0.28 respectively. Based on the empirical equivalent stiffness model of Talakokula et al. [6], the corrosion process can be identified to be in the "propagation" phase for these values of $\Delta k/k$.

4.2. Equivalent mass parameter (EMP)

The initial masses of all the rebars were measured before embedding them in concrete and after splitting all the specimens, the final mass loss of the steel bars was determined by the gravimetric method after chemically cleaning the corrosion products. The EMP was also calculated directly using Eq. (7) after mathematical calculations (refer [6], for details) during the carbonation corrosion exposure, without any destructive measure.

Knowing the actual mass and the PZT identified EMP, a non-dimensional mass model is derived as

$$\left(\frac{\Delta m}{m}\right)_{actual} = \Lambda_m \left(\frac{\Delta m}{m}\right)_{PZT} \tag{8}$$

where Λ_m is a constant relating PZT identified EMP with actual mass loss determined as 0.54 [6]. This correlation will be useful in calculating the corrosion rates of the rebar non-destructively, because measuring the actual mass loss in real life structures is not possible as rebar is inside the concrete.

Knowing the mass loss, the corrosion rate is calculated as described in ASTM G1-03 [30]using the equation

$$\Delta_c = \frac{(k \times \Delta m)}{(a \times T \times D)} \tag{9}$$

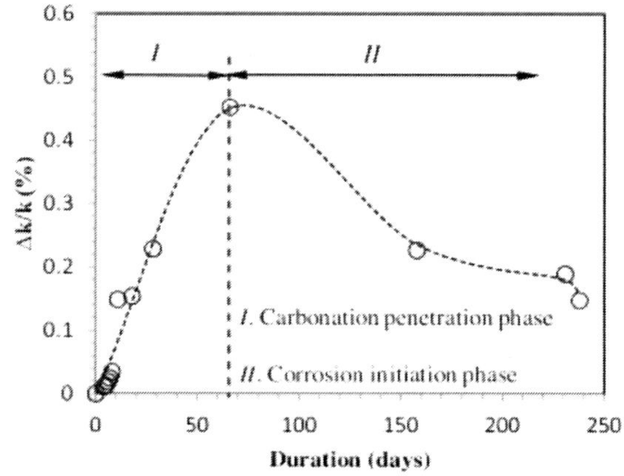

(a)

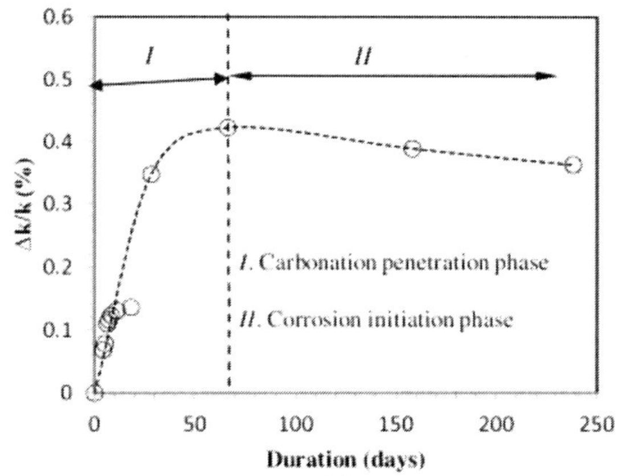

(b)

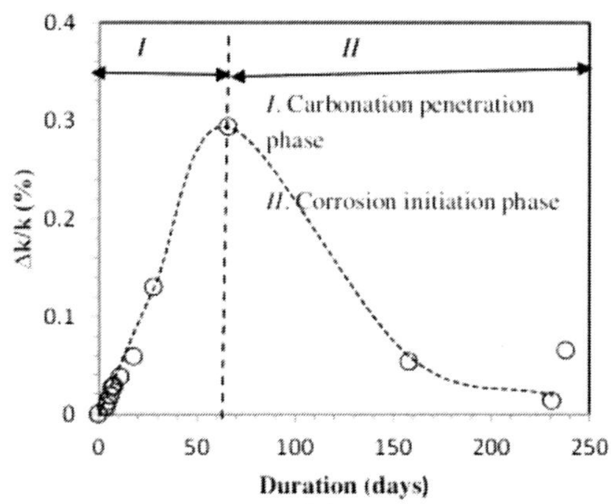

(c)

Figure. 9. Variation of Δk/k during accelerated corrosion procress. Cylinder No. 2 (b) Cylinder No. 3 (c) Cylinder No. 4.

where Δ_c is the corrosion rate (mm/year) K is a constant equal to 8.76 × 10_4, Δm is the mass loss in g, a the area in cm^2, T time of corrosion exposure in hours and D is the density of steel i.e.7.8 g/cc. The corrosion rates were calculated using both actual mass loss as well as the non-dimension mass model Eq. (8) and are listed in Table 4. Upon comparing both, it was found that using the model developed, the corrosion rates can be calculated with reasonable accuracy (average error of around 10% only). This is a very important observation, which establishes that the application of the proposed methodology enables circumventing the determination of actual mass loss, which in practice is infeasible.

5. CORRELATION WITH MICROSCOPIC IMAGE ANALYSIS AND RAMAN ANALYSIS

To verify the nature of changes highlighted by the ESP, conventionally available experimental techniques including optical microscopy, scanning electron microscopy (SEM) and Raman spectroscopy were applied to representative cross sections of the specimens. The sections were cut during the exposure period for performing these tests. In order to produce high quality polished sections and reduce the possibility of mechanical damage, the end of each cylinder was embedded in low-viscosity resin at atmospheric pressure prior to polishing. The sections were then cut perpendicularly to the cylinder longitudinal axis (parallel to the basal face) using a diamond impregnated cutting disc in dry conditions. The surface of each section was subsequently ground using progressively finer silicon carbide papers from 180 to 1200 grit size before polishing with 6, 3 and 1 μm diamond paste. A final polish was achieved using colloidal silica suspension. Optical inspection of the polished sections was carried out using a Leica DM ILM inverted metallurgical microscope equipped with an Infinity 1 digital camera (Leica Microsystems, Milton Keynes, United Kingdom). Images of the sections were acquired using Studio Capture Software version 4.0.1.

Fig. 10b shows an SEM image taken at 243 days (reference specimen) which shows calcium carbonate crystals within the pore structure which can be used to explain the increase in the ESP values on account of stiffening action (see Fig. 10a carbonation penetration phase pre-peak of ESP of specimen 2). It should be noted that calcium carbonate is a stable phase and within the time period between the peak equivalent stiffness at approximately 67 days and the end of the monitoring period when the image was taken at 243 days no significant changes would be expected. Fig. 10c shows cracking in specimen 2, which resulted in the decrease of the ESP values in post-peak phases. It is also noteworthy that such cracking would not be visible in the SEM image presented.

Table 4. Comparison of corrosion rates for specimens Nos. 2, 3 and 4 determined from gravimetric mass loss and PZT identified mass loss.

Specimen	Actual corrosion rates using gravimetric mass loss (mm/year)	Corrosion rates using PZT identified mass loss after incorporating Λ_m (mm/year)	Percentage error
2	0.059	0.067	13.5
3	0.048	0.052	8.3
4	0.012	0.013	8.3

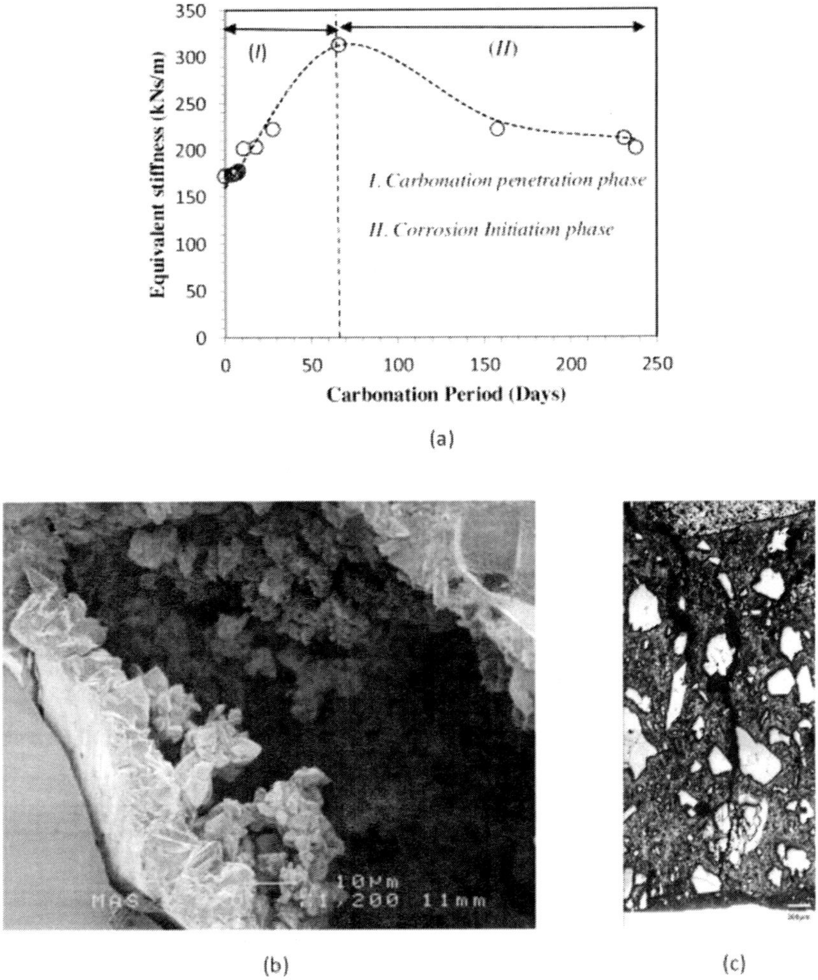

Figure. 10. Correlation of ESP values with SEM and optical images. (a) ESP values for carbon penetration phase (Region I) and corrosion initiation/cracking phase (Region II). (b) high magnification secondary electron image (scale bar 10 μm) corresponding to a typical internal crack microstructure at 243 days (c) low magnification optical image (scale bar 300 μm) of a polished section between the rebar (top) and sample edge (bottom) at 243 days.

Several cracks (black areas) are visible propagating from the rebar (upper part of the image) towards the external surface (lower part). Cracks often follow the interface between the aggregate and the cement matrix. This results in the decrease of ESP values. The relative ESP values have already indicated the corrosion to be in the propagation phase, which correlates well with the microscopic findings. Hence, the ESP provides a strong indication of the changes occurring during the carbonation process.

To verify the nature of the oxides formed on the rebar surface (which, through expansion led to the formation of the cracks identified during with the optical microscope), analysis was performed using Raman spectroscopy of the metal phase located at the cement/rebar interface. A Renishaw inVia Raman Spectrometer with a laser of wavelength 532 nm (green visible light) was used. The laser was set at 100% of the power; and scans were performed between 100 and 3200 cm^{-1} with only one accumulation and an exposure time of 10 s. A WiRE software version 3.4 developed by Renishaw plc, New Mills, Wotton-under-Edge, Gloucestershire, United Kingdom, was used to set the experiment and acquire the spectra.

Fig. 11 shows one of these spectra with the corresponding location along the interface highlighted by a red dot in the box (upper right corner). Peaks in the spectrum are located at 220, 278, 390, 486, 582, 804, 1058, 1280 cm^{-1}. Data reported in the scientific literature by Oh et al. [31] attributed the peaks at 220, 278 and 1280 cm^{-1} to hematite α-Fe_2O_3. It is already known, in fact, that depending on the crystallinity and grain size, the characteristic double at 294/225 cm^{-1} of this mineral in some cases shifts downward in wave number [31]. Peaks at 390, 486 and 582 cm^{-1} are attributed to goethite α-FeOOH[32]. Both these phases are common in the corrosion products associated with rebar corrosion.

6. CONCLUSIONS

This paper has presented a new approach for monitoring the progression of carbonation and the coupled rebar corrosion in RC structures based on the ESP and EMP parameters extracted from the admittance signatures of PZT patches surface-bonded to the rebars. This is the first ever study involving PZT sensors and utilizing the EMI technique to diagnose carbonation induced corrosion initiation and progression. The main objective was proof-of-concept demonstration of the ESP and EMP based rebar carbonation initiation, progression and quantification. A total of four specimens were tested, all of which showed consistent results, commensurate with the individual cover thickness. The ESP extracted from the admittance signatures of surface-bonded PZT patches is established as a reliable carbonation progression as well as rebar corrosion indicator. ESP correlates well with the microscopic image analysis. The average rate of corrosion has been determined using EMP and compared with the actual corrosion rates using gravimetric mass loss measurements, the error has been found to be around10% only. The most promising feature of the proposed approach is that the two main phases in the process, namely carbonation penetration through concrete cover and corrosion initiation are determined. The significance of the proposed approach is that it is non-destructive in nature, simple and fully autonomous. The empirical correlation derived from the PZT patches can be utilized in carbonation induced corrosion of rebars in real-life RC structures, where they are not exposed for direct visual check.

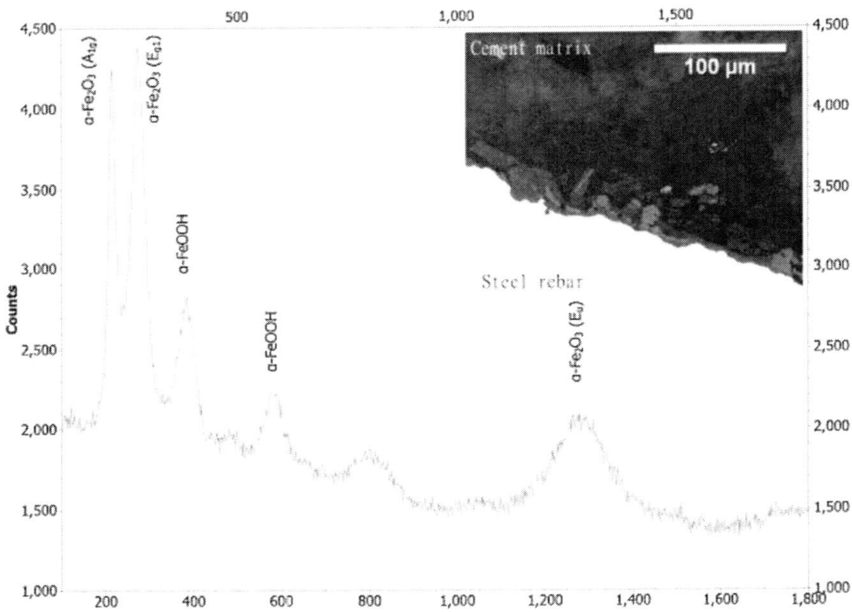

Figure. 11. Raman spectrum of α-hematite (Fe_2O_3) oxide layer on steel rebar extracted from sample 2 at 243 days. Inset shows optical image of polished section and location of analysis. Cement matrix shown in top right hand side dark area and steel re bar shown in light bottom left hand side.

ACKNOWLEDGEMENTS

Authors are thankful to the UK-India Education and Research Initiative (UKIERI-II) Thematic Partnership project "Interactive Systems and Coatings for a Sustainable Built Environment" under grant number UKUTP201100031 coordinated by the British Council, New Delhi, India, for financial support. The authors would also like to thank theEngineering and

Physical Sciences Research Council (EPSRC) for partial financial support through grant EP/K025597/1.

REFERENCES

1. M. Moreno, W. Morrism, M.G. Alvarez, G.S. Duffo **Corrosion of reinforcing steel in simulated concrete pore solutions: effect of carbonation and chloride content** Corros. Sci., 42 (11) (2004), pp. 2681–2699

2. J.P. Moreno **Corrosion of Steel in Concrete Understanding Investigation and Repair** (Second ed.)Taylor and Francis (2007)

3. H.A.F. Dehwah, M. Maslehuddin, S.A. Austin **Long-term effect of sulphate ions and associated cation type on chloride induced reinforcement corrosion in Portland cement concretes**

4. Cem. Concr. Compos., 24 (2002), pp. 17–25

5. P. Claisse, H. Elsayad, E. Ganjian **Permeability and Pore Volume of Carbonated Concrete European Concerted Action** Final Report, Brussels (1997)

6. L.J. Parrott, A review of carbonation in reinforced concrete, British Cement Association, contract report for the Building Research Establishment C/I-0987 (1987).

7. V. Talakokula, S. Bhalla, A. Gupta **Corrosion assessment of reinforced concrete structures based on equivalent parameters using electro-mechanical impedance technique** J. Intell. Mater. Syst. Struct., 25 (4) (2014), pp. 484–500

8. G. Miller, P. Gaydecki, S. Quek, B. Fernandes, M. Zaid **A combined Q and heterodyne sensor incorporating real-time DSP for reinforcement imaging, corrosion detection and material characterization** Sens. Actuator A-Phys., 121 (2) (2005), pp. 339–346

9. T. Ikeda **Fundamentals of Piezoelectricity** Oxford University Press, Oxford (1990)

10. W. Li, Y. Cho **Quantification and imaging of corrosion wall thinning using shear horizontal guided waves generated by magnetostrictive sensors** Sens. Actuators A-Phys., 232 (2015), pp. 251–258

11. C.K. Soh, K.K.H. Tseng, S. Bhalla, A. Gupta **Performance of smart piezoceramic patches in health monitoring of a RC bridge** Smart Mater. Struct., 9 (4) (2000), pp. 533–542

12. C.K. Soh, S. Bhalla **Calibration of piezo-impedance transducers for strength prediction and damage assessment of concrete** Smart Mater. Struct., 14 (2005), pp. 671–684

13. S. Bhalla, A.P.R. Vittal, M. Veljkovic **Piezo-impedance transducers for residual fatigue life assessment of bolted steel joints** Struct. Health Monit., 11 (6) (2012), pp. 733–750

14. D.M. Peairs, G. Park, D.J. Inman **IMproving accessibility of the impedance-based structural health monitoring method** J. Intell. Mater. Syst. Struct., 15 (2) (2004), pp. 129–139

15. T.G. Overly, G. Park, K.M. Farinholt, C.R. Farrar **Development of an extremely compact impedance-based wireless sensing device** Smart Mater. Struct., 17 (6) (2008), p. 065011

16. S. Bhalla, A. Gupta, S. Bansal, A. Garg **Ultra low cost adaptation of electro-mechanical impedance for structural health monitoring** J. Intell. Mater. Syst. Struct., 20 (2009), pp. 991–999

17. V. Giurgiutiu, A.N. Zagrai **Embedded self-sensing piezoelectric active sensors for on-line structural identification** J. Vib. Acoust. ASME, 124 (2002), pp. 116–125

18. C. Liang, F.P. Sun, C.A. Rogers **Coupled electro-mechanical analysis of adaptive material systems-determination of the actuator power consumption and system energy transfer** J. Intell. Mater. Syst. Struct., 5 (1994), pp. 12–20

19. S. Bhalla, C.K. Soh **Structural health monitoring by piezo-impedance transducer: modelling** J. Aerosp. ASCE, 17 (4) (2004), pp. 154–165

20. S. Bhalla, C.K. Soh **Structural health monitoring by piezo-impedance transducer: application** J. Aerosp. Eng. ASCE, 17 (4) (2004), pp. 166–171

21. E.L. Hixon **Mechanical Impedance Shock and Vibration Handbook** C.M. Harris (Ed.) (3rd edition)Mc Graw Hill Book Co., New York (1988)

22. I.S. 456 **Indian Standard Plain and Reinforced Concrete, Code of Practice** Bureau of Indian Standards, New Delhi, India (2000)

23. I.S. 1786 **Indian Standard Specification for High Strength Deformed Steel Bars and Wires for Concrete Reinforcement** Bureau of Indian Standards, New Delhi (1985)

24. PI Ceramic **Product Information Catalogue** Lindenstrabe, Germany (2012)

25. IS 383 **Indian Standard Specification for Coarse and Fine Aggregates from Natural Sources for Concrete** Bureau of Indian Standards, New Delhi, India (1970)

26. Agilent Technologies, Test and measurement catalogue (2011), www.agilent.com.

27. D. Jana, B. Erlin **Carbonation as an indicator of crack age** ACI Concr. Int. (May) (2007), pp. 61–64

28. IS: 10262-2009 **Indian standard specification for concrete mix proportions guidelines** Bureau of Indian Standards, New Delhi (2009)

29. C.L. Page **Basic Principles of Corrosion** P. Schiessl (Ed.), Corrosion of Steel in Concrete, Chapman and Hall, London (1988)

30. M. Castellote, L. Fernandez, C. Andrade, C. Alonso **Chemical changes and phase analysis of OPC pastes carbonated at different CO_2 concentrations** Mater. Struct., 42 (2009), pp. 515–525

31. ASTM G I. -03 **Standard Practice for Preparing, Cleaning and Evaluating Corrosion Test Specimens** West Conshohocken, PA (2003)

32. S.J. Oh, D.C. Cook, H.E. Townsend **Characterization of iron oxides commonly formed as corrosion products on steel** Hyperfine Interact., 112 (1998), pp. 59–66

33. Wang, L.A. Haskin, B.L. Jolliff **characterization of mineral products of oxidation and hydration by laser Raman spectroscopy—implications for in situ petrologic investigation on the surface of mars** Lunar and Planetary Institute Science Conference Abstracts, 29 (1998), p. 1819

8

BASE METAL CO-FIRED MULTILAYER PIEZOELECTRICS

Lisheng Gao *, Hanzheng Guo, Shujun Zhang and Clive A. Randall

Department of Materials Science and Engineering, Materials Research Institute, The Pennsylvania State University, University Park, PA 16802, USA

ABSTRACT

Piezoelectrics have been widely used in different kinds of applications, from the automobile industry to consumer electronics. The novel multilayer piezoelectrics, which are inspired by multilayer ceramic capacitors, not only minimize the size of the functional parts, but also maximize energy efficiency. Development of multilayer piezoelectric devices is at a significant crossroads on the way to achieving low costs, high efficiency, and excellent reliability. Concerning the costs of manufacturing multilayer piezoelectrics, the trend is to replace the costly noble metal internal electrodes with base metal materials. This paper discusses the materials development of metal co-firing and the progress of integrating current base metal chemistries. There are some significant considerations in metal co-firing multilayer piezoelectrics: retaining stoichiometry with volatile Pb and alkaline elements in ceramics, the selection of appropriate sintering agents to lower the sintering temperature with minimum impact on piezoelectric performance, and designing effective binder formulation for low pO_2 burnout to prevent oxidation of Ni and Cu base metal.

KEYWORDS

multilayer actuator; base-metal cofiring; piezoelectrics

1. INTRODUCTION

Piezoelectric applications are trending towards miniaturization, better performance, lower energy consumption, being more eco-friendly, higher reliability, and lower cost [1]. The demand for piezoelectric components has been increasing over the years [2,3]. With continued technology development, the piezoelectric business is rapidly expanding in the number and diversity of commercial products. It is estimated that the industrial and consumer piezoelectric business will reach a total of 36 billion USD in 2017, to which actuators would contribute roughly 14 billion USD [4]. Piezoelectric ceramics are becoming popular because of the ease of fabricating devices into different geometries and sizes. Piezoelectrics are experiencing growth in areas such as industrial automation, home automation, and consumer electronics. Currently, the global markets for conventional piezoelectric products are quite mature. There are market predictions of a two-orders-of-magnitude growth of the new piezoelectric products over the next five years [4]. The growth rates of piezoelectric actuators and motors are being driven primarily by automobiles (especially the piezoelectric fuel engine injectors, which are mostly diesel but with recent growth into gasoline injectors), information technology, and biomedical engineering. With the recent scandals regarding diesel emissions, though, there are further opportunities for piezoelectrics, as piezoelectric actuators can provide more sophisticated injection algorithms than the solenoid injectors and provide superior emission control. This is a very complex decision for companies, and those with joint development strategies may be technologically better-positioned to take advantage of the post-scandal era. In view of the global map of piezoelectric products, the robust, growing civilian piezoelectric products have continually diluted the market share of military applications, which were mostly sonar based. Despite the performance, reliability, and market demands, the piezoelectric industry has been continually suffering from the high costs of precious metal electrodes. Additionally, manufacturing multilayer actuators is different from that of multilayer ceramic capacitors (MLCC), which have industrial standards for size and performance providing a focus for large-scale mass production. Multilayer ceramic actuators do not have standards for production; business relationships and cooperative development in the supply change, so this diversity creates a more customer-based product. Furthermore, the economic crisis in major palladium producer Russia back in the 1990s drastically increased the market price of palladium, resulting in cost issues to manufacturers for using this metal electrode material [5]. Collectively, an effective strategy to decreasing the costs is to switch the widely used precious metal electrodes to base metal electrodes. Therefore, a reliable base metal co-firing technology would set up a solid foundation for the continually robust market growth in the future through further lowering the manufacturing costs.

2. APPLICATIONS OF MULTILAYER PIEZOELECTRICS

Multilayer piezoelectrics are widely used in actuators/motors, transformers, and sensors. The multilayer actuators can be utilized in a diesel engine to replace the electromagnetic solenoid valves. A carefully designed multilayer piezoelectric actuator injector and control system can provide a number of attractive advantages over the solenoid system, including, but not limited to, easy starting, high fuel efficiency, and rapid response speed. The response time of the nozzle needle is less than 10^{-4} s. This fast response reduces the fuel delivery rate, thus cutting the energy consumption of the fuel pump. A more sophisticated design of the fuel injector will also increase the efficiency of the diesel combustion engine. For this kind of application, piezoelectric material is required that is elastically stiffer and can charge-discharge quickly. The delicate designs of both injectors and the integrated actuators vary with each case. Specifically, geometry, capacitance, and electromechanical performance all change. All these variations cause the manufacturer to tend to be original equipment manufacturer (OEM) orientated. Other than injectors, the multilayer piezoelectrics are used for haptic applications, airbag sensors, keyless entry systems, and tire pressure monitors. Compared to the traditional electromagnetic-driven actuators/motors or transformers, the energy efficiency of multilayer piezoelectric actuators/motors is independent of the size and geometry. This independence is important because most home or office automation equipment requires tiny driven motors. Meanwhile, such advantages in controlling size offer opportunities to perfectly integrate a multilayer piezoelectric transformer into thin LCDs (liquid crystal displays) to increase the voltage for quick ignition and operation of the fluorescent backlight. Multilayer transformers are also being considered for step-up and step-down transformers in power electronic packages, and also in voltage sensors [6].

3. MULTILAYER CO-FIRING

The conventional single layer disk design and glued stacking design were preliminarily used over an extended period. Those have very limited performance, due to not only the relatively thick ceramic layers and higher drive voltage, but also to the time-consuming polishing step in manufacturing to reach the requirement for applications. However, by using the metal co-fired multilayer structures, the performance of the piezoelectrics is improved through lowering the driving voltage and increasing the overall displacement. As shown in Equation (1), the piezoelectric displacement (ΔL) can be expressed by the number of stacked layers (n) times the normalized strain coefficient (d*33) and electric field (E).

$$\Delta L = n \cdot d_{33}^{*} \cdot E \tag{1}$$

The piezoelectric actuators convert electrical energy into mechanical energy so that it can develop usable force. Blocking force is a maximum force that a piezoelectric actuator can generate under electrical load. It is represented in the form of force required to push back an electric field–loaded actuator to its original length. Equation (2) illustrates the relationship between blocking force (F_{max}) and displacement (ΔL); k is the actuator stiffness. The enhanced maximum displacement of a multilayer actuator raises the blocking force, and that improves the mechanical performance.

$$F_{max} = k \cdot \Delta L \tag{2}$$

The multilayer co-fired actuators (MLA) have thinner electrodes, requiring a relatively small driving voltage to obtain the equivalent strain. Despite the performance of the piezoelectric materials, a comprehensive understanding of material processing for multilayer co-fired actuators must address the complexity of co-firing, including optimizing each step of processing to obtain desired microstructures, density, and interaction-free interfaces simultaneously to achieve reliable electromechanical performance. The ceramic powders are dispersed and mixed with binder solution; these solutions contain constituents such as solvents, dispersants, binder polymers, and plasticizers. Various primary binder polymer systems used include materials such as poly(vinyl butyl) (PVB), poly(acrylic alcohol) (PVA), as well as, rarely, poly(propylene carbonate) (PPC). As shown in Figure 1, once a homogeneous ceramic slurry is prepared, it is then fabricated into the ceramic tapes, and metal electrodes are printed. The printed tapes are aligned and stacked together to build up multilayer actuators. The thick-film metal particle electrodes are designed with ceramic fillers to match the thermal profiles of the ceramic tapes to minimize differential shrinkage between the electrode and the tapes. The multilayer structures are binder burned-out and sintered with delicately designed profiles. There are multiple issues to consider for the process engineers, including the formulation of the binder systems, solvent attack resilience of the binder material, thermal expansion mismatch between electrodes and ceramics, sintering environments, and residual carbon after binder removal. Co-firing ceramics with metal electrodes has been challenging for decades. The noble metal electrode multilayer structures are then sintered under ambient air conditions. However, when base metal electrodes are applied, sintering conditions have to be converted from ambient air to reducing atmosphere to protect the metal electrode from oxidation. One of the common methods to obtain low pO_2 atmosphere is mixing H_2 and H_2O [7]. For the reaction $H_{2(g)} + \frac{1}{2}O_{2(g)} = H_2O_{(g)}$, the standard Gibbs formation energy is expressed in Equation (3) [8].

$$\Delta G^0 = -247,500 + 55.85T, (298K - 2000K) \quad (3)$$

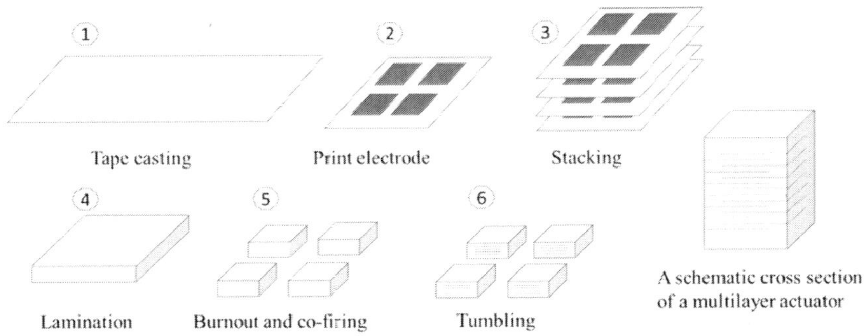

Figure 1. A schematic of processing steps toward multilayer piezoelectrics.

At the same time, the Gibbs standard formation energy can also be expressed in the form of Equation (4), where the R is the gas constant, T is the temperature, and K_p is the equilibrium constant.

$$\Delta G^0 = -RTlnK_p \quad (4)$$

The K_p can relate to the partial pressure of the H_2O, H_2, and O_2 as the relation expressed in Equation (5).

$$K_p = \frac{P_{H_2O}}{P_{H_2} \cdot (P_{O_2})^{\frac{1}{2}}} \quad (5)$$

Based on this method, almost any oxygen partial pressure can be obtained in a reducing atmosphere that necessary for controlling oxidation of copper or nickel electrodes.

4. TRENDS OF METAL ELECTRODES

Conventionally, sintering piezoceramics requires very high temperatures, over 1200 °C. At such high temperature, technically, a chemically and physically stable metal, such as Pt, is needed to co-fire with the ceramics. Costs aside, Pt is, of course, an ideal option for co-firing with ceramics; however, commercial applications mean the cost of Pt is too high. On the other hand, with the increased number of stacking layers of the actuators, the cost of electrodes dominates the total cost of the component [9]. Therefore, reduction of the costs of the electrodes would drastically lower the manufacturing costs of multilayer actuators.

Currently, there are two major electrode systems that are widely used commercially: (1) using silver (Ag) and Palladium (Pd) alloy electrodes to sinter in ambient air and (2) using base metal internal electrodes, such as nickel (Ni) or copper (Cu), to co-fire with those non-reducible piezoelectric ceramic materials in a reducing atmosphere. The related properties and price ratio of four common metals are listed in Table 1. Comparing these two systems, the later one can significantly cut the costs of the metal electrode. The Ag/Pd alloy is now widely used in multilayer actuators as internal electrodes, and these are co-fired under an air-sintering condition. Since Ag and Pd form a complete solid solution [10,11], the melting temperature can be varied by altering the composition of the alloy. The Ag-rich alloy is preferred to maintain the high conductivity and control the costs. In contrast, base metal Ni is widely used in $BaTiO_3$ MLCCs [9]; however, Cu, as an internal electrode material, is still under investigation because of the technical challenges with material processing [12,13,14]. In this paper, we will review the development of the co-firing technology of three major piezoelectric material systems and demonstrate the recent progress in the base metal co-firing technology of lead-free piezoelectric materials.

Table 1. Important properties of common metal electrode materials.

Metals	Melting Point (°C)	Resistivity ($\times 10^{-8}\ \Omega\cdot m$)	Young's Modulus (GPa)	Firing Atmosphere	Price Ratio*
Pt	1768	10.5	168	Air	16390
Ag	961	1.6	83	Air	133
Cu	1084	1.7	110	Reducing	1
Ni	1453	6.9	200	Reducing	2
Pd	1552	10.4	121	Air	5930

* The price ratio is based on the price of the investment market, and is also normalized by the factor of mass density due to the consumption amount, based on volume in real applications.

5. PB(ZR, TI)O3

$Pb(Zr, Ti)O_3$ (PZT) was discovered in the 1950s by Shirane *et al.* [15,16] at the Tokyo Institute of Technology and soon largely replaced $BaTiO_3$ as a piezoelectric material. Afterward, the Morphotropic Phase Boundary (MPB) was discovered, separating the tetragonal phase and rhombohedral phase, where it is the basic solid solution composition with the highest piezoelectric properties [16]. A more accurate MPB composition was determined at the composition with ~48 mol% of $PbTiO_3$ and ~52 mol% of $PbZrO_3$. Due to the arising equivalent energy state, it allows the optimum domain orientation during the poling process [17] for a PZT material with MPB composition, above the ferroelectric transition temperature (T_C) at ~350 °C.

The PZT was initially sintered above 1200 °C to achieve high density. However, it was reported that adding an extra 1 to 2% of excess (over stoichiometry) PbO would lower the sintering temperature. Generation of the PbO liquid phase enables a lower sintering temperature [18,19]. The sintering temperature of PZT was decreased to as low as 900 °C by adding PbO combined with Li_2CO_3 as sintering agents [20,21]. Various sintering agents for lead-based piezoelectric materials, sintering temperatures, and the associated properties are listed in Table 2.

5.1. Volatility of Pb

As is well known, lead oxide undergoes a volatilization during heat treatments. When exposed to high temperatures, lead oxide volatilizes into the surrounding atmosphere. The properties of the PZT material can be strongly influenced by the Pb vacancies in the materials. Donnelly *et al.* [27] reported Pb loss while annealing the PZT-0.75%Nb material in ambient air at 700 °C. During the heat treatment, lead (Pb) vacancies and oxygen (O) vacancies were generated at the same time as written in Kröger-Vink notation, which is shown in Equation (6).

$$Pb_{Pb}^{*} + O_{o}^{*} \rightarrow V_{Pb}^{''} + V_{O}^{\cdot\cdot} + PbO \uparrow \qquad (6)$$

$$\frac{1}{2}O_2 + V_{O}^{\cdot\cdot} \leftrightarrow O_{O}^{*} + 2h^{\cdot} \qquad (7)$$

Table 2. Sintering aids, sintering temperatures, and properties of several lead-based piezoelectric materials.

Sintering Aids (% Addition)	Ceramic	Sintering Temperature (°C)	T_c (°C)	$\varepsilon_r/\varepsilon_0$ (RT)	d_{33} (pC/N)	d_{33}^* (pm/V)	Reference
$LiBiO_2$ (1 wt%) + CuO (0.06 wt%)	PZT	880	-	~1200	-	-	Wang *et al.* [22]
Li_2CO_3 (0.2 wt%) + PbO (2 wt%)	PZT with Sr, K, Nb	900	360	1200	-	779	Donnelly *et al.* [20,21,23]
$BiFeO_3$ (5 wt%) + $Ba(Cu_{0.5}W_{0.5})O_3$ (3 wt%)	PZT with 0.5 wt% MnO_2	935	290	847	201	-	Kaneko *et al.* [24]
3ZnO-CuO (3 wt%)	PNN-PZT	900	200	3900	575	-	Ahn *et al.* [25]
$LiBiO_2$ (0.7 wt%)	PMN-PZT and 5% Ba	950	290	~1000		467	Hayashi *et al.* [26]

The controlling kinetics of the generation of oxygen vacancies and lead vacancies were different. The generation of oxygen vacancies was a bulk diffusion rate–dominated process. The dominating path of Pb volatilization was a result of Pb coupling with the oxygen adsorbed on the surface of the materials [28]. This reaction rate–limited process kinetically controlled the Pb-losing rate of the material. The generation of oxygen vacancies reached equilibrium with the oxygen in the atmosphere, as shown in Equation (7). As a consequence of the high volatilization, the PZT was often sintered with excess PbO or lead sources (PZ) to compensate for the loss of volatile PbO [23,29,30]. Figure 2 shows the ionic conductivity change over time; the gentle slope in the N_2 zone means that the Pb-losing rate is low. This evidence suggested that when air was switched to N_2 with $pO_2 < 10^{-5}$ atm, the decreased low pO_2 atmosphere suppressed the volatility of Pb through reducing the contacting chance between environmental O_2 and the materials [28,31], with the losing rate kinetics being controlled from the surface interaction of lead with oxygen at the intermediate temperature ranges of 700 °C to 1000 °C [27].

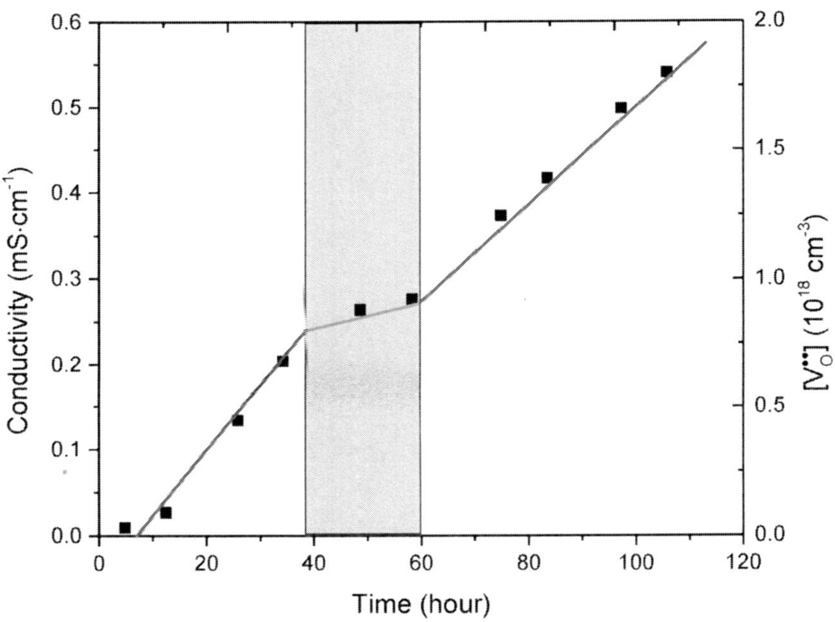

Figure 2. Ionic conductivity in a PZT-0.75%Nb buried capacitor measured over several days at 700 °C. Loss of PbO causes the increasing conductivity over time. The increasing oxygen vacancies are due to PbO evaporation: Redrawn from [27].

5.2. Low-Temperature Noble Metal Co-Firing

Co-firing PZT with the relatively cheaper metal electrodes is usually limited by the low melting temperatures of the metals. Compared with Pt, the cheaper Ag/Pd alloy has a lower melting temperature and thus is not able to undergo the high sintering temperature [10]. Various studies on PZT co-firing with different metal electrodes were reported, including the kinetics of sintering and explanations on how co-firing electrodes influence the microstructures, as well as the dielectric and piezoelectric performances. PZT has been successfully co-fired with $Ag_{1-x}Pd_x$ ($0 < x < 0.3$) electrodes into an interdigitated multilayer PZT piezoelectric structure [32]. Wersing *et al.* [33] reported their work on co-firing PZT with Ag/Pd internal electrodes. Both $Pb(Zr_{0.53}Ti_{0.47})O_3$ (PZT) and $Pb(Ni_{1/3}Nb_{2/3})O_3$-PZT (PNN-PZT) multilayer samples were prepared with Ag/Pd (75/25) alloy electrodes. The multilayer samples were sintered at various temperatures. Moreover, the methodology of lowering the sintering temperatures was engineered through various methods, including optimizing particle sizes, adding various contents of V_2O_3 as a sintering agent, and optimizing the heating rate. It was mentioned that the co-firing temperature had to be lower than 1130 °C to prevent interdiffusion and chemical reactions on the metal-ceramic interfaces. More recently, a study by Caballero *et al.* [34] on Ag/Pd (70/30) co-fired PZT and PNN-PZT further demonstrated the result from Wersing *et al.* [34] by using advanced electron microscopy. Under scanning electron microscopy (SEM), clear electrode-ceramic interfaces were observed, and energy dispersive spectroscopy (EDS) confirmed that there was no Ag/Pd or Pb interdiffusion. In Ag/Pd co-fired multilayer samples, both the EDS data and further transmission electron spectroscopy (TEM) analysis confirmed no interdiffusion of Ag and Pb. The finely controlled sintering work might not induce the Ag or Pb diffusion, but when applied with an electric field, ionized Ag would undergo electromigration along grain boundaries. Electromigration of Ag^+ ions in PZT ceramics causes catastrophic reliability issues in applications. Donnelly *et al.* [35] examined the Ag^+ electromigration in a PZT ceramic applied with high *dc* bias under extreme humidity. After a certain period, the samples exhibited metallic-like conductivity. It revealed that humidities absorbed into the grain boundaries led to the migration of Ag^+ ions under electric fields, where very small amounts of Ag ions evaporate from the electrodes and are distributed into the grain boundaries with excess PbO or other glassy phases associated with the sintering aids. The partial water solubility from the humidity into the glassy phases expedites the migration of Ag^+ ions along the grain boundaries.

5.3. Base Metal Co-Firing

Despite Ni being the major base metal electrode choice in $BaTiO_3$, paradoxically, PZT is simply not compatible with Ni co-firing because the pO_2 required to retain the metallic Ni electrodes can easily reduce PbO in PZT. As shown in Figure 3, the Ellingham diagram indicates that the NiO is less stable than PbO. Jacob and Saji [36] have conducted experiments on the interaction between Ni/NiO and $PbTiO_3$, which was part of the solid solution of the PZT. Their work displayed that direct contact between NiO and PZT did not result in chemical reactions. However, when metallic Ni had direct contact with $PbTiO_3$ while sintering in either ambient or reducing environments, it reacted with $PbTiO_3$ and generated a $NiTiO_3$ phase and a Pb-rich alloy phase, respectively. Further thermodynamic calculations on studying the coexistence between Ni and $PbZrO_3$ showed similar results. Since the change of Gibbs potential energy of oxidizing Ni to NiO is lower than that of forming PbO, the direct contact of metallic Ni and $PbZrO_3$ would also result in the generation of a Pb-rich alloy. These fatal reactions excluded metallic Ni as an appropriate electrode to co-fire with PZT.

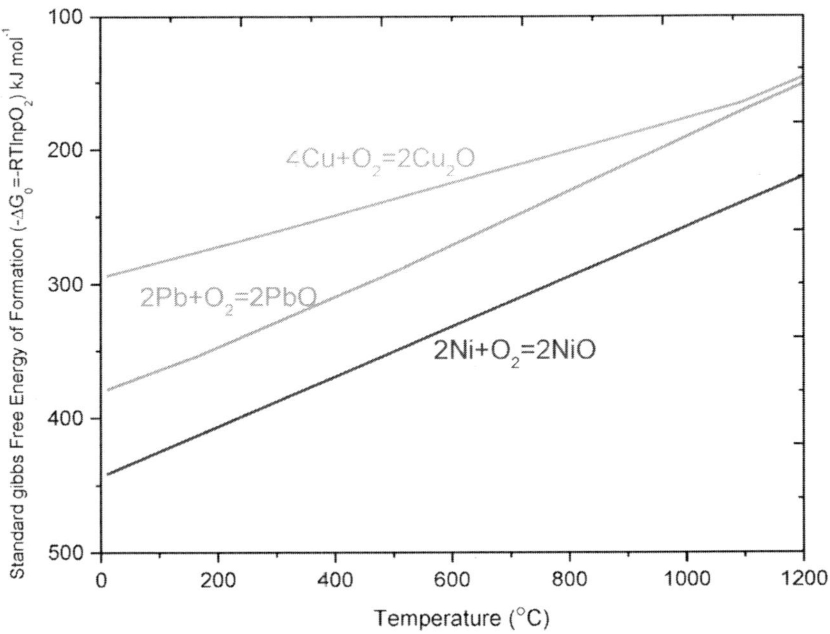

Figure 3. Standard Gibbs formation energy of oxides of Cu, Ni, and Pb, redrawn from [8].

Co-firing the kinetically more readily oxidized copper with PZT is thermodynamically feasible within a narrow pO_2window. There were published works on co-firing a Cu electrode [37] with PZT-based materials. Takayama *et al.* [38] reported the work on fabricating Cu co-fired PZT-SKN (PZT-$Sr(K_{0.25}Nb_{0.75})O_3$) [20,21,23] multilayer actuators. An acrylic binder system was used to lower the burnout temperature, instead of the more widely used PVB; however, this acrylic binder was difficult to handle. The PZT-SKN multilayer actuators with internal glass-coated copper electrodes were burned out in the ambient air at 350 °C. The function of the glass coating on the Cu particles was to prevent the oxidation of Cu during binder burnout in ambient air. Through adding excess PbO (3 wt% and 4 wt%) to the material, the actuators were sintered between 900 °C to 1000 °C under pO_2 of 10^{-12} atm. As a consequence of low pO_2 sintering, reduced piezoelectric constants (both intrinsic and extrinsic) were obtained compared with the air-sintered PZT-SKN. As shown in Figure 4, both gentle slopes and low y-intercepts are obtained from the low pO_2–sintered PZT samples. The Rayleigh analysis indicates that this special sintering process suppressed both extrinsic and intrinsic piezoelectric properties of the PZT slightly. The SEM micrographs of the cross-section of the multilayer actuators are shown in Figure 5. The results were promising, since there was no Cu diffusion detected under the EDS. A transmission electron microscopy (TEM) micrograph is presented inFigure 6. Figure 6a displays the interface between the Cu electrode and PZT ceramic. Figure 6b shows the vicinity triple points of PZT near the interface. The Cu-free vicinity triple points and grain boundaries verify that there was no Cu diffusion. However, in the Ag or Ag/Pd electrode cases, electrode materials were often observable in these places. Regarding commercial applications, the Cu co-fired PZT had been patented and mass produced by EPCOS, a TDK Group Company [39]. In this EPCOS/TDK product, the Cu co-fired PZT actuators presented a superior performance with a billion switching operations at 170 °C without failures. In contrast, under the same testing conditions, the Ag/Pd co-fired actuators already displayed 50% failures with only 145 million switching operations [40].

6. BATIO$_3$ AND BASE METAL CO-FIRING

As a widely used material for base metal co-fired multilayer capacitors, $BaTiO_3$ has been intensively studied for use in the base metal co-fired multilayer ceramic capacitors (MLCC) with ultra-thin ceramic layers [9,12,41,42,43,44,45,46,47]. The robust success of materials processing led to tremendous work on studying Ni co-fired $BaTiO_3$, accumulating a rich theoretical and technical basis for future co-firing work toward base metal co-fired multilayer piezoelectrics. Yang *et al.* [48] reported the interaction between Ni electrodes and the dielectric $BaTiO_3$ during sintering. Carbonization of binder polymers during burnout generated carbon residues, which dramatically changed the sintering condition through modifying the internal oxygen activity locally on the electrode-ceramic interfaces. The Ni-catalyzed oxidation of residual carbon resulted in the reduction reaction of $BaTiO_3$ on the interfaces. The reduced Ba and Ti dissolved into the liquid metal phase to form a metal alloy along the ceramic-electrode interface. At the same time, one of the Cu co-fired X7R multilayer capacitors was successfully fabricated by using an acrylic binder system [12]. Utilization of oxide-coated Cu as internal electrodes improved the oxygen resistivity by shielding the atmospheric oxygen from contacting the metallic Cu. Concerning the residual carbon and metal-ceramic interactions, no major issues were noticed. Base metal co-fired $BaTiO_3$-based dielectric materials have been successfully demonstrated during the past decades, but we still have to face problems from the perspective of ceramic processing, especially in designing an effective binder system that is easy to handle and free of residual carbon.

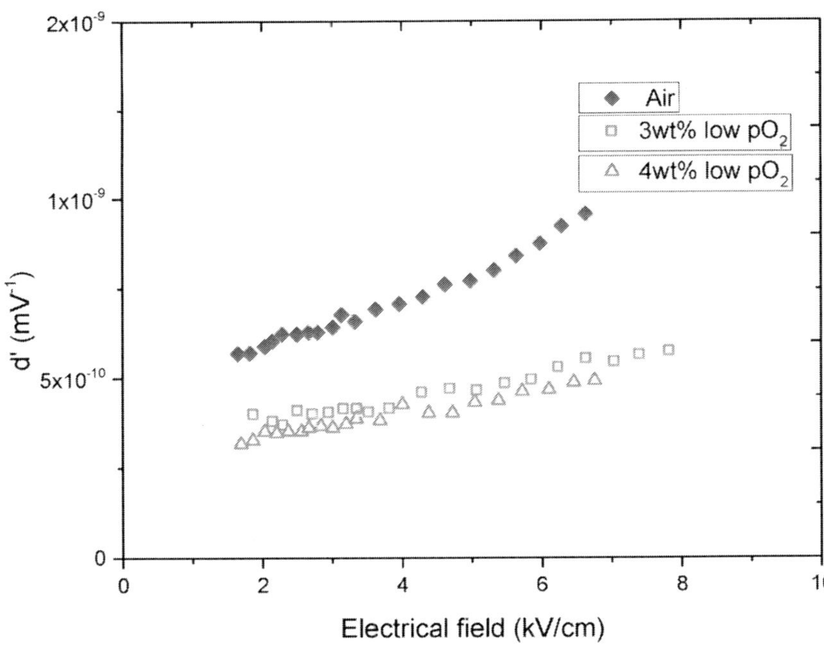

Figure 4. Real piezoelectric coefficients of the sintered PZT-SKN, redrawn from [38].

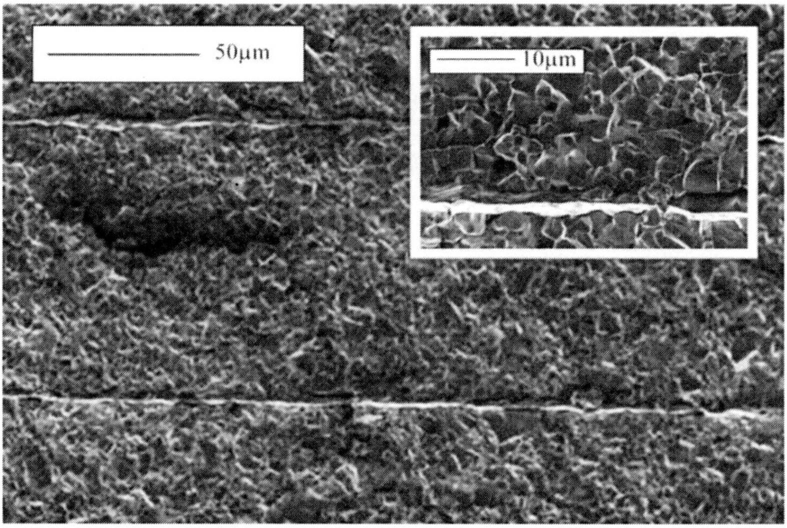

Figure 5. SEM image of multilayer sample with 3 wt% excess PbO [38].

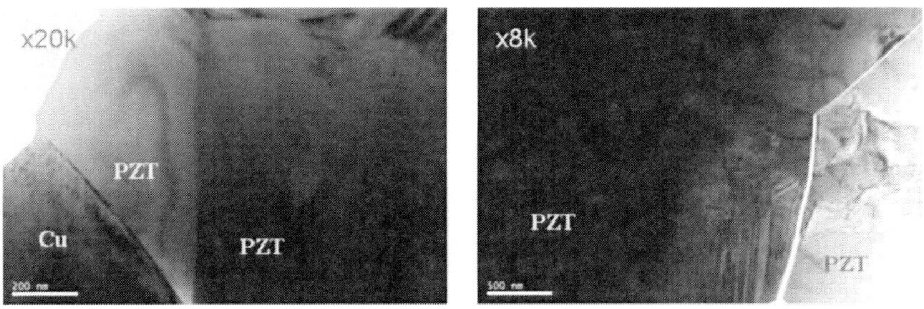

(a) (b)

Figure 6. TEM micrograph of (**a**) Cu-PZT interface, (**b**) triple points and grain boundaries near electrode [38].

7. (NA, K)NBO$_3$

Lead-based Pb(Zr, Ti)O_3 (PZT) and Pb($Mg_{1/3}Nb_{2/3}$)O_3: $PbTiO_3$ (PMN-PT) are widely used in commercial applications due to their superior piezoelectric properties. However, concerning the toxicity of lead to both the environment and human beings, the European Union has set a cap on the maximum amount of usage of lead in electronics to 1000 ppm by 2006 and 2011 [49,50]. Though the piezoelectric industries have been continually exempt from the Restriction of Hazardous Substances Directive (RoHS), tremendous effort has been put into studying the lead-free piezoelectric materials to replace the lead-based materials. The perovskite $(Na_{0.5}K_{0.5})NbO_3$ (NKN) was first reported in 1959 by Egerton and Dillon [51]. Lately, the NKN-based materials have been considered as a potential substitution for PZT, although their performance is still inferior to PZT-based materials. Saito *et al.* [52] reported that textured $(Na_{0.5}K_{0.5})NbO_3$-based piezoelectric ceramic materials possess comparable piezoelectric properties to PZT. NKN-based materials gained a large spotlight over the other lead-free piezoelectric materials. A series of studies have been conducted by modifying the composition of NKN for better piezoelectric performance and base metal electrode (BME) co-firing. Properties of different modified NKN-based piezoelectric ceramics are presented in Table 3.

7.1. Volatility of Alkaline Elements

Similar to the Pb, alkaline elements in NKN are highly volatile. The NKN materials are haunted by densification and microstructural grain morphology and bimodal grain size issues [64,65]. During the air sintering, the volatile alkaline elements could proceed a volatilization process, as described in the Equation (8), where M represents the alkaline elements.

$$M_M^* + O_O^* \leftrightarrow M_2O + 2V_M' + V_O^{\cdot\cdot} \quad (8)$$

It is similar to the volatilization path of Pb; the alkaline elements go into the atmosphere via combining with adsorbed atmospheric oxygen at the surface of particles and grains. With the effective zero atmospheric partial pressure of M_2O, vaporization of M_2O would continually proceed until no alkaline elements are left in the material. Generation of oxygen vacancies, though, is coupling with the loss of alkaline elements; such a reaction would reach equilibrium with the atmospheric O_2 following the path as illustrated in Equation (9), which is similar to the case of Pb.

$$\frac{1}{2}O_2 + V_O^{\cdot\cdot} + V_M'' \leftrightarrow O_O^* + V_M'' + 2h^{\cdot} \quad (9)$$

Table 3. Properties, additives, and sintering temperatures of several NKN-based piezoelectric ceramics.

Ceramics	Sintering Aid (mol%)	Sintering Temperature (°C)	T_c (°C)	$\varepsilon_r/\varepsilon_0$ (RT)	d_{33} (pC/N)	d_{33}^* pm/V	Reference
$0.99((Na_{0.5}K_{0.5})NbO_3)$-$0.01(SrTiO_3)$	-	1100	344	649	90	-	Guo *et al.* [53]
$0.95((Na_{0.5}K_{0.5})NbO_3)$-$0.05LiTaO_3$	-	1110	420	570	200	-	Guo *et al.* [54]
$(Na_{0.47}K_{0.47}Li_{0.03})(Nb_{0.8}Ta_{0.2})O_3$	-	~1100	~310	~1000	190	310	E. Hollenstein *et al.* [55]
$0.95[(Na_{0.5}K_{0.5})NbO_3]$-$0.05LiTaO_3$	Li_2O 1 mol%	1000	-	~450	250	267	Kim *et al.* [56,57]
$(Na_{0.5}K_{0.5})NbO_3$	LiF 5 mol%	1000	456	415	138	210	Kobayashi *et al.* [58]
$0.96(K_{0.5}Na_{0.5})NbO_3$-$0.04CaZrO_3$+0.03Zr	$MnCO_3$ 5 mol%	1080	260	1180	160	360	Kawada *et al.* [59,60]
$0.02(NaF–0.5Nb_2O_5)$-0.98 $[(Na_{0.5}K_{0.5})(Nb_{0.8}Ta_{0.2})O_3]$	NaF	1200	296	900	140	149	Liu *et al.* [61]
$0.96(K_{0.48}Na_{0.52})(Nb_{0.96}Sb_{0.05})O_3$-$0.04Bi_{0.5}(Na_{0.82}K_{0.18})ZrO_3$	-	1060–1120	227	~2200	490	-	Wang *et al.* [62]
$(K_{0.5}Na_{0.5})NbO_3$-$LiSbO_3$	-	1160	368	1380	265	-	Zhang *et al.* [63]

Kobayashi *et al.* [58,66] have demonstrated the work on studying how varying sintering atmosphere affects the microstructures and properties of NKN. The results indicated that once the NKN materials were sintered in the low pO_2atmosphere, which reduced the possibility of oxygen adsorption onto the surface of the materials, the volatility of alkaline elements decreased. When compared with the air-sintered NKN, the low pO_2–sintered NKN samples obtained higher electrical resistivity, which further indicated that a higher amount of alkaline elements was retained in the NKN. Similarly, advantages were also noted in the sintering of the end member $NaNbO_3$ [67]. These results were consistent with the first principle calculation that was conducted by Shigemi and Wada [68,69], showing that the formation energy of alkaline vacancies increased when the atmosphere changed from air to low pO_2.

7.2. Material Modification

Guo *et al.* [53,54] reported the work of $(Na_{0.5}K_{0.5})NbO_3$ materials by developing the dopant system. They doped the $SrTiO_3$ into the NKN system to study how $SrTiO_3$ would influence the piezoelectric properties of NKN [53]. With the highest d_{33} achieved in a non-doped system, the increasing doping amount of $SrTiO_3$ not only caused a drop of d_{33}, but also induced dielectric relaxor behavior. The 4 mol% $SrTiO_3$ has become a critical point separating the orthorhombic phase and tetragonal phase. Such a donor (SrO) and acceptor (TiO_2)–combined doping system solved the poor densification and deliquescence issues of NKN. Following this study, a $LiTaO_3$-doped NKN system was investigated [54]. The $LaTiO_3$-doped system gave better piezoelectric properties than that of the $SrTiO_3$-doped system. The piezoelectric coefficient d_{33}~200 pC/N was obtained with 5 mol% dopant. The ferroelectric measurement was carried out with remnant the polarization P_r~9 μC/cm^2 and coercive field ~12.5 kV/cm. The Li- and Ta-modified NKN ceramic was further studied by Hollenstein *et al.* [55] through varying the amount of Li and Ta dopant in the composition. The piezoelectric properties were compared between the only Li–doped NKN and both the Li- and Ta-modified NKN system. The 3 mol% Li– and 20 mol% Ta–modified NKN ceramic displayed the best piezoelectric properties of d∗33~310 pm/V but possessed a lower Curie temperature of about 310 °C.

7.3. Low-Temperature Co-Firing with Noble Metal

Kim *et al.* [56,57] focused on the Li- and Ta-modified system with just the 5 mol% $LiTaO_3$-doped NKN material, due to the coupling coefficient reaching its highest value at this composition. A Li_2O sintering agent was used to lower the sintering temperature, in order to meet the temperature requirement of co-firing with the Ag/Pd (70/30) electrode and decrease the volatility of alkaline elements. The sintering temperature was reduced to 1000 °C. Well-densified samples were achieved with a relative density of 95%. However, this 1 mol% additional sintering aid caused abnormal grain growth because of the change in the critical driving force of grain growth on the faceted interface and grain boundaries. Although adding more than 1 mol% Li_2O, up to 7 mol%, impeded the abnormal grain growth, the considerable sacrifice of d_{33} was observed. The sintered disk bulk samples presented a piezoelectric coefficient of 250 pC/N. A normalized strain coefficient d_{33}^* reached as high as 292 pm/V with Ag/Pd co-fired multilayer samples. The Ag electrode diffused into the ceramic and *vice versa*, forming a thin transition layer, though the Pd could have had some effect on impeding the diffusion slightly [70].

7.4. Low-Temperature Co-Firing with Base Metal

NKN has relatively poor performance compared with PZT-based materials. In order to compete with the PZT-based materials, NKN actuators have to be cheap enough to compensate for their inferior performance. Precious metal electrodes such as the Ag/Pd alloy are not reasonable in view of the costs; at the same time, the diffusion and electromigration issues of Ag could lower the reliability of the devices. Therefore, the spotlight moved to the base metal electrode in seeking low-cost substitutions, as well as looking at the reliability of base metal co-fired NKN piezoelectrics. Kawada *et al.* [59] fabricated the Ni co-fired NKN ceramic ($0.96(K_{0.5}Na_{0.5})NbO_3$-$0.04CaZrO_3$+0.03Zr) actuators and displayed a $d{*}33$ as high as ~360 pm/V.Figure 7 shows the strain-electric field behavior of the Ni co-fired multilayer actuators. The ceramic tape was fabricated with a poly(vinyl acetate) binder system. The binder was burned out at 400 °C for five hours. Samples were sintered at 1080 °C under pO_2 between 10^{-11} to 10^{-10} atm for two hours. It was reported that the Ni electrodes have a higher resistivity to electromigration than the Ag/Pd electrode, which is widely used in PZT multilayer actuators [60].

Kobayashi *et al.* [71] conducted parallel experiments on sintering NKN ceramics in both air and low pO_2 atmosphere to understand the differences that low pO_2 sintering made to the properties of NKN. The $(Na_{0.5}K_{0.5})NbO_3$ with a 5 mol% LiF sintering aid was studied. The results showed that the piezoelectric coefficient of NKN was independent of the sintering atmosphere. As shown in Figure 8, no significant changes in Rayleigh coefficients were observed. The Rayleigh equation is shown in Equation (10), where $d_{initial}$ is the intrinsic piezoelectric coefficient, d is the normalized strain coefficient, and α_d is the Rayleigh coefficient. This phenomenon was quite different from the case of PZT materials [38].

$$d = d_{initial} + \alpha_d E_0 \tag{10}$$

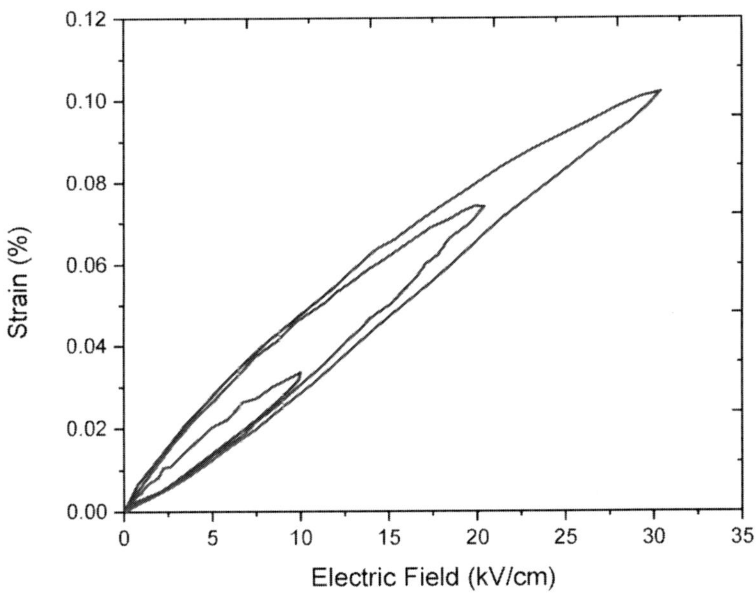

Figure 7. Strain-electric field curve of $0.96(K_{0.5}Na_{0.5})NbO_3$-$0.04CaZrO_3$+0.03Zr multilayer sample with nickel internal electrode, redrawn from [59].

As mentioned, the low pO_2–sintered NKN had higher resistivity than those that were air-sintered, due to the suppressed volatility of alkaline elements in low pO_2. Another advantage was that the NKN ceramic sintered in the low pO_2atmosphere obtained higher density (95% of theoretical density) than those sintered in the air (91% dense) at the same temperature with the same dwelling time. Afterward, further study on fabricating Ni co-fired multilayer NKN actuators was done.

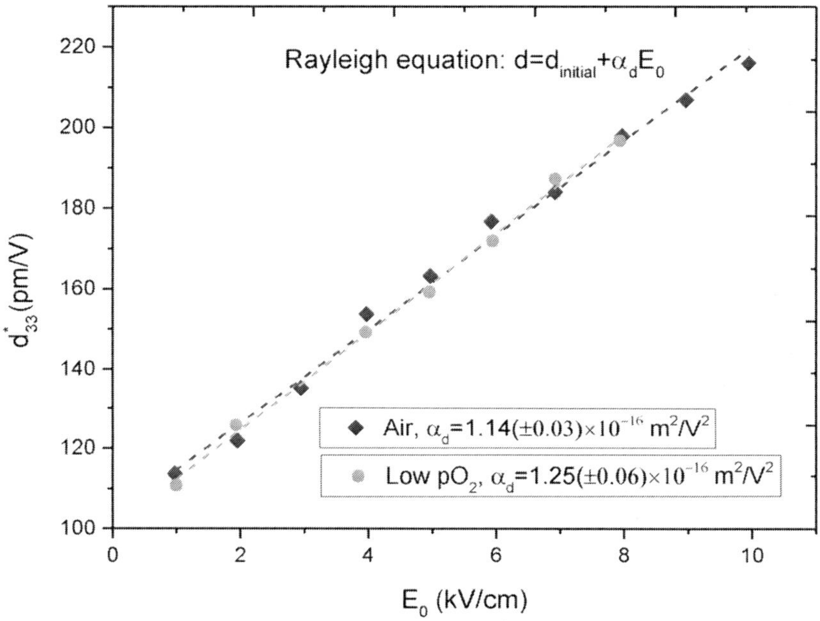

Figure 8. Piezoelectric coefficient d_{33} of a NKN system as a function of applied *ac* field, and linear fitting with the Rayleigh equation, redrawn from [71].

Kobayashi *et al.* [58,66] used the same NKN [71], then fabricated prototype multilayer actuators with Ni internal electrodes. The PVB-formed tapes were printed with a layer of Ni electrodes to prepare the multilayer actuators. The multilayer actuators were burned out at 270 °C for 12 h in N_2 atmosphere. There were no observable macroscopic defects, pores, or delaminations, which were common issues causing low reliability in multilayer actuators. Furthermore, no Ni interdiffusion was detected along the interfaces. Moreover, the Ni electrode did not affect the microstructure of NKN, unlike the alloy formation in the Ni co-fired $BaTiO_3$ system due to the reduction of $BaTiO_3$ by residual carbon. The alloy-free electrode-ceramic interfaces in Ni co-fired NKN multilayer actuators suggested the feature of working against the reduction of NKN. The following year, a NaF-Nb_2O_5 flux-doped NKN was reported by Liu *et al.* [61,72] for multilayer actuator applications. The $0.02(NaF_{0.5}Nb_2O_5)$-$(0.98)[(Na_{0.5}K_{0.5})(Nb_{0.8}Ta_{0.2})O_3]$ was synthesized and then used to fabricate Ni co-fired multilayer actuators. Similar to the majority of the reported work, a PVB binder was used. The binder burnout was performed at 270 °C for 12 h in the air. The multilayer actuators were sintered in a low pO_2 of 10^{-10} atm at 1150 °C for 2 h, and then further reoxidized at 850 °C in a pO_2 of 10^{-7} atm for eight hours. The piezoelectric performance was excellent compared with other lead-free piezoelectric materials; up to 385 pm/V of d^*_{33} was achieved. As a lead-free piezoelectric ceramic, NKN is promising for the future exploration of BME co-firing.

8. $BI_{0.5}NA_{0.5}TIO_3$

$Bi_{0.5}Na_{0.5}TiO_3$ (NBT)-based lead-free piezoelectric materials drew attention because of their moderate piezoelectric properties. The Ag/Pd co-fired NBT has already been fabricated and reported with reasonable performance and reliability by Sapper *et al.* [73]. Due to the increasing interest in base metal co-firing, recently Safari and his coworkers reported on co-firing $0.88(Bi_{0.5}Na_{0.5}TiO_3)$-$0.08(Bi_{0.5}K_{0.5}TiO_3)$-$0.04(BaTiO_3)$ (BNT-BKT-BT) with metallic Cu [74]. The NBT-based piezoelectric ceramic disks were sintered in both air and low pO_2 atmosphere. The comparison has been made on samples sintered under different conditions. Neither sintering agents nor sintering atmospheres had significant influence on piezoelectric properties; similar piezoelectric properties were obtained. The Cu electrode was pressed between two already burned-out ceramic powder disks. The co-fired samples showed a d_{33} of 77 pC/N, which was lower than the simple disk sample, having a d_{33} of ~160 pC/N. Collectively, the study presented that this NBT material was compatible for Cu co-firing; further exploration on microstructures and the interdiffusion process is needed.

9. BINDER MATERIALS

In view of the work we summarized here, we only saw a couple of works on co-firing Cu with piezoelectric ceramics. Those foreseeable challenges became roadblocks when co-firing involved metallic Cu; Cu is more readily oxidized and has a low melting temperature. Although the glass-coated copper can prevent Cu from oxidation, the influence of the glass coating on Cu particles may introduce contamination and sacrifice of the dielectric and piezoelectric performances. Therefore, a direct solution is to provide a binder system that can be cleanly burned off at low temperature and in a protective atmosphere. The traditional PVB has been widely used in both Ag/Pd and Ni co-fired multilayer actuators, due to the relatively higher oxidation temperature of these metals. Collectively, the binder system in the previous study was usually burned off in ambient air up to 400 °C. This temperature already exceeds the maximum temperature of retaining the metallic state of both Ni and Cu, though Ni electrodes are relatively feasible for burnout up to 300 °C. To effectively prevent Ni and Cu from oxidation, clean burnout has to be achieved at either a reasonably low temperature or in a low pO_2atmosphere. With the conventional PVB binder system, although well developed, clean burnout of PVB has never been reported at low temperatures. Masia *et al.* [75] reported that the amount of carbon residues of PVB varied with different ceramics. It was revealed, as a matter of fact, that usually at least 500 °C was needed for burnout due to the high activation energy of degradation [76,77]. Salam *et al.* [78] have studied the burnout behavior of PVB, serving as a binder of thermoelectric green tapes, in different kinds of atmosphere up to 450 °C. The results showed that PVB left residues in all conditions. A burnout conducted in ambient air at 450 °C showed that 19.3% of PVB was not removed. In this work, the fraction percentage of PVB left was estimated to be 2% at 600 °C and 1% between 700 °C to 800 °C. While burning out in N_2, the pyrolysis could take place and convert the PVB into carbonaceous char. Breakdown char usually requires higher temperatures than breakdown polymer. It was realized that even at 1000 °C, an appreciable amount of residual carbon could still exist under the N_2 condition [79].

Poly(propylene carbonate) (PPC) has been also studied as a potential binder polymer [80]. It was found that PPC can be cleanly burned out in both air and N_2 at relatively low temperatures. There was no detectable residue left after burnout. Unlike the PVB system, the burnout of PPC followed the depolymerization path to form a propylene carbonate monomer rather than pyrolysis of the polymer into char. The propylene carbonate vaporized without leaving residual carbon. The authors considered that lack of successful formulation of the PPC binder system hindered the utilization of PPC as a successful binder for ceramics.

10. RECENT PROGRESS ON CU CO-FIRED NKN

Recently, our group successfully fabricated prototype multilayer actuators through co-firing Cu metal electrodes with NKN ceramic [81]. We used the Li- and Ta-modified NKN, which was kindly provided by PI Ceramic. A new poly(propylene carbonate) (PPC) binder was used. The advantage of PPC is that it can be cleanly burned off in reducing atmosphere. The prototype samples were sintered at 1050 °C under a pO_2 of 10^{-12} atm for one hour, then further annealed at 1000 °C for another hour to allow grain growth. After that, the samples were reoxidized at 850 °C under a pO_2 of 10^{-5} atm for 12 h. Microstructures were compared between the low pO_2–sintered, and air-sintered pellets. The SEM surface micrographs are shown in Figure 9. Both samples were sintered at 1050 °C for 1 h. The samples sintered in low pO_2 showed a higher homogeneity of grain size with minimized abnormal growth (Figure 9a). However, lower homogeneity (Figure 9b) was obtained in air-sintered NKN ceramic with a certain amount of abnormal grain growth. Both of them have well-faceted grains.

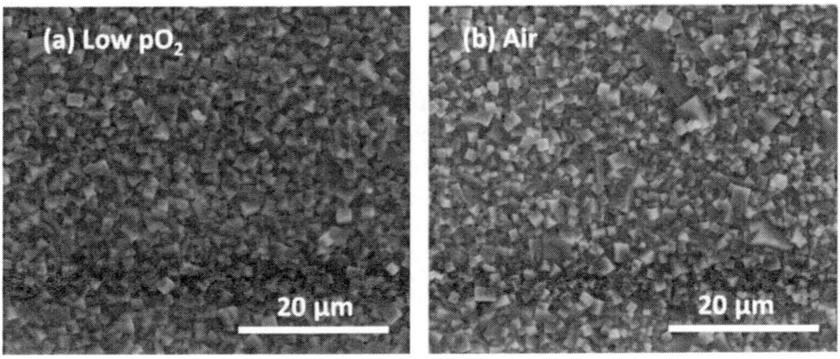

Figure 9. Surface micrograph of (**a**) low pO_2–sintered and (**b**) air-sintered pellets at 1050 °C for 1 h.

A 96% dense NKN ceramic was achieved from sintering plain tapes. The surface SEM micrograph (Figure 10) displayed well-faceted grains (Figure 10a) on the surface of the multilayer co-fired samples. The cross-section is shown in Figure 10b. The internal Cu electrode was continuous and unoxidized. EDS line scan data (Figure 10c) indicated that there was no residual carbon and Cu interdiffusion between Cu electrodes and the NKN ceramic. The microstructure was further studied by TEM (Figure 11a,b); the micrographs were taken along the <001> direction. An atomistically sharp interface between the Cu electrode and ceramic was observed (Figure 11c,d). There was no residual carbon, alloy formation, or CuO in the co-fired multilayer samples under TEM. The dielectric constant was 830, with a low dielectric loss of 3.6%. The co-fired multilayer actuators displayed a Curie temperature (T_c) of ~330 °C. The classic ferroelectric switching (Figure 12) was induced with minimal contribution from space charge and conduction [82], and a coercive field (E_c) of 10.9 kV/cm with a remanent polarization of 22.1 µC/cm^2 was obtained. The normalized strain $d*33$ was 220 pm/V (Figure 13), due to the fact that the interdigitated electrodes would have some clamping effect on the piezoelectric performance. The results are summarized inTable 4. Collectively, the data indicated that the Cu electrode is suitable and attractive for co-firing with NKN.

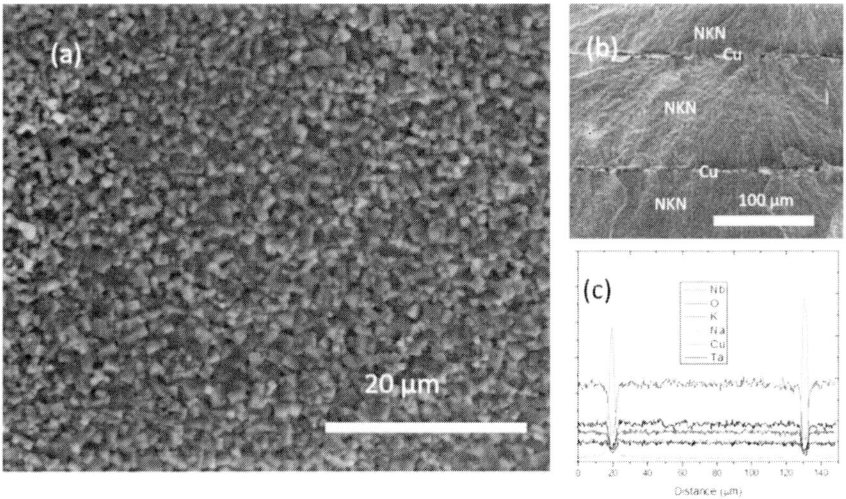

Figure 10. SEM Micrograph on (**a**) surface, (**b**) cross-section of Cu co-fired actuator, (**c**) EDS line scan on the cross-section of Cu co-fired actuator.

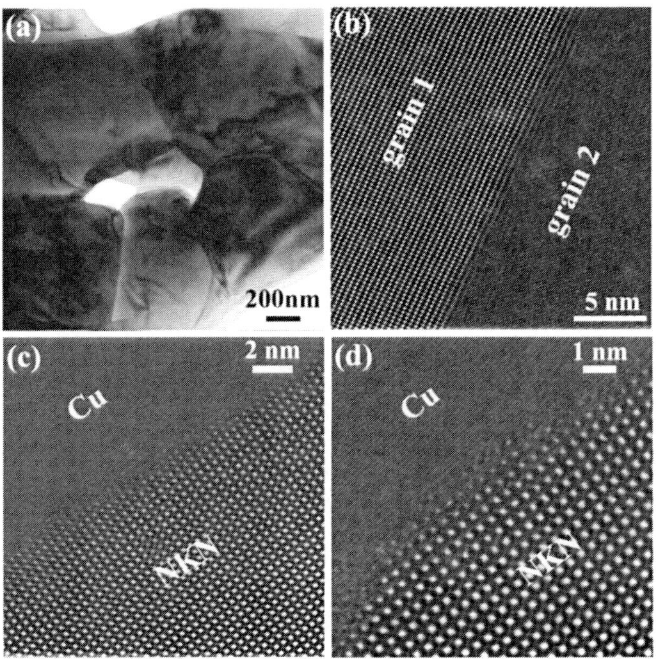

Figure 11. Cu-NKN MLCC (sintered at 1050 °C). (**a**) The typical TEM micrograph of the grains; (**b**,**c**,**d**) Atomistic STEM images of the grain boundary and Cu/NKN interface, respectively.

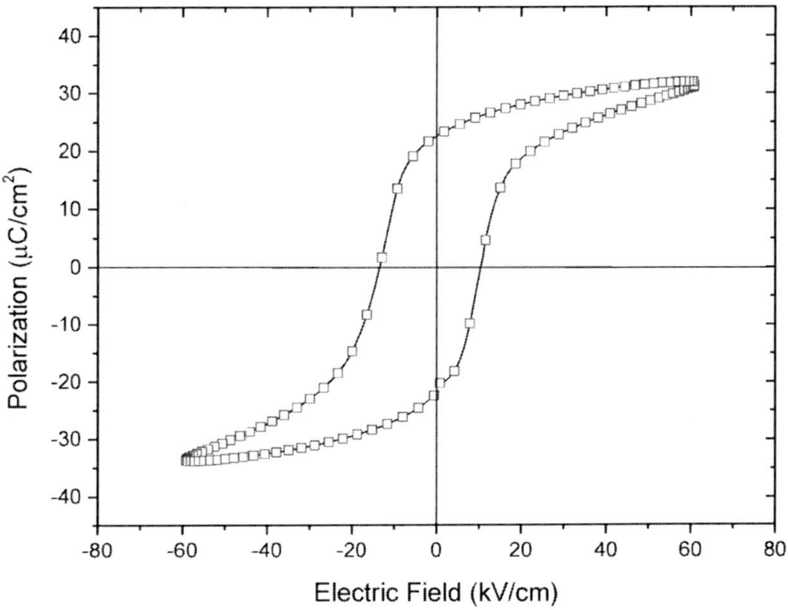

Figure 12. Ferroelectric hysteresis loop of Cu co-fired NKN actuator.

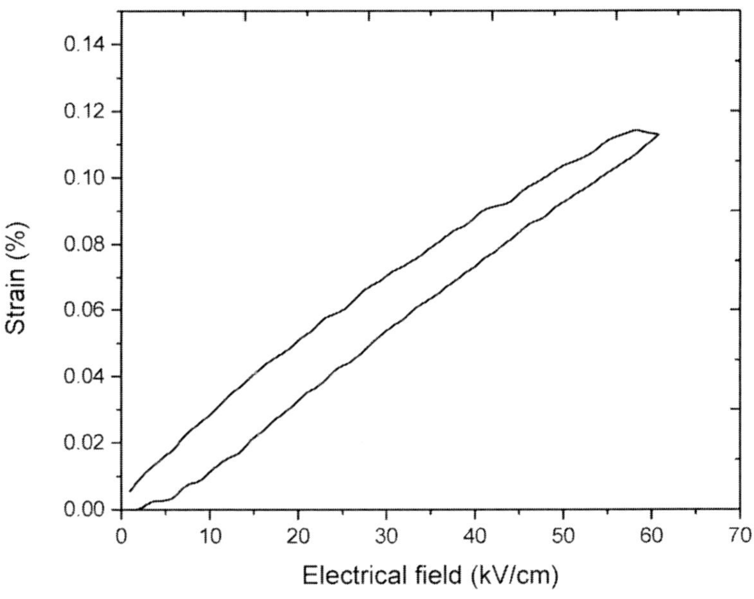

Figure 13. Unipolar electric field–induced strain of Cu Co-fired NKN sample.

Table 4. Values of electrical properties of Cu co-sintered NKN ceramics.

T_c (°C)	ε_r (1 kHz)	d^*_{33} (pm/V) @ 20 kV/cm	Tanδ (1 kHz)	E_c (kV/cm)	P_r (μC/cm^2)
330	830	220	0.036	10.9	22.1

11. SUMMARY

Recent progress on co-firing Ag/Pd, Ni, and Cu metal electrodes with both PZT and lead-free piezoelectric ceramics was presented. The sinterability of the piezoelectric ceramics was successfully improved via using sintering agents. The effects and mechanisms of adding sintering agents were carefully investigated. It was found that liquid phase sintering significantly lowered the sintering temperatures, thus widening the choices of metal electrode materials.

The volatility of Pb and alkaline elements can be suppressed through sintering in a reduced atmosphere. It was confirmed that the low pO_2–sintered samples showed higher resistivity than the air-sintered ones. The dominating path of losing volatile elements underwent coupling with surface-adsorbed oxygen. Therefore, reducing the pO_2 in the atmosphere consequently cut the chances of contact between the volatile elements and atmospheric oxygen, thereby retaining the volatile elements while sintering.

Interdiffusion of metal electrodes was investigated. The Ag/Pd internal electrode faced both diffusion during sintering and electromigration under electrical load; both resulted in the drop of reliability and lifespan of the piezoelectric applications. Both experimental and calculated results showed that PZT was intrinsically incompatible with metallic Ni electrodes, yet was eligible to co-fire with Cu internal electrodes. The Cu co-fired PZT displayed higher reliability than the Ag/Pd co-fired PZT. The Ni co-fired NKN displayed under-detectable interdiffusion along the electrode-ceramic interfaces and good electromigration resistivity.

The major roadblock toward base metal co-firing was the binder system. While burning out in the low pO_2 condition, the PVB binder could undergo a pyrolysis process, thus leaving carbon residues. However, burning out PPC polymer in the low pO_2 atmosphere follows depolymerization, generating monomers. Such a feature made it a better binder polymer for low pO_2 burnout.

Until very recently, our group utilized the PPC binder and successfully fabricated Cu co-fired prototyped NKN multilayer actuators. The whole heat treatment was conducted under low pO_2 atmosphere. There was no detectable residual carbon, interdiffusion, or alloy formation. Such success could give a general direction for future base metal co-firing investigations. Collectively, base metal (especially Cu) co-firing is largely unexplored; further deep investigations of the binder formulations, burnout conditions, and sintering profiles to realize highly reliable multilayer piezoelectrics are required. Some important information and facts are summarized in Table 5 shown below.

Table 5. Summary of important findings and progress on base metal co-firing for major piezoelectrics.

Materials	Incompatible Metal Electrode	Future Trend of Metal Electrodes	Challenges	Progress on Base-Metal Co-Firing
PZT	Ni	Cu	Binder, pO_2 control	Production in EPCOS/TDK [40]
NBT	-	Cu	Binder burnout	Pressed Cu with ceramic pellets [74]
NKN	-	Cu	Binder burnout	Cu co-fired prototyped multilayer actuators [81]

ACKNOWLEDGMENTS

The authors want to thank Amanda Baker, Steve Perini, and Jeff Long for their dedicated work on managing the lab affairs. Thanks to the Materials Chraracterization Laboratory (MCL) and Ke Wang for his assistance on TEM. The authors also want to thank PI Ceramics for providing the high-quality NKN ceramic powders. This material is based upon work supported in part by the National Science Foundation, as part of the Center of Dielectrics and Piezoelectrics under Grant No. IIP-1361571.

AUTHOR CONTRIBUTIONS

C.A.R. and S.Z. conceived and designed the experiments; H.G. provided TEM micrographs; L.G. integrated information and wrote the paper.

CONFLICTS OF INTEREST

The authors declare no conflict of interest.

REFERENCES

1. Randall, C.A.; Kelnberger, A.; Yang, G.Y.; Eitel, R.E.; Shrout, T.R. High Strain Piezoelectric Multilayer Actuators—A Material Science and Engineering Challenge. J. Electroceramics 2005, 14, 177–191.

2. Uchino, K.; Takahashi, S. Multilayer Ceramic Actuators. Curr. Opin. Solid State Mater. Sci. 1996, 1, 698–705.

3. Uchino, K. Ferroelectric Devices, 2nd ed.; CRC Press: Boca Raton, FL, USA, 2009.

4. Abraham, T. Expanding Markets for Piezoelectrics. Available online: http://www.ceramicindustry.com/articles/93845-expanding-markets-for-piezoelectrics (accessed on 22 October 2015).

5. Uchino, K. Entrepreneurship for Engineers; CRC Press: Boca Raton, FL, USA, 2009.

6. Knowles, G.J.; Bradley, W.M.; Bird, R. Galvanic Isolated Ceramic Based Voltage Sensors. WO2015,153,798 A1, 8 October 2015.

7. Imanaka, Y. Multilayered Low Temperature Cofired Ceramics (LTCC) Technology; Springer Science & Business Media: New York, NY, USA, 2005.

8. Gaskell, D.R. Introduction to the Thermodynamics of Materials, 5th ed.; CRC Taylor & Francis Group, LLC: New York, NY, USA, 2008.

9. Kishi, H.; Mizuno, Y.; Chazono, H. Base-Metal Electrode-Multilayer Ceramic Capacitors: Past, Present and Future Perspectives. Jpn. J. Appl. Phys. 2003, 42, 1–15.

10. Karakaya, I.; Thompson, W.T. The Ag-Pd (Silver-Palladium) System. Bull. Alloy Phase Diagrams 1988, 9, 237–243.

11. Wang, S.F.; Dougherty, J.P.; Huebner, W.; Pepin, J.G. Silver-Palladium Thick-Film Conductors. J. Am. Ceram. Soc. 1994,77, 3051.

12. Song, T.; Randall, C.A. Copper Cofire X7R Dielectrics and Multilayer Capacitors Based on Zinc Borate Fluxed Barium Titanate Ceramic. J. Electroceramics 2003, 10, 39–46.

13. Yonezawa, T.; Takeoka, S.; Kishi, H.; Ida, K.; Tomonari, M. The Preparation of Copper Fine Particle Paste and its Application as the Inner Electrode Material of a Multilayered Ceramic Capacitor. Nanotechnology 2008, 19, 145706.

14. Takeoka, S.; Mizuno, Y. Effect of Internal Electrode Materials in Multilayer Ceramic Capacitors on Electrical Properties. Jpn. J. Appl. Phys. 2011, 50, 1–6.

15. Shirane, G.; Takeda, A. Phase Transitions in Solid Solutions of PbZrO3 and PbTiO3 (I) Small Concentrations of PbTiO3.J. Phys. Soc. Jpn 1952, 7, 5–11.

16. Shirane, G.; Suzuki, K.; Takeda, A. Phase Transitions in Solid Solutions of PbZrO3 and PbTiO3 (II) X-ray Study. J. Phys. Soc. Jpn. 1952, 7, 12–18.

17. Jaffe, B.; Cook, W.R. Piezoelectric Ceramic, 1st ed.; Academic Press: London, UK, 1971.

18. Kingon, A.I.; Clark, J.B. Sintering of PZT Ceramics: II, Effect of PbO Content on Densification Kinetics. J. Am. Ceram. Soc. 1983, 66, 256–260.

19. Holman, R.L.; Fulrath, R.M. Intrinsic Nonstoichiometry in Single-Phase Pb(Zr0.5Ti0.5)O3. J. Am. Ceram. Soc. 1972, 55, 192–194.

20. Donnelly, N.J.; Shrout, T.R.; Randall, C.A. Properties of (1-x)PZT-xSKN Ceramics Sintered at Low Temperature Using Li2CO3. J. Am. Ceram. Soc. 2008, 91, 2182–2188.

21. Donnelly, N.J.; Shrout, T.R.; Randall, C.A. The Role of Li2CO3 and PbO in the Low-Temperature Sintering of Sr, K, Nb (SKN)-Doped PZT. J. Am. Ceram. Soc. 2009, 92, 1203–1207.

22. Wang, X.X.; Murakami, K.; Sugiyama, O.; Kaneko, S. Piezoelectric Properties, Densification Behavior and Microstructural Evolution of Low Temperature Sintered PZT Ceramics with Sintering Aids. J. Eur. Ceram. Soc. 2001,21, 1367–1370.

23. Donnelly, N.J.; Shrout, T.R.; Randall, C.A. Addition of a Sr, K, Nb (SKN) Combination to PZT(53/47) for High Strain Applications. J. Am. Ceram. Soc. 2007, 90, 490–495.

24. Kaneko, S.; Dong, D.; Murakami, K. Effect of Simultaneous Addition of BiFeO3 and Ba(Cu0.5W0.5)O3 on Lowering of Sintering Temperature of Pb(Zr,Ti)O3 Ceramics. J. Am. Ceram. Soc. 1998, 81, 1013–1018.

25. Ahn, C.W.; Nahm, S.; Ryu, J.; Uchino, K.; Yoon, S.J.; Jung, S.J.; Song, J.S. Effects of CuO and ZnO Additives on Sintering Temperature and Piezoelectric Properties of 0.41Pb(Ni1/3Nb2/3)O3-0.36PbTiO3-0.23PbZrO3 Ceramics. Jpn. J. Appl. Phys. 2004, 43, 205–210.

26. Hayashi, T.; Hasegawa, T. Piezoelectric Properties of Low-temperature Sintered Pb0.95Ba0.05[(Mg1/3Nb2/3)0.125Zr0.445Ti0.43]O3 Ceramics with Chemically-added LiBiO2 Sintering Aid. J. Eur. Ceram. Soc. 2005, 25, 2437–2441.

27. Donnelly, N.J.; Randall, C.A. Impedance Spectroscopy of PZT Ceramics—Measuring Diffusion Coefficients, Mixed Conduction, and Pb Loss. IEEE Trans. Ultrason. Ferroelectr. Freq. Control 2012, 59, 1883–1887.

28. Yoshida, T.; Nagasaka, T.; Hino, M. Effect of Oxygen on the Evaporation Rate of Lead from Liquid Copper under Reduced Pressure. ISIJ Int. 2001, 41, 706–715.

29. Helke, G.; Seifert, S.; Cho, S.-J. Phenomenological and Structural Properties of Piezoelectric Ceramics Based on xPb(Zr,Ti)O3-(1-x)Sr(K0•25Nb0•75)O3 (PZT/SKN) Solid Solutions. J. Eur. Ceram. Soc. 1999, 19, 1265–1268.

30. Atkin, R.B.; Fulrath, R.M. Point Defects and Sintering of Lead Zirconate-Titanate. J. Am. Ceram. Soc. 1971, 54, 265–270.

31. Donnelly, N.J.; Randall, C.A. Pb Loss in Pb(Zr,Ti)O3 Ceramics Observed by in Situ Ionic Conductivity Measurements.J. Appl. Phys. 2011, 109, 104107.

32. Uchino, K.; Nomura, S.; Cross, L.E.; Newnham, R.E.; Jang, S.J. Electrostrictive Effect in Perovskites and Its Transducer Applications. J. Mater. Sci. 1981, 16, 569–578.

33. Wersing, W.; Wahl, H.; Schnöller, M. PZT-based Multilayer Piezoelectric Ceramics with AgPd-Internal Electrodes.Ferroelectrics 1988, 87, 271–294.

34. Caballero, A.C.; Nieto, E.; Duran, P.; Moure, C.; Kosec, M.; Samardzija, Z.; Drazig, G. Ceramic-Electrode Interaction in PZT and PNN-PZT Multilayer Piezoelectric Ceramics with AG/PD 70/30 Inner Electrode. J. Mater. Sci. 1997, 32, 3257–3262.

35. Donnelly, N.J.; Randall, C.A. Refined Model of Electromigration of Ag/Pd Electrodes in Multilayer PZT Ceramics Under Extreme Humidity. J. Am. Ceram. Soc. 2009, 92, 405–410.

36. Jacob, K.T.; Saji, V.S. Interaction between Ni/NiO and PbTiO3: Phase Reversal with Redox Switching. J. Phase Equilibria Diffus. 2006, 27, 456–461.

37. Florian, H.; Reichmann, K.; Kainz, G. Piezoactuator and method for the production thereof. US Patent US7,304,414, 4 December 2007.

38. Takayama, Y.; Donnelly, N.J.; Randall, C.A. Piezoelectric Actuators with Cu Inner Electrode. In Proceedings of the 13th US-Japan Seminar on Dielectric and Piezoelectric Ceramics, Hyogo, Japan, 4–7 November 2007; pp. 113–116.

39. Florian, H.; Ottlinger, M.; Sedlmaier, P. Method for Producing a Multilayer Ceramic Component. US patent US8,776,364 B2, 15 July 2014.

40. Piezo Actuators for Fuel Injection Systems—Most Reliable and Cost-efficient. Available online: http://en.tdk.eu/tdk-en/373562/tech-library/articles/applications---cases/applications---cases/most-reliable-and-cost-efficient/1039300(accessed on 21 November 2015).

41. Polotai, A.V.; Yang, G.Y.; Dickey, E.C.; Randall, C.A. Utilization of Multiple-Stage Sintering to Control Ni Electrode Continuity in Ultrathin Ni-BaTiO3 Multilayer Capacitors. J. Am. Ceram. Soc. 2007, 90, 3811–3817.

42. Opitz, M.R.; Albertsen, K.; Beeson, J.J.; Hennings, D.F.; Routbort, J.L.; Randall, C.A. Kinetic Process of Reoxidation of Base Metal Technology BaTiO3-based Multilayer Capacitors. J. Am. Ceram. Soc. 2003, 86, 1879–1884.

43. Pithan, C.; Hennings, D.; Waser, R. Progress in the Synthesis of Nanocrystalline BaTiO3 Powders for MLCC. Int. J. Appl. Ceram. Technol. 2005, 2, 1–14.

44. Gui, Z.L.; Wang, Y.L.; Li, L.T. Study on the Interdiffusion in Base-Metal-Electrode MLCCs. Ceram. Int. 2004, 30, 1275–1278.

45. Wang, S.; Tsai, Y.; Lee, W. Study on (Ba, Ca)(Ti, Zr)O3 Dielectric Cofired with Copper Electrode. Jpn. J. Appl. Phys.2014, 53, 061501.

46. Yamamatsu, J.; Kawano, N.; Arashi, T.; Sato, A.; Nakano, Y.; Nomura, T. Reliability of Multilayer Ceramic Capacitors with Nickel Electrodes. J. Power Sources 1996, 60, 199–203.

47. Randall, C.A. Scientific and Engineering Issues of the State-of-the-Art and Future Multilayer Capacitors. J. Ceram. Soc. Jpn. 2001, 109, S2–S6.

48. Yang, G.Y.; Lee, S.I.; Liu, Z.J.; Anthony, C.J.; Dickey, E.C.; Liu, Z.K.; Randall, C.A. Effect of Local Oxygen Activity on Ni-BaTiO3 Interfacial Reactions. Acta Mater. 2006, 54, 3513–3523.

49. RoHS Compliance Guide: Regulations, 6 Substances, Exemptions, WEEE. Available online: http://www.rohsguide.com/ (accessed on 11 October 2015).

50. Directive 2011/65/Eu of the European Parliament and of the Council of 8 June 2011 on the Restriction of the Use of Certain Hazardous Substances in Electrical and Electronic Equipment. Available online: http://eur-lex.europa.eu/legal-content/EN/TXT/HTML/?uri= CELEX: 32011L0065&from=EN (accessed on 25 October 2015).

51. Egerton, L.; Dillon, D.M. Piezoelectric and Dielectric Properties of Ceramics in the System Potassium-Sodium Niobate.J. Am. Ceram. Soc. 1959, 42, 438–442.

52. Saito, Y.; Takao, H.; Tani, T.; Nonoyama, T.; Takatori, K.; Homma, T.; Nagaya, T.; Nakamura, M. Lead-free Piezoceramics. Nature 2004, 432, 84–87.

53. Guo, Y.; Kakimoto, K.; Ohsato, H. Dielectric and Piezoelectric Properties of Lead-free (Na0.5K0.5)NbO3-SrTiO3Ceramics. Solid State Commun. 2004, 129, 279–284.

54. Guo, Y.; Kakimoto, K.; Ohsato, H. (Na0.5K0.5)NbO3-LiTaO3 Lead-free Piezoelectric Ceramics. Mater. Lett. 2005, 59, 241–244.

55. Hollenstein, E.; Davis, M.; Damjanovic, D.; Setter, N. Piezoelectric Properties of Li- and Ta-modified (K0.5Na0.5)NbO3Ceramics. Appl. Phys. Lett. 2005, 87, 182905.

56. Kim, M.S.; Jeong, S.J.; Song, J.S. Electromechanical Properties of NKN-5LT Multilayer Actuator. Adv. Mater. Res. 2007,26–28, 263–266.

57. Kim, M.S.; Jeon, S.; Lee, D.S.; Jeong, S.J.; Song, J.S. Lead-free NKN-5LT Piezoelectric Materials for Multilayer Ceramic Actuator. J. Electroceramics 2009, 23, 372–375.

58. Kobayashi, K.; Doshida, Y.; Mizuno, Y.; Randall, C.A. Possibility of Cofiring a Nickel Inner Electrode in a (Na0.5K0.5)NbO3–LiF Piezoelectric Actuator. Jpn. J. Appl. Phys. 2013, 52, 1–5.

59. Kawada, S.; Kimura, M.; Higuchi, Y.; Takagi, H. (K,Na)NbO3-Based Multilayer Piezoelectric Ceramics with Nickel Inner Electrodes. Appl. Phys. Express 2009, 2, 111401.

60. Kawada, S.; Kimura, M.; Hayashi, H.; Ogiso, K.; Konoike, T.; Takagi, H. Study of Nickel Inner Electrode Lead-free Piezoelectric Ceramics. In Proceedings of the 2011 International Symposium on Applications of Ferroelectrics and 2011 International Symposium on Piezoresponse Force Microscopy and Nanoscale Phenomena in Polar Materials (ISAF/PFM), Vancouver, BC, Canada, 24–27 July 2011.

61. Liu, C.; Liu, P.; Kobayashi, K.; Qu, W.; Randall, C.A. Enhancement of Piezoelectric Performance of Lead-Free NKN-Based Ceramics via A High-Performance Flux-NaF-Nb2O5. J. Am. Ceram. Soc. 2013, 96, 3120–3126.

62. Wang, X.; Wu, J.; Xiao, D.; Zhu, J.; Cheng, X.; Zheng, T.; Zhang, B.; Lou, X.; Wang, X. Giant Piezoelectricity in Potassium-Sodium Niobate Lead-free Ceramics. J. Am. Chem. Soc. 2014, 136, 2905–2910.

63. Zhang, S.; Xia, R.; Shrout, T.R.; Zang, G.; Wang, J. Characterization of Lead Free (K0.5Na0.5)NbO3-LiSbO3Piezoceramic. Solid State Commun. 2007, 141, 675–679.

64. Lévêque, G.; Marchet, P.; Levassort, F.; Tran-Huu-Hue, L.P.; Duclere, J.R. Lead Free (Li,Na,K)(Nb,Ta,Sb)O3Piezoelectric Ceramics: Influence of Sintering Atmosphere and ZrO2 Doping on Densification, Microstructure and Piezoelectric Properties. J. Eur. Ceram. Soc. 2011, 31, 577–588.

65. Matsubara, M.; Yamaguchi, T.; Sakamoto, W.; Kikuta, K.; Yogo, T.; Hirano, S. Processing and Piezoelectric Properties of Lead-Free (K,Na)(Nb,Ta)O3 Ceramics. J. Am. Ceram. Soc. 2005, 88, 1190–1196.

66. Kobayashi, K.; Doshida, Y.; Mizuno, Y.; Randall, C.A. A Route Forwards to Narrow the Performance Gap between PZT and Lead-free Piezoelectric Ceramic with Low Oxygen Partial Pressure Processed (Na0.5K0.5)NbO3. J. Am. Ceram. Soc. 2012, 95, 2928–2933.

67. Shimizu, H.; Kobayashi, K.; Mizuno, Y.; Randall, C.A. Advantages of Low Partial Pressure of Oxygen Processing of Alkali Niobate: NaNbO3. J. Am. Ceram. Soc. 2014, 97, 1791–1796.

68. Shigemi, A.; Wada, T. Enthalpy of Formation of Various Phases and Formation Energy of Point Defects in Perovskite-type NaNbO3 by First-principles Calculation. Jpn. J. Appl. Phys. 2004, 43, 6793–6798.

69. Shigemi, A.; Wada, T. Evaluations of Phases and Vacancy Formation Energies in KNbO3 by First-Principles Calculation. Jpn. J. Appl. Physics 2005, 44, 8048–8054.

70. Gao, R.; Chu, X.; Huan, Y.; Zhong, Z.; Wang, X.; Li, L. Ceramic-Electrode inter-Diffusion of (K, Na)NbO3-Based Multilayer Ceramics with Ag0.7Pd0.3 Electrode. J. Eur. Ceram. Soc. 2015, 35, 389–392.

71. Kobayashi, K.; Randall, C.A. New Opportunity in Alkali Niobate Ceramics Processed in Low Oxygen Partial Pressure. In Proceedings of the Applications of Ferroelectrics held jointly with 2012 European Conference on the Applications of Polar Dielectrics and 2012 International Symp Piezoresponse Force Microscopy and Nanoscale Phenomena in Polar Materials (ISAF/ECAPD/PFM), Aveiro, Portugal, 9–13 July 2012.

72. Liu, C.; Liu, P.; Kobayashi, K.; Randall, C.A. Base Metal Co-fired (Na, K)NbO3 Structures with Enhanced Piezoelectric Performance. J. Electroceramics 2014, 32, 301–306.

73. Sapper, E.; Gassmann, A.; Gjødvad, L.; Jo, W.; Granzow, T.; Rödel, J. Cycling Stability of Lead-free BNT–8BT and BNT–6BT–3KNN Multilayer Actuators and Bulk Ceramics. J. Eur. Ceram. Soc. 2014, 34, 653–661.

74. Yesner, G.; Kuciej, M.; Safari, A. Low Temperature Sintering of Bi0.5Na0.5TiO3 Based Ceramics. In Proceedings of the 2015 Joint IEEE International Symposium on the Applications of Ferroelectric, International Symposium on Integrated Functionalities and Piezoelectric Force Microscopy Workshop (ISAF/ISIF/PFM), Singapore, 24–27 May 2015; pp. 21–23.

75. Masia, S.; Calvert, P.; Rhine, W.; Bowen, H. Effect of Oxides on Binder Burnout during Ceramics Processing. J. Mater. Sci. 1989, 24, 1907–1912.

76. Liau, L.C.K.; Yang, T.C.K.; Viswanath, D.S. Mechanism of Degradation of Poly(Vinyl Butyral) Using Thermogravimetry/Fourier Transform Infrared Spectrometry. Polym. Eng. Sci. 1996, 36, 2589–2600.

77. Liau, L.C.K.; Viswanath, D.S. Thermal Degradation of Poly (Vinyl Butyral)/ Ceramic Composites: A Kinetic Approach.Ind. Eng. Chem. Res. 1998, 37, 49–57.

78. Salam, L.A.; Matthews, R.D.; Robertson, H. Pyrolysis of Polyvinyl Butyral (PVB) Binder in Thermoelectric Green Tapes. J. Eur. Ceram. Soc. 2000, 20, 1375–1383.

79. Yan, H.; Cannon, W.R.; Shanefield, D.J. Poly(Vinyl Butyral) Pyrolysis: Interactions with Plasticizer and AIN Ceramic Powder. MRS Proc. 1991, 249, 377–382.

80. Yan, H.; Cannon, W.R.; Shanefield, D.J. Thermal Decomposition Behaviour of Poly(Propylene Carbonate). Ceram. Int.1998, 24, 433–439.

81. Gao, L.; Ko, S.-W.; Guo, H.; Hennig, E.; Randall, C.A. Demonstration of Copper Co-fired (Na, K)NbO3 Multilayer Structures for Piezoelectric Applications. J. Am. Ceram. Soc. 2016. in press.

82. Scott, J.F. Ferroelectrics Go Bananas. J. Phys. Condens. Matter 2008, 20, 021001.

9

SIMULTANEOUS SMART ACTUATING-SENSING DEVICES BASED ON CONDUCTING POLYMERS

José G. Martínez, Joaquín Arias-Pardilla and Toribio F. Otero

[1] Universidad Politécnica de Cartagena, Spain

1. INTRODUCTION

Towards the end of 1970s new artificial organic materials, conducting polymers (CP), were discovered (Chiang et al., 1977; Inzelt, 2011; Shirakawa et al., 1977). Since then, most of the scientists working on CP became interested by the fact that their conductivity can shift, in a reversible way, over several orders of magnitude by oxidation/reduction (also called doping/dedoping) processes. The availability of these new organic semiconductors has opened up possibilities to rebuilt electronics and microelectronics producing flexible devices (Guo et al., 2010; Klauk, 2006; Perepichka & Perepichka, 2009; So, 2010).

Besides conductivity other properties such as stored charge, stored chemicals, volume, porosity or colour also change during doping/dedoping, under electrochemical control, in parallel with the material composition (counterion content) along the redox reactions (Otero, 2008). Any intermediate oxidation state determines a chemical equilibrium characterized by an equilibrium potential. Any physical (temperature, pressure, applied current) or chemical (electrolyte concentration) magnitude affecting the chemical equilibrium modifies the electrical potential of the material that therefore can be used as a sensor of that magnitude (Otero, 2009).

These properties, whose value changes with the material composition, are allowing the development of different devices (Onoda et al., 1999a; Otero et al., 1992a; Otero et al., 1992c; Pei & Inganäs, 1992a). Volume variations driven by oxidation/reduction reactions are being used to generate new electrical motors having different configurations of the polymeric actuators (Alici et al., 2007; Smela et al., 1993;Wu et al., 2005).

We will present devices constructed with conducting polymers based on an electrochemical reaction and working, simultaneously, as an actuator and as several sensors of the surrounding conditions. During the movement they store and release charge working as a battery: they are multifunctional devices.

2. ELECTROCHEMICAL REACTIONS IN CONDUCTING POLYMERS

Once generated (chemical or electrochemically), conducting polymers can be considered as both, classical unreactive materials having either a constant composition and a constant value of the magnitude of its physical and chemical properties; or new reactive materials varying their composition (polymeric chains, ions and solvent in different ratios) and properties along several orders of magnitude in a reversible way under electrochemical reaction (Otero, 1999), or promoted by redox agents present in solution (Kuttel et al., 2009).

2.1. P-Doping

When the polymer chains are oxidized, consecutive electrons are removed from each chain generating an excess of positive charges (holes) along the chains. This excess of positive charges (lack of electrons) promotes the repulsion between the polymeric chains and the generation of free volume between them. This free space is occupied by anions arriving from the solution to compensate the emerging positive charges (keeping the electroneutrality) and solvent molecules to keep osmotic pressure balance (Huang et al., 1986; Otero, 1999; Tsai et al., 1988).

When the polymer is generated in the presence of small anions, they can be exchanged by other small anions present in solution by electrochemical reactions so a prevailing exchange of anion occurs during reaction:

$$\left(Pol^{0}\right)_{s}+n\left(A^{-}\right)_{sol}+m\left(Solv\right)\rightleftarrows\left[\left(Pol^{n+}\right)_{s}\left(A^{-}\right)_{n}\left(Solv\right)_{m}\right]_{gel}+n\left(e^{-}\right)_{metal} \quad (1)$$

where the different subscripts mean: *s*, solid; *sol*, in solution; *gel* indicates that the material is a gel formed by oxidized polymeric chains (Pol^{n+}) generated after the extraction of *n* electrons (e^-) through the metal (indicated by subscript *metal*) from neutral polymer chains (Pol^0), *n* anions (A^-) coming from the solution to keep the gel electroneutrality and m molecules of solvent (*Solv*) required to keep osmotic pressure balance.

When the polymer is generated in the presence of a macroanion, due to its volume and the interaction with polymer chains, this macroanion cannot be exchanged by the electrochemical reaction keeping trapped inside the polymer film. So, in order to keep the electroneutrality, smaller cations are exchanged with the solution during the reaction:

$$\left[\left(Pol^0\right)\left(MA^-\right)_n\left(C^+\right)_n\left(Solv\right)_m\right]_{gel} \rightleftarrows \left[\left(Pol^{n+}\right)\left(MA^-\right)_n\right]_{gel} + n\left(C^+\right)_{sol} + m\left(Solv\right) + n\left(e^-\right)_{metal} \quad (2)$$

where MA^- is the macroanion trapped inside the polymer film and C^+ are cations exchanged in order to keep the electroneutrality.

Usually, the real redox process is not as easy as expressed by reactions (1) and (2): anions and cations are exchanged simultaneously (Hillman et al., 1989; Inzelt, 2008). Usually one of the previous interchanges prevails supporting the greater percentage of charge balance (Kim et al., 2010; Lyutov et al., 2011; Orata & Buttry, 1987; Torresi & Maranhao, 1999).

2.2. N-Doping

Some CP such as PEDOT (Ahonen et al., 2000), polythiophene (Arbizzani et al., 1995) or polyfluorenes (Ranger & Leclerc, 1998) have an electronic affinity high enough to allow transitions from the neutral state to a reduced state, storing negative charges (by electron injection) on the chains at high cathodic potentials. In this case, very stable solvents and salts are required, as electrolytes, to perform this redox reaction:

$$\left(Pol^0\right)_s + n\left(C^+\right)_{sol} + m\left(Solv\right) + n\left(e^-\right)_{metal} \rightleftarrows \left[\left(Pol^{n-}\right)_s\left(C^+\right)_n\left(Solv\right)_m\right]_{gel} \quad (3)$$

where Pol^{n-} represents the reduced polymer chains after insertion of n electrons. Here, in an analogous way compared to reaction (1), an excess of negatives charges promotes repulsion between polymeric chains generating the free volume that will be occupied by cations (exchanged with solution to keep electroneutrality) and solvent (exchanged with solution to keep pressure and osmotic balance).

2.3. Double Doping

Some conducting polymers can be doped both, by p-doping and by n-doping. Thus, from their neutral state they can be reduced (suffering n-doping) or oxidized (suffering p-doping). In those polymers the energy difference between both processes is the electrochemical bandgap (Arias-Pardilla et al., 2010;Otero et al., 2011).

3. REVERSIBLE CHANGE OF THE ELECTROCHEMICAL PROPERTIES

Electrochemical reactions 1, 2 and 3 are reversible reactions moving through *n* consecutive oxidation states (Otero et al., 2012) having different content of the counterion. The value of the magnitude of those properties being a function of the material composition (electrochemical properties) can also be shifted in a reversible way under faradic control. As previously indicated, conducting polymers can be oxidized and/or reduced from their neutral state, with the entry/exit of ions and solvent. The most studied electrochemical properties of conducting polymers are: volume, colour, stored charge, stored chemicals, porosity or permselectivity, sensing responses or wetability, among others. The progressive and reversible variation of the value of these electrochemical properties allows the development of different devices and products. The change in volume will be reviewed in detail below: this is the most important property of conducting polymers for the development of actuators or artificial muscles.

3.1. Volume Variations

A chain of conducting polymer in solution can be considered as an electrochemical molecular motor (Balzani et al., 2005; Davis, 1999; Otero, 2011): movementsare produced by reversible conformational changes in chains originated by oxidation/reduction reactions. The reversible conformational movement from a coil like structure to a rod like structure is produced by extraction (oxidation) or injection (reduction) of *n* electrons through *n* consecutive steps of one electron per step, together with movement of balancing counterions. This results in length variation of a free polymer chain in solution but, in polymer films three dimensional changes of volume are observed. The entanglement of the polymer chains in the film gives reversible swelling or shrinking changes of volume under reversible electrochemical stimulation (Fig. 1).

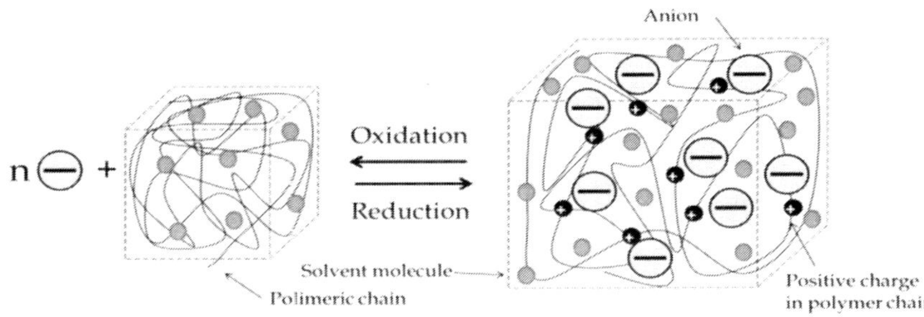

Figure 1. Schematic representation of the reversible volume change associated with the electrochemical reaction in conducting polymer chains during oxidation/reducction during p-doping exchanging anions.

Some mechanical test machines have been developed following length or thickness variations produced by submitting the film to different potential (Bay et al., 2003; Kiefer et al., 2007; Mazzoldi et al., 1998;Spinks et al., 2002) or current (Otero et al., 2006; Otero et al., 2007c) programs. In situ Atomic Force Microscopy (AFM) technique follows film thickness variation during reverse oxidation/reduction processes (Bieńkowski et al., 2011; Cho et al., 2011; Smela & Gadegaard, 2001). In this way, it has been possible to measure a volume difference between reduced and oxidized state up to 35% (Smela & Gadegaard, 1999). The volume change depends on multiple factors such as type of polymer, synthesis conditions (potential or current applied), electrolyte and solvents used.

4. SIMULTANEOUS SENSING AND ACTUATING PROPERTIES IN CONDUCTING POLYMERS

The driving reversible electrochemical reaction supports the development of simultaneous sensing and actuating properties by conducting polymer materials and by any electrochemical device based on those materials. The electrochemical working device (based on the electrochemical reaction) senses changes of any physical or chemical variable acting on the polymer reaction rate while working. This is the Le Chatelier principle applied outside equilibrium conditions (Smith, 2004). Therefore, for increasing values of the electrolyte concentrations or of the temperature working under flow of a constant current (constant reaction rate), lower values of the device potential are observed (the reaction is easier) during the transition between the same initial and final oxidation states of the materials. When a greater mechanical work is required (moving faster, applying a higher current, or a higher mechanical stress is needed to move the actuator) the reaction gives increasing potential when the device moves between the same initial and final oxidation states.

Those sensing abilities are intrinsic properties of the reaction. They are characteristics of the material reaction and of any device based on this electrochemical reaction. So, the dual and simultaneous sensing-actuation property is expected to be quantified from electrochemical equations.

The evolution of the conducting polymer film potential with time E(t) during the movement from the same initial oxidation/reduction state to the same final oxidation/reduction state driven by flow of a constant anodic current is given by a stair function (Otero et al., 2012).

$$E(t)=\sum E_n(t)p_n(t)=E_1(t)p_1(t)+E_2(t)p_2(t)+...+E_n(t)p_n(t) \tag{4}$$

where:

$$p_n(t)=u(t-t_n)-u(t-t_{n+1})=\begin{cases}1, t\in[t_n,t_{n+1}]\\0, t\notin[t_n,t_{n+1}]\end{cases} \tag{5}$$

Being t_n the time while the nth electron removed from every polymeric chain and $E_n(t)$:

$$E_n(t)=E_0+(n-1)\Delta E+\frac{RT}{(1-\alpha)F}\left\{\ln\left(\frac{i_a}{FV}\right)-d\ln\left[A^-\right]-e\ln\left(\left[Pol^*\right]_{initial}-\frac{i_a t}{FV}\right)-\ln k_{a0}\right\} \tag{6}$$

where E_0 is the standard potential, i_a is the applied current; n is the number of consecutive electrons extracted from a chain; ΔE is the increment observed in the potential when a new electron is extracted from a polymeric chain, R is the universal gas constant (R = 8.314 J K^{-1} mol^{-1}); α is the symmetry factor; F is the Faraday constant (F=96485 C mol^{-1}); V, the volume of the film; [A^-] the concentration of anions (counterion) in solution; t, the time of current flow; T is the experimental temperature; d and e are the reactions orders related with the concentration of anions in solution or to that of the active centres [Pol^*] in the film (sites of the polymer where a positive charge will be stored after oxidation) and k_{a0} is the rate constant or rate coefficient for $E=E_0$.

Therefore, Eqs. 4 and 6 are the sensing equations: the evolution of the device potential during actuation is a function of either, driving (current) and working (temperature, electrolyte concentration and film volume) variables.

$$U_a(t) = i_a \int E(t)dt = i_a t\left\{E_0 + (n-1)\Delta E\right\} + \frac{RTi_a t}{(1-\alpha)F}\left\{\ln\left(\frac{i_a}{FV}\right) - d\ln\left[A^-\right] - \ln k_{a0}\right\} + \frac{RTVc}{(1-\alpha)}\left\{\ln\left(\left[Pol^*\right]_{mass} - \frac{i_a t}{FV}\right) - 1\right\}\left\{\left[Pol^*\right] - \frac{i_a t}{FV}\right\} \quad (7)$$

Being electrical machines, by integration of Eq. 6 the evolution of the electrical energy consumed by the electrochemical device (U_a) during the actuation time is attained:

The consumed energy (U_a) after any constant time (t) of current flow is also a sensing function of the same variables. Fig. 2 shows the good agreement between experimental and theoretical results for the consumption of three different charges (from the same initial oxidation/reduction state, three different final oxidation/reduction states are obtained) at different experimental temperatures by flow of a constant anodic currentfor three different times of current flow.

5. ARTIFICIAL MUSCLES

In general, artificial muscles are devices attempting to reproduce composition, characteristics and capabilities of natural muscles. Different materials have been used as piezoelectrics (PZT), shape memory alloys (SMA) (Ouyang et al., 2008), carbon based materials as nanotubes (Baughman et al., 1999) or graphene (Huang et al., 2012) and polymeric gels. Also the pneumatic approach has been studied with contractile or extensional devices operated by pressurized air (Daerden & Lefeber, 2002).

Natural muscles transform chemical energy into mechanical energy and heat. Their actuation involves: a) aqueous media, b) an electric pulse arriving from the brain to the muscle through the nervous system, c) liberation of calcium ions inside the sarcomere, d) chemical reactions (ATP hydrolysis), e) conformational changes along natural polymeric chains (actin and myosin) with change of the sarcomere volume and f) water exchange. Among of the above mentioned materials for artificial muscles, conducting polymers are the most similar to natural devices including: electric pulses, polymeric chains, chemical reactions, aqueous solutions (ions and water), volume variations, and strain and stress changes. Two families of electrical polymeric actuators can be differentiated (Otero et al., 2007a; Otero, 2000; Otero, 2007):

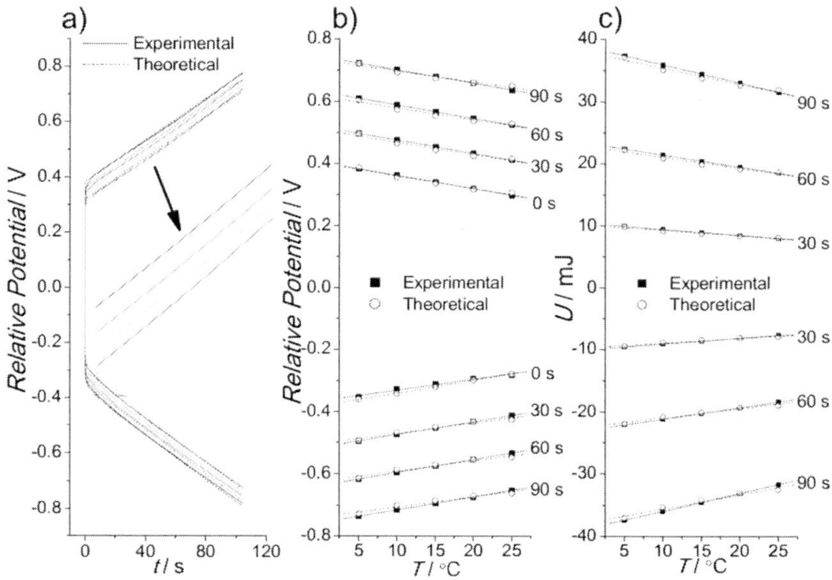

Figure 2. a) Anodic and cathodic experimental (full lines) and theoretical (dotted lines) chronopotentiograms obtained by flow of ±0.75 mA through a 1.6 mg pPy film (10.77 mm x 5.09 mm x 19 μm) at different temperatures (black line: 5ºC; red line: 10ºC; green line: 15ºC; blue line: 20ºC and cyan line: 25ºC) in 1 M $LiClO_4$ aqueous solution. b) Achieved potential after different times of anodic (positive) or cathodic (negative) current flow. c) Consumed electrical energy after the same times of current flow. Reproduced from (**Otero et al., 2012**), with permission of American Chemical Society).

- **Electromechanical actuators**: the polymer responds to electric fields (E) and the polymer chains do not participate in chemical reactions (the structure of its intramolecular chemical bonds doesn't change) during actuation. In this type of actuator the dimension variations are proportional to $E^2$for electrostrictive actuators and to E for electrostatic and electrokinetic (movement of ions and/or solvent molecules) actuators.

- **Electrochemomechanical actuators**: the polymer chains respond to electric currents and participate in electrochemical or chemical reactions, changing the structure of its chemical bonds, varying their composition and originating volume changes. In this case dimensional variations are under control of the electrochemical reaction becoming proportional to consumed charge. Conducting polymers can be used for the production of this kind of actuators. Whether carbon nanotubes and graphenes based actuators are electromechanical or electrochemomechanical is still under discussion (Gimenez et al., 2012; Mukai et al., 2011). Any electrochemically reactive material that can be laminated as stable films can be used as part of electrochemomechanical actuators.

From now on we will focus on electrochemomechanical actuators. These actuators have been built using different configurations as discussed below, always containing one or several films of conducting polymers, where reaction 1, 2 or 3 takes place. Volume variations generated by reactions 1, 2 or 3 are almost isotropic, while natural muscles are anisotropic devices. So to produce anisotropy, only volume changes following length variation of the films are used, with the consequent efficiency reduction.

5.1. Bending Artificial Muscles

Historically the first way to transduce reversible length variations in films of conducting polymers into macroscopic movements was through a bilayer, or bimorph structure, i.e. CP/passive layer (Otero et al., 1992b; Otero et al., 1992a; Otero et al., 1992c; Pei & Inganäs, 1992a) (Fig. 3). The variation of the mechanical stress gradient generated across the bilayer interface by swelling/shrinking processes induced by the electrochemical reactions in the conducting polymer film develops a macroscopic movement of the bilayer free end by the progressive bending of the device. The direction (clockwise or anticlockwise) of the movement depends on the prevalent ionic exchange (anions or cations) of the conducting polymer film. Conducting polymers with a prevalent exchange of anions swell by oxidation, pushing the bilayer free end meanwhile conducting polymers with prevalent cation exchange shrink during oxidation, trailing the device. Different materials have been used as passive layer, for example a tape (Otero et al., 1992b; Otero et al., 1992a; Otero et al., 1992c; Pei & Inganäs, 1992a), a sputtered metal (Jager et al., 2000a; Jager et al., 2000b; Smela et al., 1993), a piece of paper (Deshpande et al., 2005b), a non conducting plastic (Higgins et al., 2003), a solid state electrolyte film (Alici et al., 2011;Baughman, 1996) or a thin film of any flexible material which is metal coated (i.e. by sputtering) (Deshpande et al., 2005a).

In a similar way it is possible to obtain bending movement from asymmetrical monolayers of the same conducting polymer, having an internal asymmetry capable of producing asymmetric swelling or shrinking across the film under the same electrochemical process (Okamoto et al., 2000; Onoda et al., 1999a; Onoda & Tada, 2004; Onoda et al., 1999b; Shakuda et al., 1993; Takashima et al., 2003; Takashima et al., 1997; Wang et al., 2002). Here half of the film has a prevalent anionic exchange, while the second half experiences a prevalent cationic exchange. These asymmetrical monolayers are obtained in two separate stages of electrogeneration using different salts and the same monomer. Other ways are being studied to produce asymmetric monolayers by physical means, for example, by growing the conducting polymer on adsorbed and porous materials (Li et al., 2004), or by electrochemical means generating a film of conducting polymer with a counterion concentration gradient (Okuzaki & Hattori, 2003; Sansiñena et al., 2003), conductivity (Nakano & Okamoto, 2001; Onoda et al., 2005) or morphology gradients (Han & Shi, 2006; Okamoto et al., 2001) by crosslinked networks.

All these bilayer devices require a counter-electrode in order to close the electrical circuit allowing the current flow. In this electrode (usually a metal) different electrochemical reactions as solvent oxidation, must occur during current flow consuming a major fraction of the electrical energy, resulting in pH variations and generating new chemicals, which can deteriorate progressively the bilayer device.

Trying to avoid the counter electrode and its associated problems, a three layer structure was proposed (Otero et al., 1992c). Initially, it was produced by using a central passive film (two sides tape) each side coated with a conducting polymer film (Fig. 3). The triple layer was immersed in an electrolyte allowing the current flow. One of the conducting polymer films acts as the anode while the second film acts as the cathode (Garcia-Cordova et al., 2011; John et al., 2008; Yao et al., 2008). But using this configuration it is also possible to obtain movement outside a liquid electrolyte media using an ionic conducting membrane to separate the two films of conducting polymers. This membrane can be obtained by solvent evaporation and UV irradiation (Blonsky & Meridian, 1997; Heuer et al., 2002;Sansinena et al., 1997; Song et al., 2002), or by formation of interpenetrated networks (Cho et al., 2007;Plesse et al., 2005; Vidal et al., 2009; Vidal et al., 2003). In this case, the two conducting polymer films are generated by chemical polymerization on the external part of the membrane. Using this approach, multilayer devices were constructed and characterized (Ikushima et al., 2009; Zainudeen et al., 2008). The three-layer configuration provides greater efficiencies of the consumed energy: the same current is used two times to produce opposite electrochemical reactions and volume variations in the conducting polymer films; the anode swells and pushes the device and the cathode shrinks and trails the device.

5.2. Longitudinal Or Linear Artificial Muscles

Freestanding conducting polymer films are the simplest longitudinal actuators (DellaSanta et al., 1997). Its actuation principle is based on longitudinal expansion and contraction of the polymer during ionic exchange, although expansion and contraction occur in all three dimensions, as previously indicated. To improve the performance of these actuators overcoming problems of fragility, multilayered actuators were proposed (Hara et al., 2006; Kaneto et al., 2008; Kaneto et al., 2009), in which several thin conducting polymer films and an electrolyte (ionic liquid-soaked paper) are used to develop a compact and scalable longitudinal actuator with a high work output (Ikushima et al., 2009). Also folded films with Origami shapes provide good linear movements (Okuzaki, 2008).

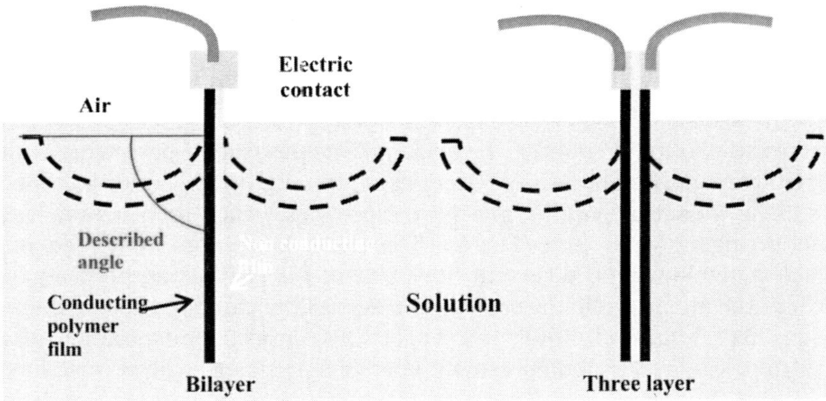

Figure 3. Bilayer and three layer devices in solution, formed by conducting polymer films and non-conductive films. Described angle is also shown.

Fibres of conducting polymer also can be considered as longitudinal actuators. Fibres can be obtained by extrusion (Mazzoldi et al., 1998) or by chemical polymerization over a fibre-shaped substrate (Ismail et al., 2011; Lu et al., 2002), or hollow fibre solid polymer electrolyte (Plesse et al., 2010) making it possible to obtain two concentric CP films separated by the electrolytic medium, allowing its movement in air (Dobbelin et al., 2010; Plesse et al., 2009; Vidal et al., 2010; Vidal et al., 2009). Microrods (Cho et al., 2011) or nanorods (Park et al., 2009; Vlad et al., 2012) of conducting polymers and bundles of films or fibres were investigated to produce vertical displacements of weights (Lu et al., 2002). Also, conducting polymer tubes were generated using springs and helical metallic wires (Ding et al., 2003;Hara et al., 2003; Hara et al., 2005; Spinks et al., 2003b) or zigzag metal wires (Hara et al., 2004;Morita et al., 2010) as substrates, looking for uniform potential and current distribution. When individual fibres, bundles or tubes are used, a counter electrode is necessary, for the same reasons given above in the case of the bilayers.

Finally, it is possible to obtain linear displacement by combination of different bending structures as bilayers (Fuchiwaki et al., 2009; Naka et al., 2010; Otero & Broschart, 2006) or trilayers (Mutlu & Alici, 2010; Otero et al., 2007b) achieving longitudinal displacements over 60% of their original length.

6. SENSING AND TACTILE MUSCLES

As mentioned above, any reactive (electrochemical) device based on conducting polymers will sense every variable influencing the electrochemical reaction rate during actuation. Following this basic principle of the chemical kinetics sensing and tactile artificial muscles have been developed.

While a current is applied to the artificial muscle, producing a mechanical work, it is possible to follow the potential achieved in the muscle at every time. Under flow of a constant current (constant reaction rate) the achieved potential is lower when a variable which favours the electrochemical reaction increases. This is the case for temperature (Garcia-Cordova et al., 2011; Ismail et al., 2011; Otero & Cortes, 2003b; Valero Conzuelo et al., 2010; Valero et al., 2010) or electrolyte concentration (Arias-Pardilla et al., 2011; Garcia-Cordova et al., 2011; Otero & Cortes, 2003b; Otero et al., 2007b; Valero Conzuelo et al., 2010). On the other hand, the potential shifts to higher values when a variable makes the reaction harder: the muscle moves larger masses (Garcia-Cordova et al., 2011; Otero et al., 2007b;Valero Conzuelo et al., 2010) or moves the mass faster by applying now rising currents (Garcia-Cordova et al., 2011; Ismail et al., 2011; Valero et al., 2011).

Being the potential evolution a sensor of the working variables, the electrical energy (U) consumed by the device during actuation and obtained by integration is also sensor:

$$U(t) = \int E(t) I dt \qquad (8)$$

Where t is the elapsed time, E(t) is the potential evolution during the actuation time and I is the constant applied current.

When a free muscle moves driven by a constant current finds an obstacle, the potential steps to higher values, trying to produce more energy and to shift the obstacle. The potential increment detects the object and its mechanical resistance to be shifted. Related with this property, artificial muscles with tactile sensitivity have been developed too (Otero & Cortes, 2003a).

7. MULTI-DEVICES: ACTUATORS-SENSORS-BATTERY-ELECTROCHROMIC

As mentioned above, different properties can be tuned simultaneously and in a reversible way during electrochemical reactions in conducting polymers. Electrochemical devices created with conducting polymers may change several of those properties, such as volume, stored charge or color, at the same time. For example, using the same configuration, changing the quantity of conducting polymer used in the device, it's possible to obtain different optical or mechanical devices (Vidal et al., 2010).

Also, as the configuration of a three layer is the same of a battery with two conducting polymer films as electrodes, electrical charge is trapped in the actuator during electrochemical reaction. A way to recover that charge may be developed in order to recover that energy, acting now as a battery and moving the actuator in the opposite directionuntil a uniform oxidation state is attained in both polymer films (vertical position in Fig. 3).

8. THEORETICAL MODELS

Several models have been proposed to characterize the electrochemomechanical behaviour of artificial muscles. At the moment, there exist different approaches from different fields as mathematics, physics or chemical-physics.

8.1. Faradic Control Of The Movement

As mentioned above, artificial muscles are electrochemomechanical machines controlled by the electrochemical reaction occurring while actuating. As any electrochemical reaction, actuation in conducting polymers is controlled by the charge consumed during actuation. Volume changes are not an exception: according to reactions 1, 2 or 3, volume variations will be given by the number of extracted (or inserted) charges from (or to) the polymeric chains, promoting the swelling/shrinking and the ionic exchange during reaction.

For bending bilayer or trilayer artificial muscles, it has been experimentally proven that the described angle (α in rad) follows a linear relationship with the consumed charge (Q in C):

$$\alpha(t) = kQ(t) \tag{9}$$

where k is a constant (rad C^{-1}) depending on every actuator system: the device (conducting polymer and isolating tape) and the electrolyte where is moving.

By definition of the angular rate of the movement (ω):

$$\omega(t) = \frac{d\alpha(t)}{dt} = k\frac{dQ(t)}{dt} = kI(t) \tag{10}$$

This expression confirms the faradic control of the movement: the angular rate of the movement is a linear function of the applied current. Any increment of the current produces (immediately, without any relaxing time) a faster movement of the actuator, by stopping the current flow the movement stops (the driving reaction and the film volume variation stops). Eq. 9 also indicates that the actuator is a positioning device: the same charge produces the same displacement and the charge consumed during description of a movement of one degree ($\alpha/Q = k$) is constant (independent of the applied current).

The above expressions can be normalized by mass unit of active conducting polymer reacting during actuation. This allows predicting the behaviour of every artificial muscle moving in a known electrolyte made of the same material whatever the geometry of the device is (shape, thickness, surface area, etc.). That means that the same change of the specific composition (according with reactions 1, 2 or 3) per unit time produce the same angular rate in devices having different geometry. This means that experiments from one muscle are only required in order to obtain this faradic characteristic of the CP.

The faradic control of the movement has been checked with different artificial muscles made of different polymers, exchanging both anions (Otero & Cortes, 2004) or cations (Valero et al., 2011).

8.2. Bending Beam Method

The bending beam method (Gere & Goodno, 2009; Timoshenko, 1925) is based on the analogy existing between a bending artificial muscle and a solid-state bending beam: the study of the forces generated at the interface between the non-conductive layer, keeping its volume constant during actuation, and the conducting polymer film, varying its volume locally.

This mechanical model assumes several characteristics related with the study of traditional mechanical bending beam: (I) the thickness of the beam is small compared to the minimum radius of curvature, (II) a linear relationship exists between stress and strain of the material and (III) the Young's modulus, Y, and the actuation expansion coefficient of the conducting polymer, α, keep constant: they do not depend on spatial location inside each layer.

The actuator curvature radius (R_∞ is the radius at the equilibrium and R_0 is the initial radius) is related to either, the Young's modulus (Y) and the thicknesses (h) of the conducting and non-conductive films (indicated by subscripts 1 and 2 respectively), and to the volume changes locally produced at the interface between both films α(t) (Pei & Inganäs, 1993b; Pei & Inganäs, 1993a; Pei & Inganäs, 1992b):

$$\frac{1}{R_\infty}-\frac{1}{R_0}=\frac{6\alpha_{(t)}}{\dfrac{\left(Y_1h_1^2-Y_2h_2^2\right)^2}{Y_1Y_2h_1h_2\left(h_1+h_2\right)}+4\left(h_1+h_2\right)} \tag{11}$$

Christophersen et al. (Christophersen et al., 2006) expanded the model by including strain and modulus variations along the direction of film thickness. Actuator's position, rate of the movement and force generated by the actuator (Alici & Huynh, 2006; Alici et al., 2006b) were simulated and applied to the design of biomimetic device (propulsion fins) (Alici et al., 2007). Du et al. (Du et al., 2010) have developed a general model for a multilayer system (N layers) to link the actuation strain of the actuator to the bending curvature.

8.3. Finite Element Method

The finite elements methodology is a well know mathematical treatment for engineering designs that can be applied to solve the movement of the artificial muscles too. Alici et al. (Alici et al., 2006a; Metz et al., 2006) developed a model based on a lumped-parameter mathematical model for trilayer actuators employing the analogy between thermal strain and the real strain (due to the insertion/extraction of ions inside the polymeric film) in polypyrrole actuators actuating in air. An optimization of the geometry was required, in order to obtain the greater output properties from a determined input voltage. Shapiro et al. (Shapiro & Smela, 2007) developed a two dimensional model (along a full area) to obtain curvature and angular moment from bilayer and trilayer actuators. Thus, they combined the results from the previous method (bending beam method) with finite element method to attain a solution. Another example of the employment of this method was carried out by Gutta et al. who applied it to the study the movement of a cylindrical ionic-polymer metal composite actuator (Gutta et al., 2011).

8.4. Equivalent Transmission Line Model

Electrochemical systems, as many other systems, can be assimilated to electrical circuits and electrochemomechanical actuators have been treated by the equivalent transmission line method. This resource is a practical tool due to the great number of facilities available to the study of electrical circuits through different steps or modules. Such treatments are employed by engineers and physicists, or electrochemists, in order to explain the claimed capacitive behaviour of CP (Albery & Mount, 1993;Bisquert et al., 2000; Paasch, 2000). Ren et al. (Ren & Pickup, 1995) proposed equivalent electrical circuits to model the electron transport and electron transfer in composite pPy-PSS films based on Albery's works. Fang et al. (Fang et al., 2008; Yang et al., 2008) have developed a scalable method including dynamic actuation performance under a given voltage input, joining three different modules for different aspects of the actuator: electrochemical dynamics, stress-generation by charge and mechanical dynamics. Shoa et al. (Shoa et al., 2011) developed a dynamic electromechanical method for electrochemically driven conducting polymer actuators based on a 2-D impedance model using an RC transmission line equivalent circuit to predict the charge transfer during actuation. Besides, a mechanical model (based on the bending beam model) is considered after the equivalent circuit that simulates ion "diffusion" through the thickness and electronic resistance along the length (Shoa et al., 2010). If the angular movement is not linear along the full geometry of the actuator, the bending beam method has to be modified, for example for cantilever type conducting polymer actuators (Alici, 2009).

From all these kind of models, it is possible to employ only one or several of them at the same time in order to obtain the best required model (Woosoon et al., 2007).

9. ACTUATORS APPLICATIONS

The investigation of these devices is mainly performed in academic laboratories. Nevertheless a rising number of applications and products are emerging with pioneering companies that are being incorporated by large multinationals. So Creganna Tactx Medical and Bayer MaterialScience recently acquired a pair of companies working in the field, indicating the potential of these technologies. Also EAMEX from Japan is developing actuators for biomedical and robotic applications. NASA and ESA space agencies consider polymeric actuators as preferential technologies, and the European Scientific Network for Artificial Muscles (ESNAM) has started funded by the European Union. Many different applications can be found in literature. The following is a summary of a few of them, both macroscopic and microscopic.

- Smart electronic textiles: Although different strategies have been used to overcome problems like short lifetimes and high response times (Carpi & De Rossi, 2005; De Rossi et al., 2009; Wallace et al., 2007), much work has to be done in order to endow textiles with efficient and reliable actuating functions.

- Braille displays: Conducting polymers is one of the materials explored to develop low-cost and efficient displays for Braille text. Actual displays are close to achieve the required specs, but they have to overcome short cycling life (Ding et al., 2003; Spinks et al., 2003a).

- Micropump: It is possible to find in literature different types of micropumps made with conducting polymer actuators able to transport fluids at a microflow rate with high precission with low energy consumption (Kiefer et al., 2008; Kim et al., 2008; Morita et al.,2010 ; Naka et al., 2010; Ramírez-García & Diamond, 2007; Wu et al., 2005).

- Propulsion or locomotion devices: Systems based on trilayer configuration were proposed in aquatic devices (Alici et al., 2007; James et al., 2007; McGovern & et al., 2009; McGovern et al., 2010) for creating enough thrust for propulsion. Also bending actuators like cilia can be used in mini-robotic devices (Alici & Gunderson, 2009).

- Microcatheters: By depositing two conducting polymer films around a passive catheter and using a fast conductor solid state electrolyte, it is possible to control the movement of the catheter tip (DellaSanta et al., 1996; Shoa et al., 2008). This kind of microcatheters can be used in Optical Coherence Tomography to enable high-resolution 3D imaging (Lee et al., 2009).

- Microactuators: The electrochemical synthesis of conducting films and their electrochemical actuation (Wilson et al., 2007) are suitable for the construction different microdevices (He et al., 2007; Jager et al., 2001; Pede et al., 1998; Roemer et al., 2002) and microtools constituted by bilayers (Jager et al., 2000a; Jager et al., 2001; Jager et al., 2000b; Jager et al., 1999; Smela, 1999;Smela & Gadegaard, 1999; Smela et al., 1995; Smela et al., 1993) or trilayers (Gursel et al., 2009;Kiefer et al., 2008) using microelectronic technologies. These microactuators can be integrated in CMOS chips containing all the elements required to transport, contain and nurture biological cells (Jager et al., 2002; Smela et al., 2007).

10. PUBLISHED PAPERS AND PATENTS EVOLUTION

An indication of the initiation and fast growth of the field is given by bibliographic and patents evolution. As can be observed, the literature related to conducting polymers actuators starts from 1990 (Fig. 4.a) reaching a maximum in 2006 with 45 papers, an excellent value for such a specific topic. After that, around 30 papers have been published every year. Fig. 4.b shows exponential evolution of citations with a value of almost 2000 in 2010. A similar evolution is observed in Fig. 4.c for patents, reaching a maximum in 2009 with 180 patents.

11. FUTURE AND CHALLENGES

Although a hard work has been performed, it does not exist yet an uncontroversial model for the description of the new reactive polymers and devices based on these reactive polymer gels whose composition and properties mimic those from artificial organs in mammals. New models should include concepts from very different fields of knowledge as electrochemistry, mechanics, polymer science and thermodynamics. In those systems, electrochemical reactions produce structural (conformational) changes in the polymeric chains and macroscopic volume changes related to the composition change of the material. Those aspects are not considered in the classical chemical kinetic models. Conducting polymer films should be used as models for the study and quantification of chemical kinetics under structural control with the aim to develop a new structural chemical model, able to quantify conformational changes and structural information in conducting polymers and in biological reactions originating life in living cells.

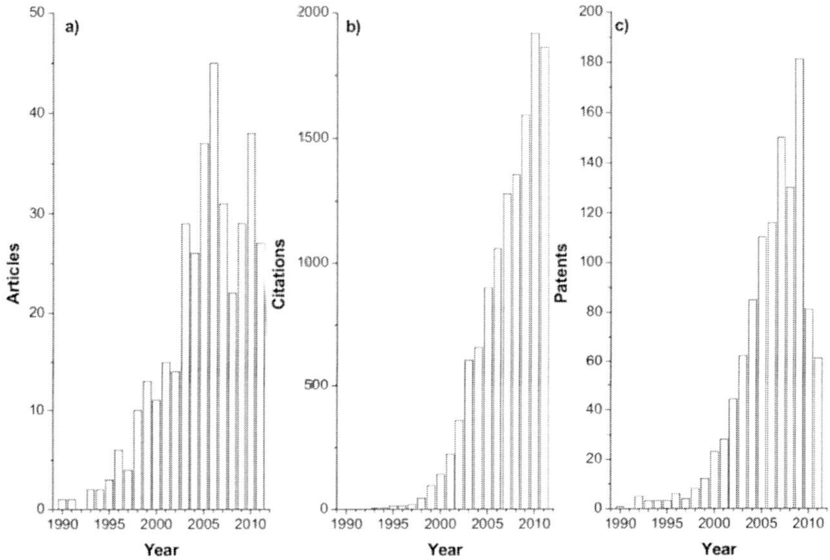

Figure 4. Evolution of a) published papers, b) citations and c) patents for "conducting polymer & actuator", from the ISI web of Knowledge® and Scopus®.

This model will allow the synthesis of new conducting polymers providing a more precise control of structural changes and intermolecular forces (polymer-ion interactions) with the reaction. Those materials should be used in a new generation of polymeric actuators able to overcome current limitations, opening possibilities for new applications.

Other factors, such as configuration, manufacture of the actuators, design of the electric contacts or electrolytic media (solid or liquid) are also very important for the improvement of the electrochemomechanical actuators. This is a multidisciplinary field, a lot of work performed by specialists from different disciplines is required in order to attain a good control and modelling of both, devices and soft robots.

ACKNOWLEDGEMENT

Authors acknowledge financial support from Spanish Government (MCINN) Projects MAT2008-06702, MAT2011-24973, Seneca Foundation Project 08684/PI/08 and to the European network for Artificial Muscles ESNAM. Jose G. Martinez acknowledges Spanish Education Ministry for a FPU grant (AP2010-3460).

REFERENCES

1. H. J. Ahonen, J. Lukkari, J. Kankare, 2000n- and p-doped poly(3,4-ethylenedioxythiophene): Two electronically conducting states of the polymer. Macromolecules, 3318678767930024-9297

2. W. J. Albery, A. R. Mount, 1993Application of A Transmission-Line Model to Impedance Studies on A Poly(Vinylferrocene)-Modified Electrode. J.Chem.Soc.Faraday T., 8923273310956-5000

3. G. Alici, N. N. Huynh, 2006Predicting force output of trilayer polymer actuatorsSens. Actuat. A-Phys, 13226166250924-4247

4. G. Alici, P. Metz, G. M. Spinks, 2006a). A methodology towards geometry optimization of high performance polypyrrole (PPy) actuators. Smart Mater.Struct., 1522432520964-1726

5. G. Alici, B. Mui, C. Cook, 2006bBending modeling and its experimental verification for conducting polymer actuators dedicated to manipulation applicationsSens. Actuat. A-Phys, 12623964040924-4247

6. G. Alici, G. Spinks, N. N. Huynh, L. Sarmadi, R. Minato, 2007Establishment of a biomimetic device based on tri-layer polymer actuators-propulsion fins.Bioinspir. Biomim., 22S18S301748-3182

7. G. Alici, 2009An effective modelling approach to estimate nonlinear bending behaviour of cantilever type conducting polymer actuatorsSens. Actuat. B-Chem, 14112842920925-4005

8. G. Alici, D. Gunderson, 2009A bio-inspired robotic locomotion system based on conducting polymer actuatorsIn: 2009 IEEE/ASME International Conference on Advanced Intelligent Mechatronics 9981004IEEE, 978-1-42442-852-6New York.

9. G. Alici, A. Punning, H. R. Shea, 2011Enhancement of actuation ability of ionic-type conducting polymer actuators using metal ion implantation Actuat. B-Chem, 157172840925-4005

10. C. Arbizzani, M. Catellani, M. Mastragostino, C. Mingazzini, 1995N-doped and p-doped polydithieno[3,4-b-3',4'-d] thiophene- a narrow-band gap polymer for redox supercapacitors. Electrochim. Acta, 4012187118760013-4686

11. J. Arias-Pardilla, W. Walker, F. Wudl, T. F. Otero, 2010Reduction and Oxidation Doping Kinetics of an Electropolymerized Donor-Acceptor Low-Bandgap Conjugated Copolymer.J. Phys. Chem. B, 1144012777127841520-6106

12. J. Arias-Pardilla, C. Plesse, A. Khaldi, F. Vidal, C. Chevrot, T. F. Otero, 2011Self-supported semi-interpenetrating polymer networks as reactive ambient sensorsJ. Electroanal. Chem., 6521-237431572-6657

13. V. Balzani, A. Credi, B. Ferrer, S. Silvi, M. Venturi, 2005Artificial molecular motors and machines: Design principles and prototype systemsSpringer-Verlag, 03401022

14. R. H. Baughman, 1996Conducting polymer artificial musclesSynth. Met., 7833393530379-6779

15. R. H. Baughman, C. X. Cui, A. A. Zakhidov, Z. Iqbal, J. N. Barisci, G. M. Spinks, G. G. Wallace, A. Mazzoldi, D. De Rossi, A. G. Rinzler, O. Jaschinski, S. Roth, M. Kertesz, 1999Carbon nanotube actuators. Science, 28454181340134400036-8075

16. L. Bay, K. West, P. Sommer-Larsen, S. Skaarup, M. Benslimane, 2003A conducting polymer artificial muscle with 12% linear strain. Adv. Mater., 1543103130935-9648

17. K. Bieńkowski, M. Strawski, M. Szklarczyk, 2011The determination of the thickness of electrodeposited polymeric films by AFM and electrochemical techniques. J. Electroanal. Chem., 66211962031572-6657

18. J. Bisquert, G. G. Belmonte, F. F. Santiago, N. Ferriols, M. Yamashita, E. C. Pereira, 2000Application of a distributed impedance model in the analysis of conducting polymer filmsElectrochem.Commun., 286016051388-2481

19. P. M. Blonsky, I. Meridian, 1997Structurally stable gelled electrolytes, US Patent 5648011.

20. F. Carpi, D. De Rossi, 2005Electroactive polymer-based devices for e-textiles in biomedicine.IEEE T. Inf. Technol. B, 932953181089-7771

21. C. K. Chiang, C. R. Fincher, Y. W. Park, A. J. Heeger, H. Shirakawa, E. J. Louis, S. C. Gau, A. G. Macdiarmid, 1977Electrical-Conductivity in Doped Polyacetylene. Phys.Rev. Lett., 3917109811010031-9007

22. M. Cho, H. Seo, J. Nam, H. Choi, J. Koo, Y. Lee, 2007High ionic conductivity and mechanical strength of solid polymer electrolytes based on NBr/ionic liquid and its application to an electrochemical actuator. Sens. Actuat. B-Chem, 128170740925-4005

23. M. S. Cho, J. J. Choi, T. S. Kim, Y. Lee, 2011In situ three-dimensional analysis of the linear actuation of polypyrrole micro-rod actuators using optical microscopySens. Actuat. B-Chem, 15612182210925-4005

24. M. Christophersen, B. Shapiro, E. Smela, 2006Characterization and modeling of PPy bilayer microactuators- Part 1. Curvature. Sens. Actuat. B-Chem, 11525966090925-4005

25. F. Daerden, D. Lefeber, 2002Pneumatic Artificial Muscles: actuators for robotics and automationEur. J. Mech. Environ. Eng., 47110211371-6980

26. A. P. Davis, 1999Nanotechnology- Synthetic molecular motors. Nature, 40167491201210028-0836

27. D. De Rossi, F. Carpi, F. Galantini, 2009Functional Materials for Wearable Sensing, Actuating and Energy HarvestingIn: Biomedical Applications of Smart Materials, Nanotechnology and Micro/Nano Engineering. 247256Trans Tech Publications Ltd, 978-3-90815-823-3Stafa-Zurich.

28. A. Della Santa, A. Mazzoldi, D. de Rossi, 1996Steerable Microcatheters Actuated by Embedded Conducting Polymer Structures. J. Intel. Mat. Syst. Str., 732923000104-5389X.

29. A. Della Santa, D. De Rossi, A. Mazzoldi, 1997Performance and work capacity of a polypyrrole conducting polymer linear actuator. Synth. Met., 902931000379-6779

30. S. D. Deshpande, J. Kim, S. R. Yun, 2005aNew electro-active paper actuator using conducting polypyrrole: actuation behaviour in LiClO4 acetonitrile solutionSynth. Met., 149153580379-6779

31. S. D. Deshpande, J. Kim, S. R. Yun, 2005bStudies on conducting polymer electroactive paper actuators: Effect of humidity and electrode thicknessSmart Mater. Struct., 1448768800964-1726

32. J. Ding, L. Liu, G. M. Spinks, D. Zhou, G. G. Wallace, J. Gillespie, 2003High performance conducting polymer actuators utilising a tubular geometry and helical wire interconnectsSynth. Met., 13833913980379-6779

33. M. Dobbelin, R. Marcilla, C. Pozo-Gonzalo, D. Mecerreyes, 2010Innovative materials and applications based on poly(3,4-ethylenedioxythiophene) and ionic liquids. J. Mater. Chem., 2036761376220959-9428

34. P. Du, X. Lin, X. Zhang, 2010A multilayer bending model for conducting polymer actuatorsSens. Actuat. A-Phys, 16312402460924-4247

35. Y. Fang, X. O. Tan, Y. T. Shen, N. Xi, G. Alici, 2008A scalable model for trilayer conjugated polymer actuators and its experimental validationMater. Sci. Eng., C, 2834214280928-4931

36. M. Fuchiwaki, K. Tanaka, K. Kaneto, 2009Planate conducting polymer actuator based on polypyrrole and its applicationSens. Actuat. A-Phys, 15022722760924-4247

37. F. Garcia-Cordova, L. Valero, Y. A. Ismail, T. F. Otero, 2011Biomimetic polypyrrole based all three-in-one triple layer sensing actuators exchanging cationsJ. Mater. Chem., 2143172651727220959-9428

38. J. M. Gere, B. J. Goodno, (2009). Mechanics of Materials 4th Ed., Cengage Learning, 0-53455-397-4

39. P. Gimenez, K. Mukai, K. Asaka, K. Hata, H. Oike, T. F. Otero, 2012Capacitive and faradic charge components in high-speed carbon nanotube actuatorElectrochim. Acta, 60177-183177 EOF183 EOF0013-4686

40. Y. Guo, G. Yu, Y. Liu, 2010Functional Organic Field-Effect TransistorsAdv. Mater., 2240442744471521-4095

41. A. Gursel, D. Valerie, R. Philippe, S. Geoff, 2009Conducting polymer microactuators operating in airJ. Micromech. Microeng., 1920250 170960-1317

42. S. Gutta, J. Realmuto, Y. Woosoon, K. J. Kim, 2011Dynamic model of a cylindrical ionic polymer-metal composite actuator. Proceedings of 8th International Conference on Ubiquitous Robots and Ambient Intelligence (URAI), Incheon.

43. G. Han, G. Shi, 2006Electrochemical actuator based on single-layer polypyrrole filmSens. Actuat. B-Chem, 11312592640925-4005

44. S. Hara, T. Zama, S. Sewa, W. Takashima, K. Kaneto, 2003Polypyrrole-metal coil composites as fibrous artificial musclesChem. Lett., 3298008010366-7022

45. S. Hara, T. Zama, A. Ametani, W. Takashima, K. Kaneto, 2004Enhancement in electrochemical strain of a polypyrrole-metal composite film actuatorJ. Mater. Chem., 1418272427250959-9428

46. S. Hara, T. Zama, N. Tanaka, W. Takashima, K. Kaneto, 2005Artificial fibular muscles with 20% strain based on polypyrrole-metal coil composites. Chem.Lett., 3467847850366-7022

47. S. Hara, T. Zama, W. Takashima, K. Kaneto, 2006Tris(trifluoromethylsulfonyl)methide-doped polypyrrole as a conducting polymer actuator with large electrochemical strain. Synth.Met., 1562-43513550379-6779

48. X. M. He, C. Li, F. G. Chen, G. Q. Shi, 2007Polypyrrole microtubule actuators for seizing and transferring microparticlesAdv. Funct. Mater., 1715291129170161-6301X.

49. H. W. Heuer, R. Wehrmann, S. Kirchmeyer, 2002Electrochromic window based on conducting poly (3,4-ethylenedioxythiophene)poly(styrene sulfonate). Adv. Funct. Mater., 12289940161-6301X.

50. S. J. Higgins, K. V. Lovell, R. M. G. Rajapakse, N. M. Walsby, 2003Grafting and electrochemical characterisation of poly-(3,4-ethylenedioxythiophene) films, on Nafion and on radiation-grafted polystyrenesulfonate-polyvinylidene fluoride composite surfacesJ. Mater. Chem., 1310248524890959-9428

51. A. R. Hillman, D. C. Loveday, M. J. Swann, R. M. Eales, A. Hamnett, S. J. Higgins, S. Bruckenstein, C. P. Wilde, 1989Charge Transport in Electroactive Polymer-FilmsFaraday Discuss. Chem. Soc., 88151-1630301-7249

52. W. S. Huang, B. D. Humphrey, Diarmid. A. G. Mac, 1986Polyaniline, A Novel Conducting Polymer- Morphology and Chemistry of Its Oxidation and Reduction in Aqueous-Electrolytes. J. Chem. Soc., Faraday Trans. I, 822385-24000300-9599

53. Y. Huang, J. Liang, Y. Chen, 2012The application of graphene based materials for actuators. J. Mater. Chem., 229367136790959-9428

54. K. Ikushima, S. John, K. Yokoyama, S. Nagamitsu, 2009A practical multilayered conducting polymer actuator with scalable work outputSmart. Mater. Struct., 18990964-1726

55. G. Inzelt, 2008Redox Transformations and Transport Processes. In: Conducting Polymers, Scholz, F. 169224 , Springer-Verlag Berlin, 978-3-54075-929-4Heidelberg (Germany).

56. G. Inzelt, 2011Rise and rise of conducting polymersJ. Solid State Electrochem., 157-8171117181432-8488

57. Y. A. Ismail, J. G. Martínez, A. S. A. Harrasi, S. J. Kim, T. F. Otero, 2011Sensing characteristics of a conducting polymer/hydrogel hybrid microfiber artificial muscleActuat. B-Chem, Vol. 0pp. 0925-4005

58. E. W. H. Jager, E. Smela, O. Inganäs, I. Lundstrom, 1999Polypyrrole microactuators. Synth. Met., 1021-3130913100379-6779

59. E. W. H. Jager, O. Inganäs, I. Lundstrom, 2000aMicrorobots for micrometer-size objects in aqueous media: Potential tools for single-cell manipulation.Science, 288547523352338 0036-8075

60. E. W. H. Jager, E. Smela, O. Inganäs, 2000bMicrofabricating conjugated polymer actuators.Science, 290549615401545 0036-8075

61. E. W. H. Jager, O. Inganäs, I. Lundstrom, 2001Perpendicular actuation with individually controlled polymer microactuators. Adv. Mater., 13176790935-9648

62. E. W. H. Jager, C. Immerstrand, K. H. Peterson, K. Magnusson, E. , I. Lundström, O. Inganäs, 2002The Cell Clinic: Closable Microvials for Single Cell Studies.Biomed. Microdev, 431771871387-2176

63. T. James, A. Patrick, F. Timothy, C. Angela, Z. Mike Del, H. Ian, 2007The application of conducting polymers to a biorobotic fin propulsorBioinspir. Biomim., 22S61748-3190

64. S. John, G. Alici, C. Cook, 2008Frequency response of polypyrrole trilayer actuator displacementProceedings of Electroactive Polymer Actuators and Devices (Eapad) 2008, San Diego (USA).

65. K. Kaneto, H. Suematsu, K. Yamato, 2008Training effect and fatigue in polypyrrole-based artificial muscles.Bioinspir. Biomim., 330350051748-3182

66. K. Kaneto, H. Suematsu, K. Yamato, 2009Conducting Polymer Soft Actuators based on Polypyrrole-Training Effect and Fatigue.In: Artificial Muscle Actuators Using Electroactive Polymers, Vincenzini, P., Bar-Cohen, Y. & Carpi, F. 122130Trans Tech Publications Inc., 16620356Zuerich, Switzerland.

67. R. Kiefer, S. Y. Chu, P. A. Kilmartin, G. A. Bowmaker, R. P. Cooney, J. Travas-Sejdic, 2007Mixed-ion linear actuation behaviour of polypyrroleElectrochim. Acta, 527238623910013-4686

68. R. Kiefer, X. Mandviwalla, R. Archer, S. S. Tjahyono, H. Wang, Donald. B. Mac, G. A. Bowmaker, P. A. Kilmartin, J. Travas-Sejdic, 2008The application of polypyrrole trilayer actuators in microfluidics and roboticsProceedings of Electroactive Polymer Actuators and Devices (Eapad).

69. J. H. Kim, K. T. Lau, R. Shepherd, Y. Wu, G. Wallace, D. Diamond, 2008Performance characteristics of a polypyrrole modified polydimethylsiloxane (PDMS) membrane based microfluidic pump Sensors and Actuators A: Physical, 14812392440924-4247

70. L. T. T. Kim, O. Sel, C. Biemme-Chouvy, C. Gabrielli, C. Laberty-Robert, H. Perrot, C. Sanchez, 2010Proton transport properties in hybrid membranes investigated by ac-electrogravimetryElectrochem.Commun., 128113611391388-2481

71. H. Klauk, Ed, 2006Organic Electronics: Materials, Manufacturing, and ApplicationsWiley-VCH Verlag, 978-3-52731-264-1Weinheim.

72. C. Kuttel, A. Stemmer, X. Wei, 2009Strain response of polypyrrole actuators induced by redox agents in solutionSens. Actuat. B-Chem, 14124784840925-4005

73. K. K. C. Lee, N. R. Munce, T. Shoa, L. G. Charron, G. A. Wright, J. D. Madden, V. X. D. Yang, 2009Fabrication and characterization of laser-micromachined polypyrrole-based artificial muscle actuated cathetersSens. Actuat. A-Phys, 15322302360924-4247

74. W. G. Li, C. L. Johnson, H. L. Wang, 2004Preparation and characterization of monolithic polyaniline-graphite composite actuators Polymer4514476947750032-3861

75. W. Lu, A. G. Fadeev, B. H. Qi, E. Smela, B. R. Mattes, J. Ding, G. M. Spinks, J. Mazurkiewicz, D. Z. Zhou, G. G. Wallace, Farlane. D. R. Mac, S. A. Forsyth, M. Forsyth, 2002Use of ionic liquids for pi-conjugated polymer electrochemical devices.Science, 29755839839870036-8075

76. V. Lyutov, V. Tsakova, A. Bund, 2011Microgravimetric study on the formation and redox behavior of poly(2-acrylamido-2-methyl-1-propanesulfonate)-doped thin polyaniline layers. Electrochim. Acta, 561348034810013-4686

77. A. Mazzoldi, C. Degl'Innocenti, M. Michelucci, D. De Rossi, 1998Actuative properties of polyaniline fibers under electrochemical stimulationMater. Sci. Eng., C, 6165720928-4931

78. S. Mc Govern, et al.2009Finding NEMO (novel electromaterial muscle oscillator): a polypyrrole powered robotic fish with real-time wireless speed and directional controlSmart Mater. Struct., 1890950090964-1726

79. S. T. Mc Govern, M. Abbot, R. Emery, G. Alici, V. T. Truong, G. M. Spinks, G. G. Wallace, 2010Evaluation of thrust force generated for a robotic fish propelled with polypyrrole actuatorsPolym. Int., 5933573640959-8103

80. P. Metz, G. Alici, G. M. Spinks, 2006A finite element model for bending behaviour of conducting polymer electromechanical actuatorsSens. Actuat. A-Phys, 1301-110924-4247

81. T. Morita, Y. Chida, D. Hoshino, T. Fujiya, Y. Nishioka, Fabrication and characterization of a polypyrrole soft actuator having corrugated structuresMol. Cryst. Liq. Cryst., 519121-127121 EOF127 EOF1542-1406

82. T. Morita, Y. Chida, D. Hoshino, T. Fujiya, Y. Nishioka, 2010Fabrication and Characterization of a Polypyrrole Soft Actuator Having Corrugated StructuresMol. Cryst. Liq. Cryst., 519121-127121 EOF127 EOF1542-1406

83. K. Mukai, K. Asaka, K. Hata, T. F. Otero, H. Oike, 2011High-Speed Carbon Nanotube Actuators Based on an Oxidation/Reduction Reaction.Chem.-Eur. J., 1739109651097115 21-3765

84. R. Mutlu, G. Alici, 2010A Multistable Linear Actuation Mechanism Based on Artificial MusclesJ. Mech. Des., 132111110011050-0472

85. Y. Naka, M. Fuchiwaki, K. Tanaka, 2010A micropump driven by a polypyrrole-based conducting polymer soft actuatorPolym. Int., 5933523561097-0126

86. T. Nakano, Y. Okamoto, 2001Synthetic Helical Polymers: Conformation and FunctionChem. Rev., 1011240134038ISSN

87. T. Okamoto, K. Tada, M. Onoda, 2000Bending machine using anisotropic polypyrrole filmsJpn. J. Appl. Phys. 1, 395A285428580021-4922

88. T. Okamoto, Y. Kato, K. Tada, M. Onoda, 2001Actuator based on doping/undoping-induced volume change in anisotropic polypyrrole filmThin Solid Films3931-23833870040-6090

89. H. Okuzaki, T. Hattori, 2003Electrically induced anisotropic contraction of polypyrrole filmsSynth. Met., 135-13645-460379-6779

90. H. Okuzaki, 2008A biomorphic origami actuator fabricated by folding a conducting paper. J. Phys. Conf. Ser., 12710120011742-6596

91. M. Onoda, T. Okamoto, K. Tada, H. Nakayama, 1999aPolypyrrole films with anisotropy for artificial muscles and examination of bending behaviorJpn. J. Appl. Phys. 2, 389ABL1070L10720021-4922

92. M. Onoda, K. Tada, H. Nakayama, 1999bPolypyrrole films with anisotropySynth. Met., 1021-3132113220379-6779

93. M. Onoda, K. Tada, 2004Anisotropic bending machine using conducting polypyrroleIEICE Trans. Electron., E87C21281350916-8524

94. M. Onoda, H. Shonaka, K. Tada, 2005A self-organized bending-beam electrochemical actuatorCurr. Appl. Phys., 521942011567-1739

95. D. Orata, D. A. Buttry, 1987Determination of ion populations and solvent content as functions of redox state and pH in polyanilineJ. Am. Chem. Soc., 10912357435810000-2786

96. T. Otero, J. Cascales, A. Fernandez-Romero, 2007aAttempting a classification for electrical polymeric actuators. Proceedings of Electroactive Polymer Actuators and Devices (EAPAD), San Diego.

97. T. F. Otero, E. Angulo, J. Rodriguez, C. Santamaria, 1992aElectrochemomechanical Properties from A Bilayer- Polypyrrole Nonconducting and Flexible Material Artificial Muscle. J. Electroanal. Chem., 3411-23693750022-0728

98. T. F. Otero, E. Angulo, J. Rodriguez, C. Santamaria, 1992bDispositivos laminares que emplean polímeros conductores capaces de provocar movimientos mecánicos, ES Patent 2 048 086.

99. T. F. Otero, J. Rodriguez, C. Santamaria, 1992cMúsculos artificiales formados por multicapas: polímeros conductores-no conductores, ES Patent 2 062 930.

100. T. F. Otero, 1999Conducting polymers, electrochemistry, and biomimicking processes. In: Modern Aspects of Electrochemistry, Bockris, J.O.M., White, R.E. & Conway, B.E. 307434Kluwer Academic/Plenum Publ, 00769924York.

101. T. F. Otero, 2000Biomimicking materials with smart polymers. In: Structural biological materials. Design and structure-properties relationships, Elices, M. & Cahn, R.W. 187220Pergamon Materials Series, 0-08043-416-9The Netherlands).

102. T. F. Otero, M. T. Cortes, 2003aArtificial muscles with tactile sensitivityAdv. Mater., 1542792820935-9648

103. T. F. Otero, M. T. Cortes, 2003bA sensing muscleSens. Actuators, B, 961-21521560925-4005

104. T. F. Otero, M. T. Cortes, 2004Artificial muscle: movement and position controlChem. Commun., Vol. 32842851359-7345

105. T. F. Otero, G. V. Arenas, J. J. L. Cascales, 2006Effect of the doping ion on the electrical response of a free-standing polypyrrole strip subjected to different preloads: Perspectives and limitations associated with the use of these devices as actuatorsMacromolecules3926955195560024-9297

106. T. F. Otero, M. Broschart, 2006Polypyrrole artificial muscles: a new rhombic element. Construction and electrochemomechanical characterization. JAppl. Electrochem., 3622052140002-1891X.

107. T. F. Otero, 2007Artificial muscles. In: Handbook of Conducting Polymers, Skotheim, T.A., Elsenbaumer, R.L. & Reynolds, J.R. 591623CRC Press, ISBN Boca Raton.

108. T. F. Otero, M. T. Cortes, G. V. Arenas, 2007bLinear movements from two bending triple-layersElectrochim. Acta, 533125212580013-4686

109. T. F. Otero, Cascales. J. J. Lopez, Arenas. G. Vazquez, 2007cMechanical characterization of free-standing polypyrrole filmMater. Sci. Eng., C, 27118220928-4931

110. T. F. Otero, 2008Artificial Muscles, Sensing and Multifunctionality. In: Intelligent Materials, Shahinpoor, M. & Schenider, H.-J. 142190Royal Society of Chemistry, 978-0-85404-335-4Cambridge (U.K.).

111. T. F. Otero, 2009Soft, wet, and reactive polymers. Sensing artificial muscles and conformational energy. J. Mater. Chem., 19681-6890959-9428

112. T. F. Otero, 2011From Electrochemically-Driven Conformational Polymeric States to Macroscopic Sensing and Tactile Muscles. In: From

Non-Covalent Assemblies to Molecular Machines, Sauvage, J.-P. & Gaspard, P. pp. Wiley-VCH Verlag & Co., ISBN Weinheim, Germany.

113. T. F. Otero, J. Arias-Pardilla, H. Herrera, J. L. Segura, C. Seoane, 2011Electropolymerization of naphthaleneamidinemonoimide-modified poly(thiophene). Phys. Chem. Chem. Phys., 133716513165161463-9076

114. T. F. Otero, J. J. Sánchez, J. G. Martinez, 2012Biomimetic Dual Sensing-Actuators based on Conducting Polymers. Galvanostatic Theoretical Model for Actuators Sensing TemperatureJ. Phys. Chem. B, 11617527952901089-5647

115. P. R. Ouyang, R. C. Tjiptoprodjo, W. J. Zhang, G. S. Yang, 2008Micro-motion devices technology: The state of arts reviewInt. J. Adv. Manuf. Technol., 385-64634780268-3768

116. G. Paasch, 2000The transmission line equivalent circuit model in solid-state electrochemistryElectrochem. Commun., 253713751388-2481

117. S. J. Park, M. S. Cho, J. D. Nam, I. H. Kim, H. R. Choi, J. C. Koo, Y. Lee, 2009The linear stretching actuation behavior of polypyrrole nanorod in AAO templateSens. Actuat. B-Chem, 13525925960925-4005

118. D. Pede, E. Smela, T. Johansson, M. Johansson, O. Inganäs, 1998A general-purpose conjugated-polymer device array for imagingAdv. Mater., 1032332370935-9648

119. Q. Pei, O. Inganäs, 1993aElectrochemical applications of the bending beam method; a novel way to study ion transport in electroactive polymersSolid State Ionics601-31611660167-2738

120. Q. Pei, O. Inganäs, 1993bElectrochemical applications of the bending beam method. 2. Electroshrinking and slow relaxation in polypyrroleJ. Phys. Chem., 9722603460410022-3654

121. Q. B. Pei, O. Inganäs, 1992aConjugated Polymers and the Bending Cantilever Method- Electrical Muscles and Smart Devices. Adv. Mater., 442772780935-9648

122. Q. B. Pei, O. Inganäs, 1992bElectrochemical Applications of the Bending Beam Method.1. Mass-Transport and Volume Changes in Polypyrrole During Redox. J. Phys. Chem., 962510507105140022-3654

123. I. F. Perepichka, D. F. Perepichka, Eds, 2009Handbook of Thiophene-Based Materials: Applications in Organic Electronics and Photonics. John Wiley & Sons Ltd.,978-0-47005-732-2Chichester (UK).

124. C. Plesse, F. Vidal, H. Randriamahazaka, D. Teyssie, C. Chevrot, 2005Synthesis and characterization of conducting interpenetrating polymer networks for new actuatorsPolymer4618777177780032-3861

125. C. Plesse, F. Vidal, D. Teyssie, C. Chevrot, 2009Conducting IPN Fibers: a new design for linear actuation in open airIn: Artificial Muscle Actuators using Electroactive Polymers Vincenzini, P., Bar-Cohen, Y. & Carpi, F. 5358Trans Tech Publicactions Ltd., 16620356Zurich (Switzerland).

126. C. Plesse, F. Vidal, D. Teyssie, C. Chevrot, 2010Conducting polymer artificial muscle fibres: toward an open air linear actuationChem. Commun., 4617291029121359-7345

127. S. Ramírez-García, D. Diamond, 2007Biomimetic, low power pumps based on soft actuators. Sens. Actuat. A-Phys, 135 1 229235 ,0924-4247

128. M. Ranger, M. Leclerc, 1998Optical and electrical properties of fluorene-based pi-conjugated polymers. Can. J. Chem., 7611157115770008-4042

129. X. Ren, P. G. Pickup, 1995Impedance measurements of ionic conductivity as a probe of structure in electrochemically deposited polypyrrole filmsJ. Electroanal. Chem., 3961-23593641572-6657

130. M. Roemer, T. Kurzenknabe, E. Oesterschulze, N. Nicoloso, 2002Microactuators based on conducting polymers.Anal. Bioanal. Chem., 37387547571618-2642

131. J. M. Sansinena, V. Olazabal, T. F. Otero, Fonseca. C. N. P. da, M. A. De Paoli, 1997A solid state artificial muscle based on polypyrrole and a solid polymeric electrolyte working in airChem. Commun., Vol. 22221722181359-7345

132. J. M. Sansiñena, J. B. Gao, H. L. Wang, 2003High-performance, monolithic polyaniline electrochemical actuators. Adv. Funct. Mater., 1397037090161-6301X.

133. S. Shakuda, S. Morita, T. Kawai, K. Yoshino, 1993Dynamic Characteristics of Bimorph with Conducting Polymer GelJpn. J. Appl. Phys. 1, 3211A514351460021-4922

134. B. Shapiro, E. Smela, 2007Bending Actuators with Maximum Curvature and Force and Zero Interfacial StressJ. Intel. Mat. Syst. Str., 1821811860104-5389X.

135. H. Shirakawa, E. J. Louis, A. G. Macdiarmid, C. K. Chiang, A. J. Heeger, 1977Synthesis of Electrically Conducting Organic Polymers- Halogen Derivatives of Polyacetylene, (Ch)X. J. Chem. Soc., Chem. Commun., Vol. 165785800022-4936

136. T. Shoa, J. D. Madden, N. Fekri, N. R. Munce, V. X. D. Yang, 2008Conducting polymer based active catheter for minimally invasive interventions inside arteries.Proceedings of 30th Annual International Conference of the IEEE Engineering in Medicine and Biology Society, 2008.

137. T. Shoa, J. D. W. Madden, N. R. Munce, V. Yang, 2010Analytical modeling of a conducting polymer-driven catheterPolym.Int., 5933433511097-0126

138. T. Shoa, D. S. Yoo, K. Walus, J. D. W. Madden, 2011A Dynamic Electromechanical Model for Electrochemically Driven Conducting Polymer ActuatorsIEEE ASME Trans. Mechatronics, 16142491083-4435

139. E. Smela, O. Inganäs, Q. B. Pei, I. Lundstrom, 1993Electrochemical Muscles- Micromachining Fingers and Corkscrews. Adv. Mater., 596306320935-9648

140. E. Smela, O. Inganäs, I. Lundstrom, 1995Controlled Folding of Micrometer-Size Structures.Science, 2685218173517380036-8075

141. E. Smela, 1999A microfabricated movable electrochromic "pixel" based on polypyrroleAdv. Mater., 1116134313450935-9648

142. E. Smela, N. Gadegaard, 1999Surprising volume change in PPy(DBS): An atomic force microscopy study. Adv. Mater., 11119539570935-9648

143. E. Smela, N. Gadegaard, 2001Volume change in polypyrrole studied by atomic force microscopyJ. Phys. Chem. B, 10539939594051089-5647

144. E. Smela, M. Christophersen, S. B. Prakash, M. Urdaneta, M. Dandin, P. Abshire, 2007Integrated cell-based sensors and "cell clinics" utilizing conjugatedpolymer actuators. Proceedings of SPIE- The International Society for Optical Engineering.

145. E. B. Smith, 2004Basic Chemical ThermodynamicsImperial College Press., 1-86094-445-0

146. F. So, Ed, 2010Organic electronics : materials, processing, devices and applicationsCRC Press, 978-1-42007-290-7Boca Raton (USA).

147. M. K. Song, J. Y. Cho, B. W. Cho, H. W. Rhee, 2002Characterization of UV-cured gel polymer electrolytes for rechargeable lithium batteriesJ. Power Sources, 11012092150378-7753

148. G. M. Spinks, L. Liu, G. G. Wallace, D. Żhou, 2002Strain Response from Polypyrrole Actuators under LoadAdv. Funct. Mater., 126-74374401616-3028

149. G. M. Spinks, G. G. Wallace, J. Ding, D. Zhou, B. Xi, J. Gillespie, 2003aIonic liquids and polypyrrole helix tubes: bringing the electronic Braille screen closer to reality. Proceedings of Smart Structures and Materials 2003: Electroactive Polymer Actuators and Devices (EAPAD), San Diego, USA.

150. G. M. Spinks, D. Z. Zhou, L. Liu, G. G. Wallace, 2003bThe amounts per cycle of polypyrrole electromechanical actuatorsSmart. Mater. Struct., 1234684720964-1726

151. W. Takashima, T. Uesugi, M. Fukui, M. Kaneko, K. Kaneto, 1997Mechanochemoelectrical effect of polyaniline filmSynth. Met., 851-3139513960379-6779

152. W. Takashima, S. S. Pandey, K. Kaneto, 2003Bi-ionic actuator by polypyrrole filmsSynth. Met., 1351-361620379-6779

153. S. Timoshenko, 1925Analysis of bi-metal thermostatsJ. Opt. Soc. Am., 1132332550093-4119

154. R. M. Torresi, S. L. D. Maranhao, 1999Anion and solvent exchange as a function of the redox states in polyaniline films. J. Electrochem. Soc., 14611417941820013-4651

155. E. W. Tsai, T. Pajkossy, K. Rajeshwar, J. R. Reynolds, 1988Anion-Exchange Behavior of Polypyrrole MembranesJ. Phys. Chem., 9212356035650022-3654

156. Conzuelo. L. Valero, J. Arias-Pardilla, J. V. Cauigh-Rodríguez, M. A. Smit, T. F. Otero, 2010Sensing and Tactile Artificial Muscles from Reactive MaterialsSensors102638-26742638 EOF2674 EOF1424-8220

157. L. Valero, J. Arias-Pardilla, M. Smit, J. Cauich-Rodríguez, T. F. Otero, 2010Polypyrrole Free-Standing electrodes sense temperature or current during reaction. Polym. Int., 59337-3421097-0126

158. L. Valero, J. Arias-Pardilla, J. Cauich-Rodríguez, M. A. Smit, T. F. Otero, 2011Characterization of the movement of polypyrrole-dodecylbenzenesulfonate-perchlorate/tape artificial muscles. Faradaic

control of reactive artificial molecular motors and musclesElectrochim. Acta, 5610372137260013-4686

159. F. Vidal, J. F. Popp, C. Plesse, C. Chevrot, D. Teyssie, 2003Feasibility of conducting semi-interpenetrating networks based on a poly(ethylene oxide) network and poly(3,4-ethylenedioxythiophene) in actuator design. J. Appl. Polym. Sci., 9013356935770021-8995

160. F. Vidal, C. Plesse, G. Palaprat, J. Juger, J. Citerin, A. Kheddar, C. Chevrot, D. Teyssie, 2009Synthesis and Characterization of IPNs for Electrochemical ActuatorsIn: Artificial Muscle Actuators Using Electroactive Polymers, Vincenzini, P., BarCohen, Y. & Carpi, F. 817Trans Tech Publications Ltd, 978-3-90815-827-1Stafa-Zurich.

161. F. Vidal, C. Plesse, P. H. Aubert, L. Beouch, F. o. Tran Van, G. Palaprat, P. Verge, P. Yammine, J. Citerin, A. Kheddar, L. Sauques, C. Chevrot, D. Teyssie, 2010Poly(3,4-ethylenedioxythiophene)-containing semi-interpenetrating polymer networks: a versatile concept for the design of optical or mechanical electroactive devices. Polym. Int., 5933133201097-0126

162. A. Vlad, C. A. Dutu, P. Jedrasik, U. Sodervall, J. F. Gohy, S. Melinte, 2012Vertical single nanowire devices based on conducting polymers.Nanotechnology23250957-4484

163. G. Wallace, T. Campbell, P. Innis, 2007Putting function into fashion: Organic conducting polymer fibres and textilesFiber. Polym., 821351421229-9197

164. H. L. Wang, J. B. Gao, J. M. Sansiñena, P. Mc Carthy, 2002Fabrication and characterization of polyaniline monolithic actuators based on a novel configuration: Integrally skinned asymmetric membrane. Chem. Mater., 146254625520897-4756

165. S. A. Wilson, R. P. J. Jourdain, Q. Zhang, R. A. Dorey, C. R. Bowen, M. Willander, Q. U. Wahab, S. M. Al-hilli, O. Nur, E. Quandt, C. Johansson, E. Pagounis, M. Kohl, J. Matovic, B. Samel, W. van der Wijngaart, E. W. H. Jager, D. Carlsson, Z. Djinovic, M. Wegener, C. Moldovan, R. Iosub, E. Abad, M. Wendlandt, C. Rusu, K. Persson, 2007New materials for micro-scale sensors and actuators: An engineering review. Mat. Sci. Eng. R, 561-611290092-7796X.

166. Y. Woosoon, L. Joonsoo, J. K. Kwang, 2007An artificial muscle actuator for biomimetic underwater propulsorsBioinspir. Biomim., 22S311748-3190

167. Y. Wu, D. Zhou, G. M. Spinks, P. C. Innis, W. M. Megill, G. G. Wallace, 2005TITAN: a conducting polymer based microfluidic pumpSmart. Mater. Struct., 146151115160964-1726

168. F. Yang, T. Xiaobo, G. Alici, 2008Robust Adaptive Control of Conjugated Polymer ActuatorsIEEE Trans. Control Syst. Technol., 1646006121063-6536

169. Q. Yao, G. Alici, G. A. Spinks, 2008Feedback control of tri-layer polymer actuators to improve their positioning ability and speed of responseSens. Actuat. A-Phys, 14411761840924-4247

170. U. L. Zainudeen, M. A. Careem, S. Skaarup, 2008PEDOT and PPy conducting polymer bilayer and trilayer actuatorsSens. Actuat. B-Chem, 13424674700925-4005

INDEX

A

Assistive Robotics, 65

B

base-metal cofiring, 227

C

Carbonation, 200, 208, 226
Chirp modulation, 177
Chloride, 200
composites, 21, 123, 124, 125, 164, 165, 166, 167, 168, 169, 170, 171, 172, 174, 175, 286
Correlation receiver, 177
Corrosion, 200, 224, 226

D

dielectric elastomer generator, 107
Double Doping, 266

E

Electro-mechanical impedance (EMI) technique, 200

F

ferroelectric, 124, 126, 128, 132, 134, 138, 154, 156, 159, 161, 174, 206, 233, 243, 249
ferromagnetic, 89, 91, 92, 94, 124, 138, 139, 140, 154, 158, 160, 169, 174
friction coupling; motor, 25

G

Gait Triggering, 78

I

Inductors, 160
inverse piezoelectric effect, 25, 26, 126

J

Janus-type structure, 124
Joints, 68

M

magnetoelectric, 123, 124, 136, 144, 161, 162, 163, 164, 165,

166, 167, 168, 169, 170, 171, 172, 173, 174, 175
magnetoencephalography, 57, 66, 87
magnetostrictive, 29, 124, 125, 127, 128, 129, 131, 133, 135, 142, 144, 145, 148, 161, 165, 166, 169, 170, 171, 224
multiferroic, 124, 125, 136, 153, 155, 159, 160, 163, 165, 166, 167, 168, 169, 172, 173, 174, 175
multilayer actuator, 42, 227, 230, 246
Muscle Toning, 65

N

N-Doping, 265
Neuroscience, 57, 59, 60, 62, 66, 67, 87, 88, 89, 91, 98

P

P-Doping, 264
Phase Shifters, 156, 174
piezoelectric, 25, 26, 28, 29, 30, 243, 244, 245, 246, 247, 249, 252
piezoelectrics, 116, 227, 228, 229, 231, 238, 244, 253, 254, 269
Position measurement, 177

R

Reinforced concrete, 200
Resonators, 158
RFID tags, 177, 178, 179, 186, 195

S

Shape Memory Alloys, 1, 60, 98, 102
SMAlinear actuators, 4
Snake-like SMA actuators, 8
Surface acoustic wave devices, 177, 184

T

thick film, 124, 134, 166
thin film, 124, 132, 133, 143, 145, 151, 166, 169, 271
Tissues, 68

W

Wearable Generators, 105